Phishing for Answers

Other related titles:

You may also like

- PBSE023 | Hakak | Fake News Detection, Mitigation and Prevention: Methods for a Safer, More Secure and Ethical Cyber World | under development
- PBPC068 | Abdel-Basset | Explainable Artificial Intelligence for Trustworthy Internet of Things | To publish October 2024
- PBSE026 | Abd El-Latif | Advances in Multi-Modal Biometrics: Enhancing Security and Identity Verification | contracted 2024
- PBSE019 | Bhadoria | Blockchain Technology for Secure Social Media Computing | 2023
- PBSE001 | Awad | Information Security: Foundations, Technologies and Applications | 2018

We also publish a wide range of books on the following topics:
Computing and Networks
Control, Robotics and Sensors
Electrical Regulations
Electromagnetics and Radar
Energy Engineering
Healthcare Technologies
History and Management of Technology
IET Codes and Guidance
Materials, Circuits and Devices
Model Forms
Nanomaterials and Nanotechnologies
Optics, Photonics and Lasers
Production, Design and Manufacturing
Security
Telecommunications
Transportation

All books are available in print via https://shop.theiet.org or as eBooks via our Digital Library https://digital-library.theiet.org.

IET SECURITY SERIES 22

Phishing for Answers

Risk identification and mitigation strategies

Terry R. Merz and Lawrence E. Shaw

The Institution of Engineering and Technology

About the IET

This book is published by the Institution of Engineering and Technology (The IET).

We inspire, inform and influence the global engineering community to engineer a better world. As a diverse home across engineering and technology, we share knowledge that helps make better sense of the world, to accelerate innovation and solve the global challenges that matter.

The IET is a not-for-profit organisation. The surplus we make from our books is used to support activities and products for the engineering community and promote the positive role of science, engineering and technology in the world. This includes education resources and outreach, scholarships and awards, events and courses, publications, professional development and mentoring, and advocacy to governments.

To discover more about the IET please visit https://www.theiet.org/.

About IET books

The IET publishes books across many engineering and technology disciplines. Our authors and editors offer fresh perspectives from universities and industry. Within our subject areas, we have several book series steered by editorial boards made up of leading subject experts.

We peer review each book at the proposal stage to ensure the quality and relevance of our publications.

Get involved

If you are interested in becoming an author, editor, series advisor, or peer reviewer please visit https://www.theiet.org/publishing/publishing-with-iet-books/ or contact author_support@theiet.org.

Discovering our electronic content

All of our books are available online via the IET's Digital Library. Our Digital Library is the home of technical documents, eBooks, conference publications, real-life case studies and journal articles. To find out more, please visit https://digital-library.theiet.org.

In collaboration with the United Nations and the International Publishers Association, the IET is a Signatory member of the SDG Publishers Compact. The Compact aims to accelerate progress to achieve the Sustainable Development Goals (SDGs) by 2030. Signatories aspire to develop sustainable practices and act as champions of the SDGs during the Decade of Action (2020–30), publishing books and journals that will help inform, develop, and inspire action in that direction.

In line with our sustainable goals, our UK printing partner has FSC accreditation, which is reducing our environmental impact to the planet. We use a print-on-demand model to further reduce our carbon footprint.

British Library Cataloguing in Publication Data
A catalogue record for this product is available from the British Library

ISBN 978-1-83953-667-0 (hardback)
ISBN 978-1-83953-668-7 (PDF)

Typeset in India by MPS Limited
Printed in the UK by CPI Group (UK) Ltd, Eastbourne

Cover Image: Just_Super/iStock via Getty Images

Contents

Preface

From zero days to phishing for answers

As a senior cybersecurity researcher, I have dedicated my career to understanding the ever-evolving landscape of cyber threats. My fascination with the intricacies of advanced persistent threats (APTs) led me to focus on a particularly elusive adversary known as the Elderwood Group. This notorious collective of hackers had gained a reputation for their sophisticated zero-day attacks, which exploited previously unknown vulnerabilities in software and hardware systems.

In the course of my research, I began to develop theories around the zero-day attack engineering lifecycle employed by the Elderwood Group. Their ability to produce zero-day exploits at regular intervals suggested a well-oiled machine, a calculated approach to infiltrating even the most secure networks. However, in June 2015, my theories were put to the test as the internet was suddenly flooded with an unprecedented number of zero-day attacks.

Frustrated and at a loss, I found myself questioning the direction of my study. The sheer volume and variety of these new attacks seemed to defy the patterns I had observed in the Elderwood Group's operations. It was at this point that one of my mentors, Matt Dinmore, encouraged me to persist in my investigation, even if the path forward was unclear. His advice resonated with me, and I resolved to dig deeper into the nature of these attacks.

As I pored over the data, a common thread began to emerge: phishing. Time and time again, the attack vector appeared to be a carefully crafted email, designed to lure unsuspecting victims into compromising their own security. The realization struck me like a thunderbolt—"It's phishing!" I repeated to myself, marveling at the simplicity and effectiveness of this age-old tactic. It was then that I stumbled upon a government agency metric that crystallized my newfound understanding. The report stated that an astonishing 96 percent of all successful cyberattacks were caused by human failures, primarily the act of clicking on nefarious emails. This revelation was a turning point in my research, as it became clear that I needed to shift my focus from the technical aspects of zero-day exploits to the human element of cybersecurity.

From that moment on, I embarked on a new journey, one that would lead me to explore the psychology behind phishing and the reasons why individuals, despite their best intentions, continue to fall victim to these attacks. I was no longer just studying the Elderwood Group or the zero-day attack lifecycle; I was phishing for answers to a much larger question—how can we effectively combat cyber threats by addressing the human factor?

During the course of these studies, I managed to incessantly bend my co-author's ear. With Larry's background in law enforcement, and subsequent cyber-security career he found the exploration of the human component too alluring to resist. And so he joined me on this journey into the world of human vulnerabilities, and the ever-present threats posed by online interactions.

To that end, in the pages that follow, we invite you to join us on this exploration of the complex interplay between technology, psychology, and the ongoing battle for cybersecurity. By understanding the human vulnerabilities that lie at the heart of successful phishing attacks, we can begin to develop more effective strategies for protecting ourselves, our organizations, and our digital world as a whole.

Terry R. Merz, DCS, CISSP, CISM and Lawrence E. Shaw, CISSP, CHFI, CEH

With thanks and gratitude

My career has been blessed with the guidance and interests of some of the most amazing people. To that end, I would like to offer special thanks and gratitude to the following people, people who at some point in time took interest in my career and pointed me into the right direction. While I may have "dropped off the radar," their influence has remained with me, and served as a reminder that I should "stay the course":

- Vice Admiral M. McConnell (Ret, USN), current Executive Director of Cyber Florida and former Director of the U.S. National Security Agency, and former Director of National Intelligence
- Brigadier General J. Jaeger, (Ret., USAF), former Assistant Director of Operations, U.S. National Security Agency
- Dr. Dan J. Ryan, Esq., Former Director of Information Systems Security for the Office of the Secretary of Defense and Executive Assistant to the Director of Central Intelligence. Author, academic, teacher and husband to Dr. Julie Ryan.
- Major General M.D. Dinmore, USAF. Dr. Dinmore is faculty member of Johns Hopkins University's Engineering for Professionals program, and is also principal Professional Staff Member at the Johns Hopkins University Applied Physics Laboratory (JHU APL). Accomplished academic, researcher and military leader.
- Dr. David Manz, Chief Cybersecurity Scientist at the Pacific Northwest National Laboratory. Recognized expert in the area of cyber-physical systems security, is also an author, and accomplished researcher.

Lastly, but certainly not least, a special thank you to our research assistant, Mark Lautenschlager. Mark's enthusiasm and unsatiable quest for knowledge were key in contributing to the research that went into the sections involving the anthropomorphizing of technologies.

Acronyms

AI (Artificial Intelligence): A branch of computer science dealing with the simulation of intelligent behavior in computers.

APWG (Anti-Phishing Working Group): An international coalition focused on eliminating the identity theft and fraud that result from phishing and email spoofing.

ATT&CK (Adversarial Tactics, Techniques, and Common Knowledge): A globally accessible knowledge base of adversary tactics and techniques based on real-world observations.

BS7799: A standard created by the British Standards Institution (BSI) that outlines how information security should be managed.

CEO (Chief Executive Officer): The highest-ranking person in a company or other institution, ultimately responsible for making managerial decisions.

CERT (Computer Emergency Readiness Team): Teams responsible for receiving, reviewing, and responding to computer security incident reports and activities.

CISO (Chief Information Security Officer): An executive responsible for an organization's information and data security.

C|EH (Certified Ethical Hacker): A certification obtained by individuals skilled in understanding and finding weaknesses and vulnerabilities in computer systems.

C|HFI (Certified Hacking Forensics Investigator): A certification for professionals specializing in detecting hacking attacks and properly extracting evidence to report the crime and conduct audits to prevent future attacks.

CISSP (Certified Information Systems Security Professional): A globally recognized certification in the field of IT security provided by ISC2 (International Information System Security Certification Consortium).

COBIT (Control Objectives for Information and Related Technology): A framework created by ISACA for information technology management and IT governance.

CVSS (Common Vulnerability Scoring System): A free and open industry standard for assessing the severity of computer system security vulnerabilities.

CVE (Common Vulnerabilities and Exposures): A list of publicly disclosed computer security flaws.

DHS (Department of Homeland Security): A United States federal agency responsible for public security.

DMARC (Domain-based Message Authentication Reporting and Conformance): An email authentication, policy, and reporting protocol designed to detect and prevent email spoofing.

DKIM (DomainKeys Identified Mail): An email authentication method designed to detect forged sender addresses in emails by verifying the domain name of the sender.

DNS (Domain Name System): The system that translates human-friendly domain names into IP addresses.

DSL (Digital Subscriber Line): A family of technologies providing internet access by transmitting digital data over the wires of a local telephone network.

EDR (Endpoint Detection and Response): Security solutions that monitor and respond to cyber threats in real-time, focusing on detecting suspicious activities on endpoint devices.

EPP (Endpoint Protection Platforms): Comprehensive security solutions designed to detect, investigate, block, and contain malicious activities on endpoint devices.

HCI (Human–Computer Interaction): The study and planned design of human and computer activities.

IC3 (Internet Crime Complaint Center): A partnership between the FBI and the National White Collar Crime Center (NW3C) to receive and develop complaints related to cybercrime.

ISO (International Organization for Standardization): An international standard-setting body composed of representatives from various national standards organizations.

ISO/IE 1799: An international standard for information security management.

IT (Information Technology): The use of computers to store, retrieve, transmit, and manipulate data or information.

MFA (Multi-Factor Authentication): A security measure requiring two or more verification factors to gain access to an account or system, significantly reducing the risk of unauthorized access.

MITRE (MITRE Corporation): An American not-for-profit organization that manages federally funded research and development centers supporting government agencies.

ML (Machine Learning): A type of AI that allows software applications to become more accurate in predicting outcomes without being explicitly programmed to do so.

MR (Mixed Reality): The merging of real and virtual worlds to produce new environments where physical and digital objects co-exist and interact in real-time.

MITRE ATT&CK: A globally accessible knowledge base of adversary tactics and techniques based on real-world observations.

NFC (Near-Field Communication): A set of communication protocols for communication between two electronic devices over a distance of 4 cm (1 1/2 in) or less.

NIST (National Institute of Standards and Technology): A physical sciences laboratory and non-regulatory agency of the United States Department of Commerce.

NLP (Natural Language Processing): A field of AI that gives computers the ability to understand text and spoken words in much the same way human beings can.

OCTAVE (Operationally Critical Threat, Asset, and Vulnerability Evaluation): A risk-based strategic assessment and planning technique for security.

RMF (Risk Management Framework): A framework developed by NIST consisting of steps including preparation, categorization, selection, implementation, assessment, authorization, and monitoring of security controls.

SE-CMM (Systems Engineering Capability Maturity Model): A model for evaluating and improving systems engineering capabilities.

SSE-CMM (Systems Security Engineering Capability Maturity Model): A framework for evaluating the maturity of security engineering processes.

SETA (Security Education, Training, and Awareness): Programs designed to educate and train employees about their information security responsibilities.

SPF (Sender Policy Framework): An email authentication protocol designed to detect and block email spoofing by verifying sender IP addresses.

SSH (Secure Shell): A cryptographic network protocol for operating network services securely over an unsecured network.

TARA (Threat Assessment and Remediation Analysis): A framework developed by MITRE for identifying and assessing cyber vulnerabilities and deploying countermeasures.

TPB (Theory of Planned Behavior): A theory which suggests that behavior is directly influenced by attitudes, subjective norms, and perceived behavioral control.

UBA (User Behavior Analytics): Tools that monitor for unusual activity that could indicate a security threat by understanding normal user behavior and alerting on deviations.

UI/UX (User Interface/User Experience): The process of enhancing user satisfaction by improving the usability, accessibility, and pleasure provided in the interaction between the user and the product.

U.S. (United States): A country primarily located in North America.

UX (User Experience): The overall experience of a person using a product such as a website or computer application, especially in terms of how easy or pleasing it is to use.

RBI (Remote Browser Isolation): Technology that directs web activities through an isolated external browser, ensuring that potential malware from compromised websites does not reach the organization's internal networks.

SIEM (Security Information and Event Management): Systems that offer a comprehensive approach to detecting, analyzing, and responding to cybersecurity incidents by aggregating and analyzing log data across an organization's technology infrastructure.

About the authors

Terry R. Merz is a senior research scientist at the Pacific Northwest National Laboratory (PNNL), USA and an adjunct professor at the State University of New York (SUNY) College of Emergency Management, Homeland Security and Cybersecurity (CEHC). Her work in the field of cybersecurity spans two decades and involves security engineering, testing, and incident response. Since 2014, she has focused exclusively on research and is researching self-healing, situationally aware architectures for power systems, long-term exploratory research on data collected from commercial industrial control systems, the development of theoretical models for the testing and evaluation of binary analysis tools, behavioral information security, and advanced persistent threats. Her current area of research involves the integration of AI in asymmetric cyber operations and analytics. She holds a Doctor of Computer Science from Colorado Technical University, USA.

Lawrence E. Shaw is the former lead of the CSSP Cyber Forensic Team at the Missile Defense Agency (MDA), USA. His cybersecurity experience covers the areas of incident response, forensics, and malware analysis and cybersecurity testing (Blue/Red Team testing). In his role as a senior incident responder with Microsoft Cyber Defense Operations Center, he focused on cyber incidents related to the Microsoft enterprise. In addition to incident response, he has spent several years in law enforcement-related cyber forensics with the Florida Department of Law Enforcement. He holds a master's degree in computer science, with a concentration in cyber security from the Western Governors University, Utah, USA.

Chapter 1

The phish: an introduction to the nature of phishing

Phishing is a form of cyberattack in the class of social engineering. These attacks masquerade as legitimate forms of electronic communications designed to compel a user into taking a specific action that will benefit the attacker. Most experts would agree Phishing is a "Human" problem. As such, traditionally, organizations have responded to phishing threats using management techniques such as developing policies relative to human behaviors in the work environment and training programs.

While each of these mitigation strategies has been a key element of cybersecurity programs for the past two decades, phishing attacks not only continued successfully but also have escalated in terms of frequency and sophistication. Details issued in a report by Cisco SECURE for 2021 found in the United States:

Eighty-six percent of organizations had at least one user attempt to connect to a phishing site.
Seventy percent of the organizations in the study had users served with malicious ads.

A 2021 report from the U.S. Federal Bureau of Investigation's Internet Crime Complaint Center (IC3) reported Phishing, Vishing[1], Smishing[2], and Pharming[3] as the leading attack types with 323,972 incidences reported in 2021 in the United States and a total loss of $44,213,707 in revenues.

Given the continuing success of phishing attacks, assumptions can be made that current approaches to reducing the number of phishing incidences are not developing phishing resistant user behaviors.

With policies, training, awareness campaigns and technologies falling short, invariably questions around the user's experience, disposition, context, and psychology have taken center stage in the research community since 2009.

---

[1]Fraudulent phone calls used to induce victims to divulge personal information.
[2]Fraudulent use of SMS text messages to induce victims to divulge information.
[3]Pharming is like phishing; however rather than using email as an attack vector, it uses malicious code executed on the victim's device to redirect to an attacker's website.

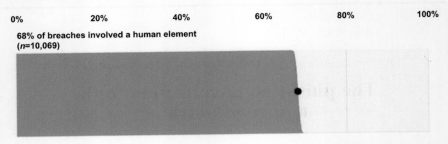

Figure 1.1   *Verizon Data Breach Investigations (DBIR) 2023 Human Element Breaches*

However, before taking a broader look at the human experience relative to phishing, understanding the types of phish users are faced with is an important first step (Figure 1.1).

Case: spear phishing attacks against Crimean activists

In a series of three case studies conducted between April 2019 and March 2020, researchers from The Engine Room, a non-profit organization studying global social justice issues, investigated digital attacks against human rights groups.

In an interview with The Engine Room researchers, the Digital Security Lab (DSL) in Ukraine discussed how in December 2019 four groups of activists became targets of a spear phishing campaign.

In this case, the spear phish sent to the group targeted individuals interested in music. The spear phish masqueraded as a communication from a music group and was requesting the activists submit articles for a music column.

By February 2020, the attack was well underway, with over 15 individuals receiving spear phishing emails.

DSL further reported that once this group discovered it had been spear phished, it was unable to discern the malicious email from an authentic email due to the personalized nature of the malicious emails.

While DSL was unable to make a 100 percent attribution of the origins of the spear phishing emails, DSL suggested the most likely source was local law enforcement. These conclusions were drawn based upon a set of targets that had local law enforcement as a common concern, IP addresses used to send the emails, data used, and email features.

The likely purpose of the spear phishing campaign was to sow discord and distrust among the activists' groups. For example, as the activists became aware of the spear phishing attack, uncertainty evolved around who had been compromised within the group. When such uncertainty in a trusted group arises, it can breed distrust, which in turn results in a cohesive group becoming fractured.

To assist the group in maintaining its cohesion, particularly with partners in remote areas requiring digital communications, DSL published the discovery of the attacks such that remote groups coordinating with the victim group could strategize on developing a trusted digital communications approach.

1.1   Understanding the nature of a phishing attack

The first and foremost characteristic of a phishing attack is its focus: the human user. The methods used to exploit the user involve deception techniques, both technical and non-technical. The purpose of a phishing attack could center around theft, information gathering, or gaining access to sensitive systems.

Phishing is often executed as a coordinated and phased campaign by motivated adversaries wishing to achieve specific goals. For example, assume a particular company is targeted by an adversary seeking to acquire company trade secrets.

Using any number of effective tools (such as the Harvester), the initial phase of a phishing campaign could involve gathering email addresses of employees working at this company. Additional tools can be used to cross reference email addresses to employee names, positions (LinkedIn accounts for example), and locations.

Once an adversary has identified potential targets, the phishing campaign can take on a more focused approach. For example, if the adversary is interested in gaining access to a system and has identified the systems administrators by name and email address, the adversary can now engage in personal information gathering such as social media scraping.

Social media scraping techniques also leverage tools and involve collecting information about individuals from various social media platforms (such as Facebook, X, and Instagram). Since gaining access to a victim's assets is central to the adversary's motives, perpetrators often look for things like:

- Dates of birth
- Personal information such as old addresses, former pets, spouse names, places of marriage (These are data points that could possibly help a perpetrator guess the responses to secret questions used to secure an account when a password is forgotten.)
- Discussions about one's place of employment
- Intellectual property
- Vacation pictures (signals to a perpetrator people may not be at home)

Once such information is gathered, the victim can be specifically targeted. This targeting can result in victims receiving various forms of communications that appear to come from someone who either knows the victim or had enough information to ensure the victim responds to the perpetrator.

Communications with victims can assume a variety of forms. These can include telephone calls, which are referred to as vishing. Or smishing, which involves victims receiving text messages. Pharming is like phishing; however, it involves the execution of malicious code on a victim's machine to direct the victim's network traffic to a perpetrator's web site.

Other phishing campaigns can involve criminal elements, often executed by international criminal actors. These campaigns tend to focus on theft and begin with an initial bombardment of phishing emails targeting a series of email addresses. This initial first wave of phishing emails is often crudely worded and obvious to a sharp-eyed victim.

Invariably someone does eventually click on the link provided in the phishing email, thus finding themselves redirected to the perpetrator's website where the victim machine is infected with a malicious payload.

Another approach using the initial bombardment of phishing emails involves botnets.[4] Again, the initial email sent in mass quantities to potential victims is an obvious phish to the keen eye. However, victims that do interact with the phish are reported to the perpetrator. This allows the perpetrator to target individuals specifically, increasing the chances the individual will yield to the demands of the perpetrator.

1.2   Phishing anatomy

Understanding the nature of the phish, i.e., the anatomy, is key to understanding the perpetrator, and potential targets said perpetrator may focus on. To describe the anatomy of a phish, and develop a policy framework around mitigation strategies, a comprehensive definition of phishing is needed.

Since phishing activities can vary significantly, to date, numerous researchers, and organizations (government and commercial alike) have presented the cybersecurity community with various definitions, most capturing aspects of phishing, but lacking in comprehensiveness. While this may initially seem like a trivial oversight, detailing that the full scope of phishing potential unnecessarily narrows the discussion around phishing, and thus potentially limits effective responses.

In a 2021 reported study conducted by Alkhalil, Hewage, Nawaf, and Khan from the Cardiff School of Technologies at the Cardiff Metropolitan University, a new phishing anatomy was explored.

This phishing anatomy differed from previous anatomies in that it is not limited to the phishing mechanisms and corresponding countermeasures. In this study conducted by the authors [1], the phishing anatomy was investigated relative to:

- Attack phases
- Attacker types

---

[4]A network of computers infected with malicious software and controlled as a group by a perpetrator (bot-herder) and often without the computer owner's knowledge. Among other nefarious activities, botnets can be used for phishing campaigns.

- Vulnerabilities
- Threats
- Targets
- Attack mediums
- Attack techniques

In addition to offering an expanded evaluation of the attack phases, Alkhalil *et al.* [1] offers an understanding around the phishing process life cycle. To wit, Alkhalil *et al.* [1] define phishing as:

> A socio-technical attack in which the attacker targets specific valuables by exploiting an existing vulnerability to pass a specific threat via a selected medium into the victim's system, utilising social engineering tricks or some other techniques to convince the victim into taking a specific action that causes various types of damages.

As with phishing definitions, the actual attack phases are described differently by various groups. While the phishing definition provided by Alkhalil *et al.* [1] is comprehensive and can be applied to nearly all phishing attacks, formulating an overarching framework relative to attack phases that is equally applicable to all phishing attacks is trickier.

To that end, the MTIRE Corporation developed a framework [2] consisting of overarching attack categories with each containing attack techniques. The information gathered to generate this comprehensive framework was derived from globally available, open sources and consists of observed in "the wild" tactics and techniques. The resultant product(s) of this effort consists of matrices, which in turn drills down into the specific details of an attack technique along with potential defensive techniques.

For the purposes of this book, we will use the Enterprise Matrix of the MITRE ATT&CK framework when discussing and referring to the anatomy of a phish (attack.mitre.org). In this framework, 14 phases of an attack are described. The attack phases are not necessarily in a prescribed order relative to an attack.

1.2.1 MITRE ATT&CK: reconnaissance

While any one of the reconnaissance techniques identified in the MITRE ATT&CK Enterprise framework could be included in a phishing attack, the ATT&CK framework addresses a specific technique as Phishing for Information. The sub-techniques attributed to the Phishing for Information Technique include:

- Spearphishing service: a spearphishing communications sent via a third party and designed to deceive users into divulging sensitive information.
- Spearphishing message: a spearphishing communications which includes a malicious attachment. The attachment can involve requesting sensitive information from a user, information that can be used in targeting.
- Spearphishing link: a spearphishing communications can include a malicious link. This may direct the user to a website (often referred to as "Watering Holes) that captures sensitive information to include account log in information.

1.2.2   MITRE ATT&CK: resource development

Resource development as understood under the MITRE ATT&CK framework involves gathering resources that will aid an attacker in executing an attack. The focus of these activities can include (but not limited to) purchasing domains, gathering email accounts, or acquiring code signing certificates.

Under this set of techniques, phishing is mostly identified with the Compromise Accounts technique. While phishing can certainly be a technique used in during the execution of other techniques found during this phase, phishers are often interested in gaining access to legitimate accounts. Gaining access to legitimate accounts allows an attacker to evade defensive controls. The sub-techniques found under the Compromise Accounts technique include:

• Compromising social media accounts
• Compromising email accounts
• Compromising Cloud accounts

Compromising accounts may include not only attacks against the actual target person but also friends or contacts of the primary target.

1.2.3   MITRE ATT&CK: initial access

In the first quarter of 2022 phishing attacks designed to grant initial access to a perpetrator jumped by 54 percent [3]. Although mostly phishing attacks focused on Initial Access to gain an ongoing foothold in a system, in 2022 the sharp increase was attributed to activities associated with the Emotet and IceID malware[5]

The types of phishing attacks that the MITRE ATT&CK framework identifies in the Initial Access phase are:

• Spearphishing with attachment
• Spearphishing link
• Spearphishing via service

1.2.4   MITRE ATT&CK: execution

The Execution phase characterizes those actions associated with perpetrator-controlled code or actions executing on a victimized system. These actions can vary widely, depending upon the perpetrator's objectives. In the case of an advanced adversary that seeks to hold a system at risk over a lengthy period, the Execution phase could include additional reconnaissance and account compromise. The Execution phase can also involve setting up infrastructure to support an adversary's command and control.

The phishing angle during the Execution phase can be found under the User Execution technique. The MITRE ATT&CK framework identifies activities in this phase as follow-on activities from a phishing attack. These activities can include a perpetrator compelling a user to execute specific actions that facilitate the attack

---

[5]Emotet and IceID are banking trojans designed to steal sensitive and private banking information.

(malicious code execution for example) or opening a malicious file or link. The phishing sub-techniques associated with the User Execution technique include:

- User executes on a malicious link
- User executes a malicious file
- User executes a malicious image

1.2.5   MITRE ATT&CK: persistence

Adversaries, in particular advanced adversaries, often execute techniques that afford them a persistent presence in a system. Persistence can become a significant problem for an organization. On average, perpetrators remain undetected in a system for up to 277 days [4]. It takes on average another 75 days to contain the breach [4]. The average cost of a data breach of less than 200 days amounts to $3.74 million with an average cost of $4.86 million for breaches remaining undetected for over 200 days [4].

The Persistence techniques identified in the MITRE ATT&CK framework that are associated with phishing relate directly to the accounts compromised during a phishing attack and are referred to in the framework as Account Manipulation. Account Manipulation includes the following sub-techniques:

- Additional Cloud credentials
- Additional Email Delegate Permissions
- Additional Cloud roles
- Secure Shell (SSH) Authorized Keys
- Device Registration

In each of these instances, an adversary expands the reach of a compromised account to include account features controlled by the adversary.

1.2.6   MITRE ATT&CK: privilege escalation

Within the context of privilege escalation, the exploitation of valid accounts is often associated with phishing attacks. In many respects, the very intention of the phishing attack could be understood as an attempt to assume control over valid accounts, thus gaining access to the target systems and avoiding detection by defensive protections.

Accounts targeted by phishing attacks include:

- Default accounts
- Domain accounts
- Local accounts
- Cloud accounts

Once an attacker gains access to an account on the system, to include accounts with limited privileges, attackers can engage in privilege escalation activities. The primary objective of most privilege escalation activities is to gain administrative control (root) over the system. Presumably, gaining administrative control over the system will enable an attacker to leverage trusted relationships between systems, and pivot from one system to the other within a network, enjoying privileged access on each system.

Privilege escalation is often achieved through configuration and/or software flaws on the system. These flaws are exploited by an adversary, even if the exploit is being driven from a compromised account with lessor privileges.

The very nature of privilege escalation enables an attacker to mature the attack. Since an attacker who has escalated privileges is now recognized by the system as a privileged user, the attacker can evade system defenses, compromise additional credentials, move laterally through the system and ultimately collect sensitive information, develop a command-and-control structure, exfiltrate information, and inflict damage on the system.

1.3   Phishing mitigation attempts

In response to the ongoing and prevalent phishing attacks, organizations have responded with a wide variety of management controls, training programs, and technologies.

Management controls are involved with the implementation of policies, standards, procedures, and processes. These represent many layers of controls to include controls implementing laws, acts, and rules.

Standards play a particularly significant role in managing cybersecurity in that standards prescribe how information systems are used, operated, and maintained. Often cybersecurity policies will reference a standard for implementation across an organization to ensure organizational uniformity. Among the standards created for these purposes is the ISO/IE 1799, which incorporates the British BS7799.

Even though organizational cybersecurity policies represent multiple layers of control, organizations often struggle with user compliance with organizational policies, and by extension compliance with the implementation of cybersecurity controls.

To that end, in 2018, 35 percent of Chief Cyber Security Officers declared security education, training and awareness (SETA) as the highest priority expenditure for their respective organizations [5]. While the focus on SETA has continuously increased, at least since 2009, several studies have found that populations in receipt of formal cybersecurity training are often less likely to correctly identify a phishing email versus a legitimate email [5]. Additionally, studies conducted by Pattison *et al.* [6], Merz [7], and Merz *et al.* [8] found that individuals who received formal cybersecurity training were less likely to maintain an awareness of cybersecurity risks.

To understand the underlying causes of such findings, several studies since 2009 have focused on user behaviors and the underlying psychology driving these behaviors.

In a 2012 study, Vance *et al.* [9] combined the Protection Motivation Theory (PMT), and the Theory of Habit into the areas of Habits, Cognitive Mediation, and Coping Mode to explain user's lack of compliance with security policies.

This study found that user's habits toward compliance with security policies affected all six factors of the PMT. The results of this study indicated that user habits played a significant role in compliance with security policies [9].

The conclusions drawn by Vance *et al.* included a theory that increased user training and awareness should yield a higher rate of compliance with cybersecurity policies.

However, despite having numerous mandatory cybersecurity training programs throughout the U.S. Federal Government for example, ("FISMA," 44 U.S.C. § 3541, et seq.), cybersecurity training in its current form appears to be ineffective [5].

In the case of the U.S. Federal Government users, an online training program is provided annually. This training is compulsory for all U.S. Federal Government personnel, to include active-duty military, government civilians and contractors alike. While this course is interactive and has been provided to Federal users for many years, phishing attacks have not been reduced. These data points beg the question, if traditional training methods are not working, what would work?

In a 2011 study, Jansson and von Solms [10] found that experiential training coupled with embedded training positively affected user behavior relative to phishing attacks (in other words, users became more resistant to phishing attacks).

However, questions remained around the formation of cybersecurity habits of experiential training over conventional training. Additional questions around the formation of cybersecurity habits of experiential training over conventional training remained around the degree to which habits were formed by experiential training.

Among the first researchers to address developing security training and awareness programs were Thomson and von Solms. These researchers integrated elements from social psychology, into training programs, thus creating a holistic training development approach [11].

Another early group of training researchers was Dodge *et al.* [12], who in 2007 addressed the obfuscating nature of phishing attacks and the need to train users relative to these types of attacks. The approach these researchers suggested involved using un-announced training exercises to evaluate a user's susceptibility to phishing attacks.

In 2010, Puhakainen and Siponen [13] tied the efficacy of security training to user compliance with information security policies. These researchers sought to create and validate a theory-based training program based on the universal constructive instructional theory and the elaboration likelihood model, which they validated through an action research study.

The results of their study demonstrated that information security training programs should motivate students to a systematic cognitive processing of the material, and that continuous communication processes were required to improve the student's compliance with information security policies [13].

However, despite having a wide selection of computer security training options, phishing attacks continued to successfully exploit systems to this day. In fact, the Federal Bureau of Investigation's 2022 Internet Crime report documented 300,497 reported phishing attacks in the United States alone for 2022. Additionally, phishing remained by far the most frequently used attack with the next most frequent attack for 2022 being Personal Data Breaches coming in at 58,859. In other words, phishing attacks occurred 5.1 times more often than then the next highest reported cyberattack [14].

With traditional training failing, researchers began looking at a variety of training approaches. Experiential training using phishing tools offered one approach, giving users a direct experience of a phishing attack.

However, Ashenden [15] found if users had concerns around the disposition of results coming from experiential training programs, users might not derive a true learning experience from such an approach.

In all, whether security training and awareness is effective or not the topic remains a point of significant debate to this day.

As mentioned earlier, organizational policy responses to phishing attacks have included compliance requirements with various standards.

Within the United States, Federal and many states require compliance with the National Institute of Standards and Technologies Special Publication series 800. This series of standards addresses not only processes and procedures for implantation but also technological solutions (such as multi-factor authentication in the NIST SP 800-63 Digital Identity Guidelines, 2022).

The technological countermeasures designed to resist phishing attacks were developed to address semantic attacks. Phishing belongs to this attack family [16].

Semantic attacks circumvent the security of a computer system leveraging user facing interfaces and deceive users into performing an action that breaches the computer's security controls [16]. A variety of sematic attacks belonging to the Phishing Attack Family can be mitigated technologically. Among these are Bluetooth phishing (Snarfing), Instant Message phishing, NFC phishing, phishing websites, spam phishing (botnet-generated), spear phishing, and spear phishing with advanced persistent threats.

Earlier technology solutions (which are often in continuing use today) involve various filtering schemas.

These earlier techniques involved email filtering. One approach involves content-based phishing filters, which review the content, and structural properties of an email [17]. In a 2010 study, Bergholz *et al.* crafted email filters that incorporated statistical models for email topics. These models included sequential analysis of email text and external links, a detection mechanism for embedded logos and hidden "salting."

In a 2015 study, Laszka *et al.* tailored email filters modeled against the Stackelberg's Security game. Spear phishing attacks were used to test this approach as spear phishing attacks are targeted attacks, and better suited for the Stackelberg's Security game [18]. At the end of the experiment, the researchers found that modeling an attacker's approach was not difficult. However, modeling the defender's best strategy was considerably more challenging [18].

Since phishing is often associated with email, and as such, mitigation solutions are often designed accordingly. However, these solutions are frequently not suited for mitigating attacks such as Domain Name Service (DNS) Poisoning [19].

When someone navigates to a website using a poisoned DNS server, the user's web browser is directed to a perpetrators phishing website. One schema tested by

Kim and Huh in 2011 attempted to train technological classifiers with a heuristic model, which used historical traffic data between a client and website to profile the connection path's traffic [19].

Whitelisting and blacklisting is another type of phishing technology-based mitigation approach. This approach is designed to alert users to malicious sites [20]. In 2014, Li *et al.* conducted a use-ability study of whitelisting and blacklisting tool bars. These tool bars were designed to alert users as to whether a website was authentic or spoofed. The researchers discovered the performance of various tools did not vary to a significant degree. However, they did find that important information was missing from the toolbars.

Website logos can function as a countermeasure in that these help users recognize whether a website is authentic or malicious [21]. This approach uses a two-step process involving a machine learning (ML) technique designed to analyze an extracted website logo image. It validates the logo against Google's image search to retrieve the associated domain name [21].

An area that began gaining interest involves systems that adapt and "learn" [22]. As opposed to static black-listing or heuristic techniques, neural networks and fuzzy systems can be merged to provide a neuro-fuzzy model. This model calculates the value of heuristics from membership functions training the system through the neural network using adaptive learning capabilities and weighted values. This approach detected 99 percent of the phishing sites used during a study [22].

Individual users also have technology options for phishing protection. Among these are cloud-based spam filters or high-grade internal spam filters. Regular maintenance can also support user protections such as maintaining current versions of web browsers or varying email addresses. A common technique includes using a disposable email address.

Biometrics can also play a role in thwarting phishing attacks. However, biometrics is vulnerable to two types of attacks: sensor bypass and sensor spoofing. Each attack can be executed with relative ease. Additionally, biometrics may introduce a host of ethical concerns as spoofing biometrics is a form of identity theft. Requiring biometrics as a function of organizational policy could expose users to phishing attacks, as well as identity theft involving their biometric data [23].

ML and neural networks represent the most recent technological advances in countermeasures toward phishing. With many phishing websites have a short-lived online presence, traditional methods of avoiding known phishing websites such as blacklisting are rendered useless. By relying on phishing identification algorithms, using ML and neural networks, phishing websites are more readily identified [24].

In summary, the human factor remains the most persistent vulnerability relative to phishing attacks, regardless of the technological solutions. This is particularly true when sophisticated technological countermeasures are deployed. There is a marked increase in user targeting when technology countermeasures are matured. With users accessing the World Wide Web using the application layer, legacy countermeasures such as firewalls are bypassed. This bypassing effect places the user directly on the frontline of cybersecurity attacks.

As the limits of technological solutions are continuously identified, a stronger focus on factors influencing the user's vulnerabilities has been taking shape for several years. A marked research interest in user psychology has emerged since 2009 and continues to this day.

However, in numerous studies, to include two studies specifically focused on phishing, the user's psychology only accounts for the ultimate behavior to 28 percent to 40 percent of the time [7].

If a user's psychology can only explain roughly 40 percent of their decision-making process, what is happening with users during that unexplained 60 percent of the time?

Case: spear phishing attacks against Crimean activists

In a series of three case studies conducted between April 2019 and March 2020, researchers from The Engine Room [25], a non-profit organization studying global social justice issues, investigated digital attacks against human rights groups.

In an interview with The Engine Room researchers, the DSL in Ukraine discussed how in December 2019 four groups of activists became targets of a spear phishing campaign.

In this case, the spear phish sent to the group targeted individuals interested in music. The spear phish masqueraded as a communication from a music group and was requesting the activists submit articles for a music column.

By February 2020, the attack was well underway, with over 15 individuals receiving spear phishing emails.

DSL further reported that once this group discovered it had been spear phished, it was unable to discern the malicious email from an authentic email due to the personalized nature of the malicious emails.

While DSL was unable to make a 100 percent attribution of the origins of the spear phishing emails, DSL suggested the most likely source was local law enforcement. These conclusions were drawn based upon a set of targets that had local law enforcement as a common concern, IP addresses used to send the emails, data used, and features of the emails.

The likely purpose of the spear phishing campaign was to sow discord and distrust among the activists' groups. For example, as the activists became aware of the spear phishing attack, uncertainty evolved around who had been compromised within the group. When such uncertainty in a trusted group arises, it can breed distrust, which in turn results in a cohesive group becoming fractured.

To assist the group in maintaining its cohesion, particularly with partners in remote areas requiring digital communications, DSL published the discovery of the attacks such that remote groups coordinating with the victim group could strategize on developing a trusted digital communications approach.

1.4   How phishing is evolving in the world of AI

The realm of cybersecurity is witnessing a significant paradigm shift, ushered in by the rapid advancements in artificial intelligence (AI). A particularly concerning aspect of this shift is the increased sophistication of AI-generated phishing attacks. These have significantly altered the landscape of digital communication and cybersecurity. The evolution from rudimentary text-based scams to sophisticated AI-generated content poses unique challenges and opportunities in the realms of cybersecurity, digital forensics, and AI ethics.

1.4.1   Technological advancements

At the heart of this transformation is the incredible progress made in the realm of generative AI technologies. Gone are the days of simple text-based scams; today's AI systems are adept at crafting emails that are not only grammatically precise but also contextually tailored to their targets. For instance, modern AI can effortlessly mimic the tone and style of legitimate corporate messages, making these fraudulent communications alarmingly convincing [26].

Central to this progress in generative AI lies natural language processing (NLP) and ML, which enables these technologies to create AI systems capable of generating contextually relevant, grammatically accurate, and stylistically convincing text. These systems leverage vast datasets and complex algorithms to understand and mimic human-like writing styles, making them potent tools for both legitimate and malicious uses. The challenges facing cybersecurity professionals include (but are not limited to) the following:

Advanced natural language processing (NLP): Modern AI models, especially those utilizing deep learning approaches like transformers (e.g., GPT-3), have revolutionized NLP. These models are trained on vast datasets comprising diverse text sources, enabling them to learn intricate patterns of human language and replicate them with high accuracy [27].

- Context-awareness: These AI models are designed to understand and generate text based on context. They analyze preceding text to maintain thematic and stylistic consistency, allowing them to create messages that are contextually aligned with the target's expectations or existing communication threads.
- Personalization techniques: By leveraging data about the target, such as past interactions, publicly available information, and typical communication patterns, AI models can tailor messages to individuals' specific characteristics. This personalization makes the messages more convincing and increases the likelihood of a successful phishing attacks.
- Grammatical precision: Traditional phishing detection often relies on identifying poor grammar and spelling as indicators of illegitimacy. AI-generated texts, however, exhibit high grammatical accuracy, making them indistinguishable from legitimate communications on this basis.
- Stylistic mimicry: AI models can mimic writing styles specific to individuals or organizations. This capability makes it challenging for recipients to distinguish

AI-generated emails from those written by legitimate sources. The stylistic mimicry extends to tone, formatting, and vocabulary, further complicating detection.

- Content sophistication: AI models can generate contextually rich and sophisticated content that aligns closely with the target's interests or concerns. Such content sophistication helps evade filters that screen for simplistic or generic phishing indicators.

1.4.2   Case studies and examples

A compelling instance of this was an organization tricked by emails purportedly from a senior executive, which were, in fact, generated by an AI system. This AI, having analyzed previous legitimate emails, crafted messages with specific project references and similar language patterns, enhancing their credibility. Statistically, there's been a 30 percent uptick in such AI-generated phishing attempts over the past year, further emphasizing the urgent need for robust countermeasures [28].

The following is an example of an AI-generated phishing email created by ChatGPT4:

Subject: Urgent: Update Your Account Information

Dear [User's Name],

I hope this message finds you well. I'm reaching out from the [User's Bank Name] security team. We've recently upgraded our online banking system to enhance your banking experience. As part of this upgrade, we require all our esteemed customers to update their account information to ensure uninterrupted service.

To update your information, please click on the link below and follow the prompt instructions:

[Hyperlinked text that appears legitimate but leads to a fraudulent site]

Your security is our top priority. It's essential to update your information within 24 hours to avoid temporary suspension of your online banking services. We appreciate your cooperation and understanding in this matter.

If you have any questions or need assistance, please contact our support team at [Fraudulent Customer Support Number].

Thank you for choosing [User's Bank Name].

Warm regards,
[AI-generated Name]
Customer Service Team
[User's Bank Name]

Key features of this phishing email that are often linked to phishing success include:

Urgency and call to action: The email creates a sense of urgency, prompting the user to act quickly, often bypassing rational judgment.

Personalization: AI enables the inclusion of specific user details (like the user's name and bank), making the email appear more legitimate.

Legitimate-looking link: Phishers often use hyperlinked text that appears genuine but redirects to a fraudulent site designed to capture user information.

Authority impersonation: The email impersonates a reputable authority (the bank's security team) to gain the user's trust.

Plausible scenario: The pretext of a system upgrade is a common and believable scenario that many users wouldn't question.

Pressure tactics: Mentioning account suspension creates a fear of losing access, pressuring the user into taking hasty actions.

1.4.3 Exploitation of AI-generated images and audio in phishing

Deepfake technology, manipulating audio and video to create realistic imitations of real people, has become another tool in the phishing arsenal. This technology, powered by advanced ML algorithms like generative adversarial networks (GANs), produces audio and video content that can deceive even the most discerning eyes and ears. A notorious example is the $35 million heist executed using deepfake audio of a corporate director [28]. The technical underpinnings of Deepfake Technology include:

GANs: Central to deepfake technology are GANs, introduced by Goodfellow *et al.* A GAN consists of two neural networks: the generator, which creates images, and the discriminator, which evaluates them. These networks are trained in opposition to each other, with the generator striving to produce increasingly realistic images and the discriminator aiming to better distinguish between real and fake images. This adversarial process results in highly convincing fakes [29].

Autoencoders and encoders in deepfake creation: Autoencoders are also crucial in deepfake technology. They are neural networks designed to encode an input into a lower-dimensional representation and then decode it back into the original format. In deepfakes, they're used to compress and reconstruct facial images, allowing for the swapping of faces between individuals in videos [30].

Voice synthesis and cloning: Beyond images, AI has also made significant strides in audio manipulation. Voice synthesis technologies, including Tacotron by Google, enable the generation of highly realistic synthetic speech. Recent advancements have made it possible to clone an individual's voice with just a few seconds of sample audio, leading to convincing audio deepfakes [31].

The growth of deepfake material has been exponential, with the number of deepfakes online doubling every six months. According to DeepMedia, in 2023, roughly 500,000 video and voice deepfakes were shared on social media around the world. By 2025, we can expect to see 8 million deepfakes shared online, consistent with doubling deepfakes every six months. The ease of access to powerful AI tools and the large quantity of publicly available data contribute to the spread of deepfakes.

Social media platforms, where the distribution of information is rapid and widespread, play a crucial role in the rise of deepfakes. A study conducted by iProov found that in 2022, less than one-third of global consumers knew what a deepfake was. This highlights the potential for misinformation to spread rapidly through these platforms.

The rise of deepfakes poses significant challenges to various parts of society, from politics to business. Here are some key ways in which deepfakes are already impacting our lives.

- Deepfakes have already disrupted political landscapes by showing political figures saying or doing things they never did. This can lead to misinformation, influence elections, and undermine public trust in political institutions.
- Deepfakes can be used to create compromising media without an individual's consent, leading to severe privacy violations. Faces may be imposed into inappropriate content, causing emotional distress and reputational damage.
- As deepfakes become more convincing, there is a growing risk of losing trust in digital content. People may become skeptical of the authenticity of any video or image, leading to a general atmosphere of doubt. This can have major implications in high-stakes industries like law enforcement and justice, where evidential integrity is paramount.
- Organizations are often targeted by deepfakes, which can be used to create fraudulent content, scams, and other activities that compromise their privacy and trust. This can lead to financial losses and damage to the organization's reputation.

1.4.4 Exploitation in phishing: techniques and examples

Phishing with deepfake video: The most direct application of deepfake technology in phishing is the creation of video clips where trusted individuals appear to say or do things that are entirely fabricated. This method can be used to manipulate stock prices, sway public opinion, or even incite geopolitical conflict [32].

Deepfake audio in corporate espionage: A notable example of deepfake audio used in phishing is the $35 million heist, where fraudsters used AI-generated audio to mimic the voice of a corporate director, instructing a subordinate to transfer funds. The employee, believing the request to be legitimate, complied, resulting in a substantial financial loss [33].

Combining deepfakes with traditional phishing tactics: Deepfakes can be combined with more conventional phishing tactics for more sophisticated attacks. For example, an email purportedly from a senior executive, accompanied by a deepfake video clip, can be more persuasive than text alone, increasing the likelihood of the recipient taking the desired action.

In a high-profile case, scammers used deepfake video technology to impersonate a company's CFO during a video call, convincing an employee to transfer $25.6 million. The deepfake was so convincing that the employee did not suspect any foul play until it was too late. This incident illustrates the potential for deepfake technology to be used in high-stakes financial fraud, where the realism of the deepfake can override the usual skepticism of the victim.

Another example involved a UK-based energy company where fraudsters used deepfake audio to mimic the voice of the CEO, resulting in a $243,000 loss. The attackers called the company's CEO, pretending to be the CEO of the parent company, and demanded an urgent wire transfer, which was initially complied with before suspicions arose. This case highlights the vulnerability of voice communications to deepfake technology and the significant financial impact that such attacks can have.

LastPass reported an attempted deepfake phishing attack where an employee received a series of calls and messages featuring an audio deepfake of the company's CEO. The employee's suspicion and adherence to security protocols prevented any financial loss, but the incident underscores the growing threat of deepfake phishing. The use of deepfake audio in this case demonstrates how attackers can exploit the trust placed in familiar voices to attempt to bypass security measures.

In a politically charged environment, deepfake videos have been used to create false narratives about candidates, potentially swaying voter opinions and election outcomes. For example, a deepfake video of a candidate making controversial statements can be released close to an election, causing significant damage to their campaign. The ability of deepfake technology to create realistic but entirely fabricated content poses a significant threat to the integrity of democratic processes.

Financial institutions are particularly vulnerable to deepfake phishing attacks due to the high value of transactions involved. Attackers can use deepfake audio to impersonate senior executives and authorize large transfers, exploiting the trust placed in voice communications within the industry. The potential for deepfake technology to be used in this way underscores the need for robust security measures and vigilance in financial transactions.

By understanding the techniques and examples of deepfake phishing, organizations can better prepare and protect themselves against this evolving cyber threat. The combination of technological solutions, employee vigilance, and robust security protocols can help mitigate the risks associated with deepfake phishing and safeguard against potential financial and reputational damage.

1.4.5   Challenges in detection and prevention

Detecting deepfakes: Detecting deepfakes remains challenging. Traditional digital forensics techniques, which look for anomalies in images or videos, are increasingly insufficient as deepfake technology evolves. ML models are being developed to detect deepfakes, but they often engage in a cat-and-mouse game, with detection capabilities constantly trying to catch up with deepfake generation technologies [34]. These models are trained on large datasets of both real and fake images or videos to learn the subtle differences between them. Convolutional neural networks (CNNs) and recurrent neural networks (RNNs) are commonly used architectures in this domain. CNNs are particularly effective at analyzing spatial features in images, while RNNs can capture temporal dependencies in video sequences.

However, the detection of deepfakes often turns into a cat-and-mouse game. As detection capabilities improve, so do the techniques for generating more convincing deepfakes. This ongoing battle necessitates continuous updates and improvements to detection algorithms. For instance, the use of GANs in creating deepfakes has led to the development of adversarial training techniques, where detection models are trained alongside generative models to improve their robustness.

One of the primary challenges in deepfake detection is the generalization capability of the models. Many detection models are trained on specific types of deepfakes and may struggle to detect those generated using different techniques. This lack of generalization can be problematic when new deepfake methods emerge. Researchers are exploring hybrid models that combine the strengths of different architectures, such as CNNs and vision transformers (ViTs), to improve generalization. These models leverage the local–global attention mechanisms of ViTs and the feature extraction capabilities of CNNs to detect a wider range of deepfake manipulations.

Ethical and legal considerations: The rise of deepfakes has triggered ethical and legal debates. Issues include the consent of individuals whose likenesses are used, the potential for defamation, and the broader impacts on trust in digital media. Legal frameworks are struggling to keep pace with the rapid development of this technology [35]. The ethical principle of autonomy dictates that individuals should have control over their own image and how it is used. However, deepfake technology often bypasses this control, leading to unauthorized and potentially harmful representations.

One of the primary ethical concerns with deepfakes is the use of an individual's likeness without their consent. This can lead to severe privacy violations, especially when deepfakes are used to create explicit or defamatory content. The ethical principle of autonomy dictates that individuals should have control over their own image and how it is used. However, deepfake technology often bypasses this control, leading to unauthorized and potentially harmful representations.

Deepfakes have the potential to cause significant reputational harm. For instance, a deepfake video showing a public figure engaging in inappropriate or illegal activities can spread rapidly on social media, leading to public outrage and damage to the individual's reputation. This misuse of deepfakes for defamation and misinformation poses a serious threat to the integrity of information in the digital age.

With that said, while deepfakes are often associated with malicious uses, they also have potential ethical applications. For example, deepfakes can be used in the entertainment industry to create realistic special effects or in education to create engaging learning materials. The key is to ensure that these uses are transparent and consensual, with clear guidelines and regulations to prevent misuse.

Preventive measures and public awareness: Preventing the misuse of deepfakes in phishing requires a combination of technological solutions, legal measures, and public awareness. Educating the public about the existence and capabilities of deepfake technology is crucial in building resilience against such attacks. Additionally, organizations need to implement robust verification processes to confirm the authenticity of seemingly legitimate requests, especially those involving financial transactions or sensitive information.

Technological solutions play a crucial role in detecting and preventing deepfake attacks. Advanced deepfake detection tools, such as those using ML and AI, can help identify manipulated content. Additionally, blockchain technology can be used to verify the authenticity of digital content, creating a tamper-proof record of its origin and modifications.

Collaboration between governments, technology companies, and academia is essential to develop effective solutions for deepfake detection and prevention. Research initiatives, such as the Deepfake Detection Challenge, bring together experts from various fields to address the technical and ethical challenges posed by deepfakes. By sharing knowledge and resources, these collaborations can accelerate the development of robust detection technologies and create a more secure digital environment.

1.4.6   Social engineering

In the realm of social engineering, AI-generated images are now used to fabricate credible social media profiles. These profiles, often adorned with AI-generated faces, can bypass traditional detection methods like reverse image searches. A cybersecurity firm's discovery of an extensive network of such LinkedIn profiles, connected to over a thousand real professionals, showcases the frightening effectiveness of AI in social engineering [28].

A recent investigation by the Stanford Internet Observatory uncovered a network of over 1,000 LinkedIn profiles that appeared to be using AI-generated profile pictures. These profiles were connected to thousands of real professionals, raising concerns about the potential misuse of such technology for social engineering purposes.

The researchers discovered that many of these profiles exhibited telltale signs of being AI-generated such as perfectly centered eye alignment and minor inconsistencies in facial features. However, these subtle anomalies are often difficult for the average user to detect, making it easier for malicious actors to create convincing fake personas.

One of the key advantages of using AI-generated images for social media profiles is their ability to bypass traditional detection methods like reverse image searches. Unlike traditional profile pictures, which can be traced back to their original sources, AI-generated images are unique and cannot be easily linked to existing online content.

This makes it challenging for social media platforms and users to identify and flag these profiles as potentially fraudulent, allowing cybercriminals to operate more freely and gain the trust of their targets. Cybercriminals can leverage AI-generated profiles in various ways to execute social engineering attacks. Some common tactics include:

- Impersonation: AI-generated profiles can be used to impersonate real individuals, organizations, or authorities, lending credibility to phishing attempts or other scams.
- Targeted attacks: By creating profiles tailored to specific industries or organizations, cybercriminals can gain access to closed groups or networks, enabling them to gather sensitive information or establish trust with potential targets.
- Spreading misinformation: AI-generated profiles can be used to disseminate false or misleading information, contributing to the spread of disinformation campaigns and eroding trust in online sources.

• Reconnaissance: Cybercriminals can use AI-generated profiles to gather intelligence on individuals or organizations, collecting data that can be used for future attacks or extortion attempts.

1.4.7 Legal and ethical implications

The rise of AI-generated phishing opens a Pandora's box of legal and ethical challenges. As AI becomes a potent tool in the hands of cybercriminals, legal systems worldwide grapple with the nuances of holding AI creators accountable. The ethical considerations are equally daunting, raising questions about the responsible use of AI and the thin line between technological advancement and misuse [36].

One of the primary legal challenges posed by AI-generated phishing is determining accountability and liability. Traditional legal frameworks are often ill-equipped to address the complexities introduced by AI. In cases of AI-generated phishing, it can be difficult to pinpoint who is responsible: the developers of the AI, the users who deploy it maliciously, or the organizations that fail to implement adequate security measures.

In the European Union, the AI Act aims to address some of these issues by categorizing AI systems based on their risk levels and imposing stringent requirements on high-risk applications. The Act mandates rigorous testing, monitoring, and transparency measures to ensure AI systems are safe and reliable. However, the Act also raises questions about the extent to which AI developers can be held liable for the misuse of their technologies by third parties.

In the United States, the approach is less prescriptive. The Executive Order on AI issued by President Biden emphasizes principles of safety, innovation, and ethical considerations but lacks specific legal requirements and enforcement mechanisms. This not only creates a more flexible regulatory environment but also leaves gaps in accountability, particularly when it comes to AI-generated phishing.

AI-generated phishing often involves the use of personal data to create convincing and targeted attacks. This raises significant privacy and data protection concerns. In the EU, the General Data Protection Regulation (GDPR) provides a robust framework for protecting personal data, and the AI Act integrates these principles to ensure AI systems respect individuals' privacy. The GDPR imposes strict requirements on data processing, including obtaining explicit consent and ensuring data minimization and security.

In contrast, the United States lacks a comprehensive federal privacy law, relying instead on a patchwork of state laws and sector-specific regulations. This fragmented approach can make it challenging to enforce consistent data protection standards across the country. However, there is growing pressure from industry stakeholders and consumer advocacy groups for the federal government to adopt more comprehensive privacy legislation.

The global nature of AI-generated phishing complicates enforcement efforts. Cybercriminals can operate from anywhere in the world, making it difficult for national authorities to investigate and prosecute these crimes. International cooperation is essential to address this challenge, but differences in legal frameworks and enforcement capabilities can hinder effective collaboration.

The EU's AI Act seeks to establish a uniform regulatory framework across member states, reducing fragmentation and facilitating cross-border enforcement. The Act also emphasizes the need for international dialogue and cooperation on AI governance, recognizing the importance of a coordinated global response to AI-related threats.

In the United States, the lack of a unified federal approach to AI regulation can lead to inconsistencies in enforcement. However, federal agencies such as the Federal Trade Commission (FTC) and the Department of Justice are increasingly focusing on AI-related issues, including privacy violations and discriminatory practices. These agencies are working to develop guidelines and enforcement strategies to address the unique challenges posed by AI-generated phishing and other AI-enabled cybercrimes.

The ethical considerations surrounding AI-generated phishing are multifaceted. One of the key issues is the responsible development and deployment of AI technologies. Developers must ensure that their AI systems are designed with robust security measures to prevent misuse. This includes implementing safeguards to detect and mitigate malicious activities, such as phishing attacks.

The EU's AI Act emphasizes the importance of ethical AI development, requiring developers to adhere to principles of transparency, accountability, and human oversight. The Act also mandates the use of high-quality datasets and rigorous testing to ensure AI systems are accurate and reliable. These measures aim to prevent the misuse of AI technologies and promote trust in AI systems.

In the United States, the focus is on encouraging voluntary compliance with ethical guidelines and industry standards. The Executive Order on AI promotes the development of safe and trustworthy AI technologies but does not impose specific legal requirements. This approach relies on the commitment of industry stakeholders to self-regulate and adopt best practices for ethical AI development.

Another ethical challenge is finding the right balance between fostering innovation and implementing effective regulations. Overly stringent regulations can stifle innovation and hinder the development of beneficial AI technologies. On the other hand, insufficient regulation can lead to the proliferation of harmful AI applications such as AI-generated phishing.

The EU's AI Act seeks to strike this balance by adopting a risk-based approach to AI regulation. The Act categorizes AI systems based on their risk levels and imposes varying degrees of regulation accordingly. High-risk applications are subject to strict requirements, while low-risk applications face minimal regulatory burdens. This approach aims to protect public safety and fundamental rights without unduly hindering innovation.

In the United States, the emphasis is on creating a flexible regulatory environment that encourages innovation while addressing potential risks. The Executive Order on AI promotes the development of industry-led standards and best practices, allowing for a more adaptive and responsive regulatory framework. However, this approach also relies heavily on the willingness of industry stakeholders to prioritize ethical considerations and self-regulate effectively.

Raising public awareness about the risks and capabilities of AI-generated phishing is crucial for building resilience against such attacks. Educating individuals and organizations about the signs of phishing and the importance of verifying the authenticity of digital communications can help mitigate the impact of these attacks.

Both the EU and the United States recognize the importance of public awareness and education in addressing the challenges posed by AI-generated phishing. The EU's AI Act includes provisions for promoting transparency and informing users about the capabilities and limitations of AI systems. The Act also encourages the development of educational programs and public awareness campaigns to help individuals recognize and respond to AI-generated threats. In the United States, federal agencies such as the FTC and the Cybersecurity and Infrastructure Security Agency (CISA) are actively involved in raising public awareness about AI-related risks. These agencies provide resources and guidance to help individuals and organizations protect themselves from AI-generated phishing and other cyber threats.

1.4.8   Advanced analytical tools for phishing detection

In a twist of fate, the key to combating AI-generated phishing might lie within AI itself. Emerging analytical tools, leveraging the same AI principles used in phishing, are being developed to identify and neutralize these threats. These include AI-driven anomaly detection systems that can spot irregular patterns in communication, flagging potential phishing attempts before they cause harm [26]. Some specific details:

Anomaly detection: Advanced ML models are being developed to detect subtle anomalies in communication patterns such as unusual sentence structures or deviations from an individual's typical communication style.

Behavioral analysis: Some systems focus on analyzing user behavior, such as how they interact with an email, to identify potential phishing attempts. This approach can help spot even well-crafted AI-generated emails.

Continual learning systems: Cybersecurity systems equipped with continual learning capabilities can adapt to evolving AI-generated threats. By continuously learning from new data, these systems stay updated with the latest phishing tactics.

1.4.9   Case studies: successful defense against AI-phishing

There are glimmers of hope in this ongoing battle. For instance, a financial institution successfully thwarted an AI-phishing attack by employing cutting-edge ML-based email filtering systems. These systems, trained on vast datasets of legitimate and phishing emails, were able to detect subtle anomalies in the AI-generated messages, preventing a significant financial loss [28].

1.4.10   Future trends and predictions in AI-phishing

Looking ahead, the landscape of AI-phishing is likely to grow more complex and perilous. Cybercriminals are expected to harness increasingly sophisticated AI models, potentially leading to hyper-personalized phishing attacks. These future threats underscore the necessity for organizations to adopt a proactive stance in their cybersecurity practices, staying abreast of technological advancements and fortifying their defenses accordingly.

1.4.11   Conclusion and future considerations

In conclusion, the advent of AI-generated phishing represents a significant shift in the cybersecurity domain. While technological advancements have brought immense benefits, they also present new, formidable challenges in the form of AI-enabled cyber threats.

Without a doubt, this brief discussion highlights the urgent need for continuous evolution in cybersecurity strategies, emphasizing the critical role of advanced AI-based detection tools and the importance of legal and ethical frameworks to guide the responsible use of AI technologies. As the landscape of cyber threats evolves, so too must our defenses, requiring ongoing vigilance and adaptability in the face of these ever-changing challenges.

1.5   Phishing and QR (quick-response) codes

QR codes are images that have encoded information embedded in them. The image is created such that an electronic device can read the encoded information and ingest the information for processing. The visual interface on scanning devices such as smartphone cameras is designed to quickly scan such an image functioning as an input device.

Information embedded in QR codes can contain links to websites, text, or commands to send emails or SMS messages. Cybercriminals exploit QR codes by embedding malicious links into the image. When scanned, these links can lead to phishing websites or directly install malware on the scanning device.

The compact and opaque nature of QR codes obscures the destination URL, making it difficult for users to discern the legitimacy of the link, thus enhancing the effectiveness of phishing attacks.

As is so often the case with emerging technologies, designing security protections for the vulnerabilities they introduce is often the last thing that is taken into consideration. The expanded use of QR codes is no exception to that trend.

As such, the security of QR codes is a rapidly evolving field, as their use becomes increasingly ubiquitous in various sectors, including payment systems, marketing, and even personal identification.

This widespread adoption has made them a lucrative target for cybercriminals, necessitating advancements in security measures to protect against phishing and other malicious activities. Here are some key aspects that outline the future of QR code security:

1.5.1   Emerging technologies to secure QR codes

Innovative solutions are being developed to enhance the security of QR codes. One approach involves the integration of encryption directly into the QR code. This method ensures that even if a QR code is tampered with, the information it directs to remains secure and inaccessible to unauthorized users.

Another promising technology is the use of blockchain to create a secure and immutable record of QR codes, which can prevent the creation of fraudulent codes.

Additionally, there's ongoing research into dynamic QR codes, which change periodically to prevent reuse by malicious actors after the initial scan.

1.5.2   *Predictions on the evolution of phishing attacks*

As security measures become more sophisticated, so too do the tactics of cyber-criminals. It is predicted that phishing attacks involving QR codes will become more complex, incorporating elements of AI to create highly personalized and convincing phishing campaigns. These attacks may use data collected from various sources to tailor malicious QR codes to individual targets, increasing the likelihood of successful deception.

In conclusion, while QR codes have become a ubiquitous tool in the digital age, their security implications cannot be overlooked. The future of QR code security looks promising, with ongoing developments in technology and an increasing emphasis on user education. As we navigate this landscape, the key to mitigating phishing attacks lies in a balanced approach that incorporates techno-logical solutions, vigilant practices, and continuous learning.

Key takeaways

- Phishing is a cybercrime that uses social engineering and technical deception to compromise user access credentials and gain unlawful entry to user systems (to include computers, online system, and mobile devices).
- Phishing is often used as the initial entry point for sophisticated attacks targeting assets in a critical infrastructure sectors.
- The Anti-Phishing Working Group reports that 2022 was a record year for phishing.
- APWG logged more than 4.7 million attacks.
- Since 2019, the number of phishing attacks has grown by 150 percent per year [37].
- Attacks against the financial sector remained the highest for the 4th quarter of 2022, accounting for 27.7 percent of the attacks [1].
- Technological advancements: The evolution of phishing in the AI era is predominantly driven by advancements in generative AI technologies, particularly NLP and ML algorithms [26]. These advancements enable AI to craft emails that are not only grammatically correct but also con-textually relevant, making them significantly more convincing than tra-ditional phishing attempts.
- Case studies and examples A striking example of AI's capabilities in phishing was witnessed in a recent incident where an organization

received emails purportedly from a senior executive, generated by an AI system [28].

- Exploitation of AI-generated images and audio in phishing deepfake technology, which manipulates audio and video to create convincing imitations of real people, is another frontier in AI-powered phishing [28].
- Social engineering: In social engineering, AI-generated images are being used to create fake social media profiles [28].
- Legal and ethical implications of AI in phishing: The integration of AI in phishing raises significant legal and ethical concerns [36,38].
- Advanced analytical tools for phishing detection: In response to the growing sophistication of AI-phishing, advanced analytical tools are being developed to detect and prevent these attacks [26].
- Case studies: Successful defense against AI-phishing real-world examples of successful defenses against AI-phishing attacks provide valuable insights into effective strategies and tools [28].
- Future trends and predictions in AI-phishing: As AI technology continues to evolve, so too will the methods employed in AI-phishing [26].

Questions to reflect on

1. How do you believe the human factors contribute to the success of phishing attacks?
2. If you were to explain the significance of the user's experience, disposition, context, and psychology in combating phishing, how would you articulate that?
3. Do you believe you are equipped to spot an AI generated phishing attack?

Chapter 2

The user and context

While a user's psychological dispositions can account for a portion of the decision to behave contrary to security best practices, a recent study found only 40 percent of a user's response can be attributed to psychological factors involving threat perceptions and vulnerabilities [7]. This data would suggest users are reacting to a cybersecurity event relative to the context in which they find themselves. When speaking of context, consider the sample case below:

A conundrum: understanding the context of a decision

Emma and Felix work in downtown Washington D.C. They are married, and work at the same Federal agency. As with most families working in the District of Columbia, they live in the suburbs outside Washington D.C. In their case, they live in Rockville, Maryland. Their two young children attend daycare in Rockville.

While Rockville is only 17.8 miles from downtown Washington D.C., the length of time it could take to reach the daycare and pick up their children could exceed a couple hours on Fridays during rush-hour (which begins around 3:30 pm). Since they must pick up their children by 4:30 pm, it is important they leave work by 3:00 pm on Fridays.

On this Friday, Emma and Felix clear their desk at 2:50 pm and prepare to close for the weekend. When it comes time to shut down their respective computers, their computers display a message that 37 Windows updates were being processed, and the computer should not be shut down.

By now it is 3:00 pm. Windows processing 37 updates could take over an hour and would preclude them from reaching the daycare on time to pick up their children.

Leaving the office while logged into their computer is in violation of policy. Not permitting their computer to install updates is in violation of policy. One of them could call an Uber and pick up the children, but that would require the other to access the computer of their spouse, while their spouse is still logged in and shut it down when the updates are finished. That too is in violation of policy.

The only legal option Emma and Felix have is to stay in the office and forgo picking up their children on time. While a daycare worker would surely remain with the children, the daycare has a firm policy that states any failure to respect pickup times could result in children being expelled from daycare in the future.

Felix and Emma have no doubt about their course of action. They both shut down their computers without allowing the updates to finish and hope for the best as they rush out the door to pick up their children on time.

Apparently, at this agency, numerous people opted for the same course of action as the Information Technology Department felt late afternoon on Friday was a reasonable time to push the Windows updates to each machine. As fate would have it, when the updates attempted to reinitiate the following Monday, many employees eager to get on with the week prematurely halted the updates thus leaving the systems unpatched.

Shortly thereafter, Chinese-backed cyberattackers launched a large-scale cyberattack on this agency. The attackers exploited an unpatched Microsoft Office zero-day vulnerability known as "Follina" and executed malicious code remotely on these systems.

2.1   The human at the center

If an individual's psychology plays into the decision-making process (to comply or not to comply with a security policy) less than 40 percent of the time, what else is influencing an individual's decision? What factors play into an individual's choice when training and policy are purposely set aside, and a different course of action is taken? Since this happens often, are we to assume individuals are simply prone to breaking rules or is something else at play? The answer may lie in the context individuals find themselves in. Context can be multilayered as the following story details.

While this is a fictitious scenario, the sequence of events and the elements involved are quite real and happen frequently. In the case of U.S. Federal employees, cybersecurity training is mandated, and required before receiving access to Federal systems. But Emma and Felix were confronted with an impossible choice: take care of their children or comply with a cybersecurity policy. This is their context. The context variables involved with their decision included:

- Employment position
- Marital status
- Socio-economic status
- Computer access
- Training and education

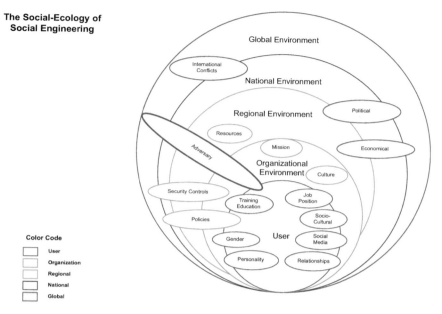

Figure 2.1   Social engineering domains affecting users [7]

It is tempting to limit our focus on the actions and context of the employees. However, in truth, the context variables involved with the final attack are far more distributed. In Figure 2.1, the context model depicts five context domains:

- User context
- Organizational environment
- Regional environment
- National environment
- Global environment

Each bubble within the diagram represents a variable that is applicable to the context domains. Bubbles that intersect one or more domains affect each domain they intersect.

The bubbles highlighted in blue represent the variables in play during the Emma and Felix scenario now the vulnerabilities were created. The adversary's actions are represented by the bold red enclosed bubble.

2.2   Perception and behavior in cyberspace

Given the prevalence of phishing attacks, and the frequency trained user populations succumb to phishing attacks, an assumption could be made that patterns in behavior potentially predict certain actions. Phishing resistance (or lack thereof) may be related to specific behavioral settings a given population finds themselves in.

The term behavioral setting was identified by the developmental psychologist Roger G. Barker (1903–1990) and refers to the geographical, physical, and social situation as it affects relationships and behavior.

In turn, ecological psychology is the analysis of behavior settings designed to predict patterns of behavior that can be expected within certain settings.

Ecological psychology is often applied to the study of human behaviors relative to their experience of their environments. This term often invokes the work of J. Gibson and E.J. Gibson in the area of perception and to R. Barker's studies of social interactions [39].

K. Lewin introduced the term "psychological ecology" in 1943. In his work, he attempted to shift the focus in psychology away from mental processes and toward environmental relationships. This focus centered around an individual's interpretations of external environments relative to goals, barriers, and boundaries.

A study-centered ecological psychology would start by characterizing the non-psychological data first. The point of this approach is to determine the conditions surrounding an individual or group's life. Once the non-psychological conditions are known, the focus of the study pivots to the psychological dimensions of the individual or group.

From this approach, two lines of research evolved: one emphasizing the perceptions of individuals/groups and the other the behavioral adaptations.

Understanding the cyber experience from the perspective of ecological psychology offers some interesting challenges. To wit, the theory of direct perception as formulated by Gibson [40] and Michaels and Carello [41] posited that direct knowledge/contact with the environment must be possible. This theory stands in contrast to more conventional cognitive theories which posit that perceptions of phenomenon are mediated by way of inferential means and mental representations.

However, whether we subscribe to the notion that we have a direct experience of cyberspace, or the perception of cyberspace is inferential and is apprehended through a cognitive mediation process, the context of cyberspace should be described first.

An earlier work by John Suler, called the Psychology of Cyberspace, describes cyberspace as a "psychological space." Suler elaborates on this description by inviting us to examine the user experience beginning with powering on a device and interacting with it to accomplish ordinary day-to-day tasks. Suler further states that whether consciously or subconsciously, users have an experience of entering a place or a space [42].

Since Suler's work in 1996, the language describing cyberspace as a place has further evolved with various forms of social media and virtual reality. Today, we speak of things like the "Metaverse," rooms, groups, chats, etc. We enter each of these places without necessarily moving our bodies to a different location. The experience of a space is strictly a psychological experience.

Suler further describes cyberspace (as experienced by many) as a type of "transitional space," an extension of an individual's internal, psychological processes. In other words, cyberspace becomes a transitional zone bridging one's connection from oneself to another, who is then perceived as having characteristics that are often like our own.

Regardless of how an individual processes this experience, invariably, in the absence of additional information such as the other person's expressions, tone of voice, body language, the virtual experience of a user is largely a projection of created by their own mind.

2.3   The personal computing experience

While the trajectory of interpersonal interactions in cyberspace had been evolving and increasing since the early 1990s, the COVID-19 pandemic (2019 onward) accelerated our reliance on cyber interactions to a degree hardly anticipated in previous years.

Where before our cyber exchanges may have been limited to work product development, email, text messages, and social media platforms, almost overnight many of our interpersonal communications were transformed to online-focused experiences.

Zoom took the place of family gatherings, our professional interactions, classrooms, and even religious worship. Working remotely is now as common as traveling to an office. A person's entire educational experience can unfold exclusively online. Dating applications dominate how people meet, and self-expression is now a function of Instagram and TikTok.

With these rapid changes in how we, as humans, perceive our interpersonal context today, understanding cyberspace as a place with specific features is imperative. It is particularly imperative if we wish to understand existing threats such as the phish.

Whether we are communicating via text message, or a video conference call, the full range of sensory input that would normally inform our perception of another is no longer available to us.

Additionally, when we are involved with interactions in cyberspace, chances are we are not physically located in the same location as the individual's we are communicating with.

While this may appear trivial, our physical location does inform our experience relative to our physical context. If I am comfortably sitting in my living room, surrounded by comforting and familiar décor, the chances of feeling threatened by an incoming communication are less likely to occur. My living room is not a dark alley, that would trigger my normal fear arousal processes. However, the possibility of "getting mugged" in my living room on my computer is just as real as being in a dangerous place physically.

Considering these points, the inevitable question arises: what is our psychological context when interacting in cyberspace?

To examine the cyberspace context, Suler's 1996 work offers the following psychological features.

2.3.1   Reduced sensations

Whether we communicate via video conference, text, or email, the interaction will not include other sensory inputs we would typically have during an in-person exchange.

In cyberspace, the term "reduced sensations" refers to the diminution of sensory stimuli that one would normally experience in direct human interactions. Unlike in-person encounters, digital communications via video conferences, texts, or emails lack many sensory inputs—such as touch, smell, and the nuanced aspects of body language. This reduction in sensory feedback can fundamentally alter the nature of interpersonal exchanges.

Communications are particularly impacted by the reduction in sensations. The absence of these sensory cues can decrease the emotional depth of exchanges. Facial expressions and gestures are often compressed or entirely absent in text-based communication. While video calls can convey visual cues, they still fall short of the full spectrum of human expression.

With reduced sensory inputs, there is an increased risk of misinterpretations. Subtle cues like sarcasm, humor, and empathy may not translate well without the full range of human expressions and intonations, leading to misunderstandings.

The concept of reduced sensations underscores the physical distance between communicators in cyberspace. This distance can contribute to a sense of isolation or detachment from others, even when engaging in what feels like direct communication.

Over time, technological advances have sought to bridge this sensory gap. Innovations in virtual reality, augmented reality, and haptic feedback technologies are examples of attempts to reintroduce some of the missing sensory dimensions to digital communications.

To mitigate the effects of reduced sensations, individuals may need to develop heightened social and emotional intelligence. This can involve learning to read subtler cues within the constraints of digital communication or choosing words more carefully to convey tone and emotion.

The blending of digital and in-person interactions, known as hybrid communication, is another approach to overcoming the limitations posed by reduced sensations. By combining the convenience and reach of digital communication with the richness of face-to-face interactions, individuals can enjoy a more holistic communication experience.

2.3.2   Texting

Typing/texting employs a different set of mental processes than speaking and listening. Unlike spoken language, which flows more naturally and allows for immediate feedback and adjustment, texting requires the sender to carefully choose words and structure sentences to convey meaning. This can lead to more thoughtful communication but may also introduce delays and reduce spontaneity.

Texting's reliance on written language strips away the nuanced emotional cues in voice inflection and body language. Emojis and other digital markers of emotion attempt to fill this gap, but they cannot fully replicate the depth of face-to-face interaction. This can make it harder to establish an emotional tone, leading to misunderstandings.

The absence of vocal tone and body language in texting can lead to a higher incidence of misinterpretation. What was intended as a joke or sarcasm can be taken literally, potentially leading to conflicts or hurt feelings. This necessitates a heightened level of digital literacy and empathy, where individuals learn to read between the lines or seek clarification when in doubt.

While texting allows for instant and constant communication, it can also foster a sense of emotional distance. The physical act of typing on a device creates a barrier, both literally and metaphorically, that can detach individuals from the immediacy and emotional engagement of conversation.

Over time, users have developed creative ways to express themselves through texting, using language, emojis, GIFs, and memes to convey tone, emotion, and personality. These adaptations showcase the human capacity to innovate within constraints, bringing a new layer of richness to digital conversations.

Recognizing texting's limitations, many now employ hybrid communication strategies, switching between texting, voice messages, and video calls based on the context and needs of the conversation. This approach allows individuals to leverage the strengths of each medium, improving the overall quality of communication.

2.3.3   Identity flexibility

Online communications provide an opportunity to be whoever you wish to be. You can be yourself, reveal only a part of yourself, be entirely invisible, or be someone entirely different. With the use of Avatars, adopting a completely different persona allows one to reinvent oneself on many levels.

This phenomenon of digital identity transformation, facilitated by the anonymity and flexibility of online platforms, has profound implications for individual psychology and social interactions.

Online environments liberate individuals from the physical constraints and societal expectations that govern real-world interactions. This freedom can be particularly empowering for those who feel marginalized or misunderstood in their offline lives, providing a space for exploration of identity, interests, and desires without fear of judgment.

The use of avatars and online personas enables individuals to explore aspects of their identity in a safe and controlled environment. This can be a valuable process for self-discovery and expression, allowing users to experiment with different facets of their personality, including those they may feel uncomfortable or unable to explore in real life.

Online platforms can foster community and belonging, connecting individuals across geographical and cultural boundaries. The ability to adapt and interact through diverse personas can encourage openness and understanding, as interactions are judged more on the basis of shared interests and ideas than on physical appearance or social status.

While the anonymity of online interactions can encourage honesty and openness, it can also lead to a disconnect between one's online and offline selves. This disjunction can impact real-life relationships and one's sense of identity, potentially leading to feelings of isolation or confusion about one's true self.

Furthermore, the veil of anonymity can sometimes facilitate negative behavior, such as cyberbullying or deceit, which can harm others and degrade trust in online communities.

An important aspect of engaging with online personas is finding a balance that integrates one's digital and real-world identities. This involves reflecting on how online experiences and explorations inform and influence one's sense of self and ensuring that digital interactions complement rather than replace real-life connections and activities.

Identity fluidity in texting also creates a normalized experience of exchanging personas, i.e. lending the experience legitimacy. If users perceive online interactions as "truthfully real," why would a user suspect an incoming email personalized to cause the user to believe the email was written by a friend? This aspect of online communications could certainly be a contributing factor to a user's phishing risk profile.

The opportunity to reinvent oneself online presents both empowering possibilities and complex challenges. As digital spaces continue to evolve, so too does the concept of identity. Navigating this landscape requires a conscious effort to maintain authenticity, integrity, and a sense of caution.

2.3.4   Altered perceptions

Suler suggests that while being absorbed in cyberspace, one can experience altered perceptions. He further likens these altered perceptions to a dream-like state of consciousness. Recent studies, such as those conducted by Hipp *et al.* validated these conclusions, and in this study in particular, visual perceptions were found to be altered during online interactions [43].

The findings of Hipp *et al.*, which demonstrate alterations in visual perceptions during online interactions, suggest that the way we process visual information can change significantly in digital contexts. This may be due to the unique visual cues and interfaces of digital platforms, which require our brains to adapt to different modes of processing information compared to the physical world. This can have both cognitive and emotional implications, potentially affecting attention span, memory recall, and emotional responses to visual stimuli.

Altered perceptions in cyberspace also extend to our sense of presence—how we perceive ourselves and others within a digital space. Virtual environments can evoke a sense of immersion that makes users feel as though they are physically part of a digital world. This can lead to a more intense engagement with digital content and interactions, mirroring the engrossment one might feel in a vivid dream. But it can also introduce cybersecurity risks.

The malleability of identity online, facilitated by the use of avatars and the ability to curate or completely reinvent oneself, further contributes to altered perceptions. In cyberspace, the boundaries of self are fluid, allowing individuals to explore aspects of their identity in ways that might be impractical or impossible in the physical world. This can lead to a transformative experience, akin to the profound self-reflection and realization one might experience in dream states.

Just as dreams can distort our perception of time and space, so too can online environments. Users may experience a sense of timelessness while engaged in digital activities, where hours can feel like minutes. Similarly, the digital realm collapses geographical distances, allowing instant communication across the globe, which can alter our perception of space and proximity to others.

These altered perceptions can have profound emotional and psychological effects. On one hand, they can provide a sense of escapism and freedom, offering a space for experimentation and discovery. On the other, they can lead to disorientation and a blurring of the lines between the self in the virtual world and the self in the physical world, raising questions about authenticity, alienation, and the impact on mental health.

Suler's analogy of cyberspace to a dreamlike state highlights digital environments' profound impact on human perception and interaction. The validation of these concepts through studies like that of Hipp *et al.* [43] underscores the need for further research into how digital interfaces influence our cognitive and emotional lives. Understanding these altered perceptions is crucial for designing digital spaces that foster healthy, meaningful, and enriching interactions while also acknowledging and mitigating the potential disorienting effects of our increasingly digital lives.

2.3.5   Equalized status

Posting a tweet on Twitter or uploading a humorous video to Instagram affords everyone the same opportunity to reach viewers. Cyberspace does not require a publisher, or any other means to vet content before it is offered to millions of viewers. A single proprietor can present a web page as detailed and complicated as that of a multi-million dollar corporation. In short, cyberspace "levels the playing field" to a certain degree. The measure of one's online success is driven largely by one's talent, ideas, and delivery. Not by access to an agent.

Context-aware systems can significantly enhance the user experience by providing personalized content and recommendations. For instance, social media platforms can use context-aware algorithms to suggest relevant content based on a user's location, time of day, and past interactions. This personalization can help users discover content that resonates with them, regardless of their social or economic status.

Moreover, context-aware systems can simplify complex tasks, making technology more accessible to a broader audience. For example, voice-activated assistants like Siri and Alexa use context awareness to understand and respond to user commands more effectively. This ease of use can empower individuals who may not have the technical skills to navigate traditional interfaces, thereby promoting digital inclusion.

The democratization of content creation is another significant aspect of equalized status in cyberspace. Context-aware tools and platforms enable individuals to create high-quality content without requiring extensive technical knowledge or expensive equipment. For instance, mobile apps with built-in editing tools allow users to produce professional-looking videos and images with minimal effort.

Additionally, context-aware platforms can provide real-time feedback and suggestions to content creators, helping them improve their work and reach a wider audience. This support can be particularly beneficial for amateur creators who may not have access to professional training or resources.

While context awareness can enhance equalized status in cyberspace, it also presents several challenges and considerations that need to be addressed. One of the primary concerns with context-aware systems is privacy. These systems often rely on collecting and analyzing large amounts of personal data to provide personalized experiences. This data collection can raise significant privacy issues, particularly if users are not fully aware of how their data is being used or if it is not adequately protected.

To address these concerns, it is essential to implement robust privacy policies and data protection measures. Users should be informed about what data is being collected, how it will be used, and who will have access to it. Additionally, context-aware systems should provide users with control over their data, allowing them to opt-out of data collection or delete their data if they choose.

The ethical implications of context-aware systems also need to be considered. For instance, the algorithms used to personalize content and recommendations can inadvertently reinforce existing biases and inequalities. If these algorithms are not carefully designed and monitored, they can perpetuate stereotypes and limit the diversity of content that users are exposed to.

To mitigate these risks, it is crucial to develop context-aware systems that are transparent and fair. This includes regularly auditing algorithms for bias, ensuring diverse representation in training data, and involving a wide range of stakeholders in the design and development process.

While context-aware systems provide a fulfilling user experience, by adapting to the user's environment, preferences, and activities, they can inadvertently increase vulnerability to phishing and other forms of social engineering. These systems leverage contextual information such as location, time, device type, and user behavior to provide personalized services. However, the same contextual data that enhances user experience can be exploited by cybercriminals to craft more convincing and targeted attacks.

Context-aware systems collect and analyze vast amounts of data to understand user behavior and preferences. This data can include browsing history, location data, device information, and interaction patterns. Cybercriminals can exploit this information to create highly personalized phishing attacks, known as spear phishing. By tailoring the content of phishing emails or messages to the specific interests and behaviors of the target, attackers can significantly increase the likelihood of success.

For example, an attacker who gains access to a user's contextual data might know that the user frequently shops online at a particular store. The attacker could then craft a phishing email that appears to be a promotional offer from that store, complete with personalized details that make the email seem legitimate. The user, recognizing the familiar context, is more likely to click on the malicious link or provide sensitive information.

Context-aware systems can also be manipulated to trigger social engineering attacks based on specific contextual cues. For instance, an attacker could use location data to send a phishing message when the user is in a particular place, such as a coffee shop or airport, where they might be less vigilant. Similarly, time-based attacks can be launched when the user is likely to be busy or distracted such as during work hours or late at night.

These contextually triggered attacks exploit the user's environment and state of mind, making them more susceptible to manipulation. For example, a phishing email sent during a busy workday might claim to be an urgent request from a colleague or supervisor, prompting the user to respond quickly without thoroughly scrutinizing the message.

Traditional phishing detection methods, such as spam filters and URL blacklists, may struggle to identify context-aware phishing attacks. These attacks are often highly personalized and may not exhibit the typical characteristics of generic phishing emails such as poor grammar or suspicious links. Instead, they blend seamlessly into the user's normal digital environment, making them harder to detect.

Context-aware phishing attacks can also bypass traditional detection methods by using legitimate-looking URLs and email addresses. For instance, an attacker might create a phishing website that closely mimics a legitimate site the user frequently visits, using a URL that differs by only a single character. The user, recognizing the familiar context, may not notice the subtle difference and proceed to enter their credentials.

Users often place a high level of trust in context-aware systems, believing that these systems are designed to enhance their security and convenience. This trust can be exploited by attackers who use contextual information to create convincing phishing scenarios. For example, a context-aware system might notify the user of a security alert based on their location or device activity. An attacker could mimic this notification, prompting the user to take immediate action such as clicking a link or entering their password.

The inherent trust in context-aware systems can make users less skeptical of unexpected requests or alerts, increasing their vulnerability to social engineering attacks. This is particularly concerning in environments where users are accustomed to receiving context-based notifications such as corporate networks or smart home systems.

One of the most effective ways to mitigate the risks associated with context-aware systems is to enhance user awareness and training. Users should be educated about the potential dangers of context-aware phishing and social engineering attacks and trained to recognize the signs of such attacks. This includes being cautious of unexpected requests for personal information, scrutinizing URLs and email addresses, and verifying the authenticity of messages through alternative channels.

Phishing simulations can be a valuable tool in this regard, providing users with hands-on experience in identifying and responding to phishing attempts. By simulating context-aware phishing scenarios, organizations can help users develop the skills needed to detect and avoid these sophisticated attacks. Figure 2.2 from the

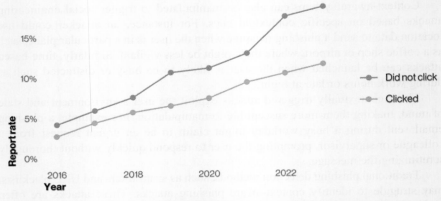

Figure 2.2   Verizon DBIR 2023 click and no click test results

Verizon Data Breach Investigations Report (DBIR) in 2023 illustrates how awareness around clicking on unknown emails can improve a user's likelihood of clicking on a nefarious link.

In addition to training, implementing robust security measures should always be a consideration. These should be designed to protect against context-aware phishing and social engineering attacks. This includes deploying advanced threat detection systems that can analyze contextual data and identify anomalies indicative of phishing attempts. Machine learning algorithms can be particularly effective in this regard, as they can learn to recognize patterns and behaviors associated with phishing attacks.

Multi-factor authentication (MFA) is another critical security measure that can help mitigate the risks of context-aware phishing. By requiring multiple forms of verification, such as a password and a biometric factor, MFA makes it more difficult for attackers to gain unauthorized access, even if they manage to obtain the user's credentials through a phishing attack.

2.3.6   Transcended spaces

Interacting with someone halfway around the globe is no longer as daunting as it was in previous times. While time zones can still pose a problem, collaboration and interactions online have bridged many gaps. The reduction of geographical barriers through online interactions has far-reaching implications for how we work, learn, create, and connect. While challenges remain, particularly in ensuring equitable access and managing the nuances of cross-cultural communication, the potential for innovation, understanding, and collaboration is unprecedented.

However, the ability to communicate and interact with anyone around the world instantly increases the potential targets for phishing attacks. Cybercriminals can launch campaigns on a global scale, targeting individuals and organizations across different regions without the need for physical presence.

A variety of communication platforms used for global interactions—email, social media, instant messaging, and professional networking sites—provides multiple channels for phishing attempts. This diversity makes it challenging for users to consistently identify and guard against phishing attempts across all platforms.

Phishers can tailor their attacks to specific regions, cultures, or even individuals, making them more convincing. This could involve using local languages, mimicking local institutions, and crafting messages that resonate with specific cultural or societal concerns. Online interactions often rely on a foundational level of trust, especially in professional and collaborative contexts. Phishers exploit this trust, posing as colleagues, business partners, or reputable organizations to deceive victims into divulging sensitive information or executing unauthorized transactions.

The global nature of online interactions complicates the legal framework for combating phishing. Cybercriminals operating across borders can exploit differences in regulatory environments, making it difficult for law enforcement to pursue and prosecute offenders.

The wide range of potential victims, from individuals to large organizations, means that the effectiveness of cybersecurity measures varies greatly. While larger organizations may have sophisticated defenses and protocols in place, individuals and smaller entities may lack the resources or knowledge to effectively protect themselves.

Frequent and sophisticated phishing attacks can erode trust in digital communications, making individuals and organizations more cautious and potentially hindering the open exchange of information that characterizes the transcended digital space.

Efforts to combat phishing, involving monitoring and analyzing communications for potential threats, must be balanced against concerns for privacy and data protection. This is especially pertinent in a global context, where expectations and legal standards for privacy vary widely.

Thus, the global nature of online interactions complicates the legal framework for combating phishing. Cybercriminals operating across borders can exploit differences in regulatory environments, making it difficult for law enforcement to pursue and prosecute offenders. Jurisdictional issues arise when the perpetrator, victim, and the servers used in the attack are located in different countries. Coordinating international efforts to track, apprehend, and prosecute cybercriminals requires robust legal frameworks and cooperation agreements that are often lacking or insufficiently enforced.

Different countries have varying legal standards and regulations concerning cybersecurity and data protection. This variability can be exploited by phishers who operate from regions with lax enforcement or inadequate legal provisions. For instance, some countries may not have stringent laws against cybercrime, making it a safe haven for cybercriminals. The lack of harmonized legal standards complicates efforts to create a unified global response to phishing and other cyber threats.

Additionally, frequent and sophisticated phishing attacks can erode trust in digital communications, making individuals and organizations more cautious and potentially hindering the open exchange of information that characterizes the transcended digital space. When users become wary of every email or message, it can slow down communication and collaboration, impacting productivity and innovation. The fear of falling victim to phishing can lead to a more guarded approach to digital interactions, which contradicts the open and collaborative nature of the internet.

The psychological impact of phishing attacks on individuals and employees can be significant. The constant vigilance required to identify and avoid these attacks, along with the potential consequences of falling victim, can contribute to increased anxiety and decreased productivity. Employees may experience stress and fear, knowing that a single mistake could lead to severe repercussions for themselves and their organization. This stress can affect overall job satisfaction and performance, leading to a less engaged and more cautious workforce.

Efforts to combat phishing involve monitoring and analyzing communications for potential threats, which must be balanced against concerns for privacy and data protection. This is especially pertinent in a global context, where expectations and legal standards for privacy vary widely. For instance, the General Data Protection Regulation in the European Union imposes strict requirements on data processing and user consent, which can complicate efforts to implement comprehensive monitoring systems. Organizations must navigate these regulatory landscapes carefully to ensure that their anti-phishing measures do not infringe on user privacy rights.

And as always, there are ethical considerations. The ethical considerations of monitoring and data analysis are significant. While it is essential to protect users from phishing attacks, it is equally important to respect their privacy and autonomy. Overly intrusive monitoring can lead to a surveillance culture, where users feel constantly watched and mistrusted. Striking the right balance between security and privacy requires transparent policies, user consent, and the implementation of privacy-preserving technologies that can detect threats without compromising personal data.

International collaboration is essential to address the global nature of phishing. Governments, law enforcement agencies, and cybersecurity organizations must work together to share information, coordinate responses, and develop unified strategies to combat phishing. Initiatives such as the Anti-Phishing Working Group facilitate collaboration and information sharing among stakeholders, helping to create a more coordinated and effective response to phishing threats. By fostering international cooperation, the global community can enhance its ability to detect, prevent, and respond to phishing attacks.

2.3.7   Social multiplicity

Social multiplicity refers to the ability to interact with a wide variety of people from all walks of life. These types of interactions would normally have been reduced by more limited means of connectivity such as chance meetings in person. However, from a cybersecurity perspective, social multiplicity also introduces complex challenges, particularly concerning phishing attacks and broader cyber-security concerns.

Phishing attackers exploit social multiplicity by leveraging the wide range of social interactions that individuals engage in online.

Phishers use social engineering tactics to manipulate users into divulging confidential information or taking certain actions. By exploiting the diverse social connections individuals maintain, attackers can craft more convincing and perso-nalized phishing messages. For example, by mimicking the communication style of a distant acquaintance or a professional contact met through online networking, attackers can trick recipients into clicking malicious links or sharing sensitive information.

Social multiplicity often involves sharing personal information across plat-forms, making it easier for attackers to gather personal data and impersonate legitimate contacts. This impersonation can be particularly effective in phishing attacks, as recipients are more likely to trust and respond to requests that appear to come from known contacts.

The dynamics of social multiplicity can amplify the reach and impact of phishing attacks in several ways.

With individuals engaging in a broader array of social interactions online, the attack surface for phishers expands. Each connection represents a potential entry point for attackers, multiplying the opportunities for successful phishing expeditions.

Trust is a foundational element of social interactions. In a digitally connected environment characterized by social multiplicity, trust can be both a strength and a vulnerability. Phishers exploit the inherent trust within social networks to propa-gate attacks more effectively, counting on the fact that individuals are more likely to open messages and click on links from someone within their network.

Social multiplicity complicates the implementation of effective cybersecurity measures in several ways.

The vast and varied nature of online interactions makes it challenging to verify the authenticity of every communication. Traditional security measures may not be sufficient to detect sophisticated phishing attempts that exploit personal and pro-fessional connections.

Raising awareness and educating users about the risks associated with social multiplicity and phishing requires tailored approaches. Given the diverse ways in which individuals engage online, cybersecurity education must account for a broad range of behaviors and threat recognition capabilities.

2.3.8   Recordability

Online interactions can easily be recorded, saved, and can be reviewed in an almost on-demand manner. The format in which interactions can be saved is also trans-ferrable, i.e. one can easily share interactions with others.

The ability to record, save, and share online interactions with ease presents a double-edged sword in the realm of cybersecurity, particularly concerning phishing attacks. This characteristic of digital communication not only transforms the way we preserve and disseminate information but also introduces specific vulnerabilities that phishers exploit to conduct their attacks. Understanding the implications of this capability sheds light on the nuanced challenges faced in protecting individuals and organizations in the digital space.

The ease with which digital interactions can be saved and shared allows for rapid dissemination of phishing attacks. Once a malicious link or deceptive message is crafted, it can be spread widely with little effort, reaching a vast audience in a short period. This amplifies the potential impact of phishing campaigns, as each shared interaction can potentially recruit more victims.

Phishers can hijack legitimate interactions, such as email threads or social media conversations, and insert malicious links or attachments. When these tampered interactions are saved and shared, the malicious content rides on the credibility of the original conversation, making it more likely to deceive recipients.

The transferrable nature of digital interactions complicates efforts to trace the origin of phishing attacks and hold perpetrators accountable. Phishers can mask their identity and location, exploiting the global nature of digital communications to evade detection and legal repercussions.

Cybercriminals continuously refine their strategies to exploit the latest technological advancements and social behaviors. The ability to save and share interactions means that phishing attempts can be more elaborately crafted, incorporating elements of legitimate interactions to create highly convincing scams.

To counter the threats posed by the ease of recording, saving, and sharing online interactions, cybersecurity defenses must be adaptable and dynamic. This includes deploying advanced threat detection algorithms capable of identifying subtle signs of tampering or malicious content within shared interactions.

Educating users about the risks associated with saving and sharing digital interactions is crucial. Awareness programs should emphasize the importance of scrutinizing shared content, verifying the credibility of links and attachments, and recognizing signs of manipulation or unauthorized alterations.

Organizations must implement robust data protection policies that regulate how interactions are recorded, saved, and shared. This includes establishing clear guidelines for data retention, access controls, and the secure transmission of sensitive information.

2.4   The experiential context versus the actual context

Assuming users do formulate a mental experience that is deeply personal, and proprietary to themselves, how does this affect elements found in their actual physical context? Do users lose sight of their physical context when embedded in their virtual, mental space?

2.4.1   A hypothetical example

Austin and his team develop complex, detailed media for communications purposes. Often this team pushes technology to its limits. Additionally, since the core of this team's activities involves creativity, Austin encourages his team to express themselves creatively, personalize their workspace, and enjoy their computing experiences.

Given the performance needs of Austin's team coupled with their use of esoteric software, Austin's team frequently requires the assistance of the tech desk. They communicate with the tech desk mostly via text message. Given the frequency of their communications, they've established a jovial report with the tech desk.

Recently Austin's team received a contract to develop the promotional material for a long-awaited sports vehicle that includes cutting-edge engineering and design. The material Austin's team is receiving to create this promotion is very sensitive and tightly guarded by their client, Lambor My Jin.

Shortly after receiving the closely held materials from their client, Alex, one of Austin's graphic designers, receives a text from Becky at the tech desk. She ribs him as usual about his odd choice of lunch sandwiches, and they banter on text back and forth before Becky texts:

"Hey, before you guys go too crazy with your new project, I need to install some patches on your system. I need to remote into your machine, but I'm working from home today, and need to run the remote connect from a personal account. Are you ok with that?"

Alex, who's interacted with the tech desk before, has always had a great experience. Therefore, he is relying on his projection of who Becky is when he agrees to let her run the updates from a personal account. In fact, Alex looks at his watch, and decides he could pick up in the morning rather than wait while she runs the updates. He tells her to "go for it," granting her machine access. In fact, Alex tells her he's leaving and requests she log out when she is done.

Of course, the text message didn't come from Becky, but rather from a phisher working on behalf of Lambor My Jin's competition, Masqueradi.

Within one afternoon, Masqueradi collected enough proprietary Lambor My Jin information they could extrapolate what the new design would be, and what engineering modifications would allow Lambor My Jin to achieve the performance benchmarks they were attaining.

While this example is fictitious, these types of interactions are happening daily. Austin's team is not operating with an awareness of their actual physical context (namely that they are working for a high-browed media firm handling very sensitive information on behalf of their clients).

Austin's team is operating within their personal contexts, which involves a highly interactive, personalized, creative environment that thrives on trust, and free-flowing, independent interactions. In other words, they are "in their own world."

Even if Austin's team were to work remotely, chances are the team's surroundings, and technology would not create an environment that projects a sense of potential threats.

These conclusions leave us with many questions. Clearly, if we were to explain the nature of phishing to end-users (much the way a training program would), users would understand that phishing is a problem. However, that understanding does not appear to create an experience of caution when users are perched behind a computer screen. Is the problem the technology?

This phenomenon of user groups experiencing a context that is different from their organization's context is known as subgroup formation. It can lead to divergent perceptions, behaviors, and norms among different subgroups within the same organization. Several factors contribute to this inconsistency in context.

Demographic differences, such as age, gender, ethnicity, or educational background, can lead to the formation of subgroups within an organization. People tend to categorize themselves and others based on these demographic attributes, creating an "us vs. them" mentality. As a result, subgroups may develop their own shared experiences, values, and norms that differ from those of the broader organization.

In large or geographically dispersed organizations, physical separation can contribute to the formation of subgroups. Employees who work in different locations or departments may have limited interaction with others, leading to the development of distinct subcultures and contexts within each subgroup. This separation can reinforce the perception of "us vs. them" and hinder the sharing of information and experiences across the organization.

Organizations often have different functional areas or teams responsible for specific tasks or projects. These functional or task-based differences can lead to the emergence of subgroups with distinct goals, priorities, and ways of working. For example, the sales team may prioritize revenue generation, while the engineering team focuses on product development. These divergent priorities can create distinct contexts and norms within each subgroup. Even in the absence of clear demographic or functional differences, the mere perception of subgroup formation can shape the context experienced by employees. If individuals perceive that subgroups exist within the organization, they may adjust their behaviors and attitudes accordingly, reinforcing the perceived subgroup boundaries and creating divergent contexts. Power dynamics and status inconsistencies within an organization can contribute to the formation of subgroups and divergent contexts. Individuals with higher status or power may form subgroups that enjoy privileges or access to resources that are not available to others. This can create a sense of "us vs. them" and lead to different experiences and norms within each subgroup.

The inconsistency in context experienced by subgroups within an organization can have significant implications for organizational functioning and performance. It can lead to communication breakdowns, conflicts, and a lack of cohesion and alignment across the organization. Additionally, subgroup formation can hinder the development of a shared organizational culture and impede the effective transfer of knowledge and best practices.

In addition to organizational challenges, subgrouping within organizations can inadvertently create fertile ground for phishing opportunities. When different groups within an organization develop their own distinct contexts, it can lead to communication silos, varying levels of security awareness, and inconsistent adherence to security protocols. These factors collectively increase the vulnerability of the organization to phishing attacks.

When subgroups operate in isolation, they often develop their own communication norms and channels. This isolation can result in a lack of information sharing and coordination across the organization. For instance, if one department receives a phishing alert and does not effectively communicate it to other departments, those other departments remain unaware and vulnerable to similar attacks. Cybercriminals can exploit these silos by targeting less informed or less prepared subgroups within the organization.

Additionally, subgrouping can lead to varying levels of security awareness and training. Different departments or teams may receive different levels of cybersecurity training based on their perceived risk or importance. For example, the IT department might undergo rigorous phishing awareness training, while administrative staff might receive only basic training. This disparity creates weak links within the organization that phishers can exploit. Attackers often target these less trained subgroups, knowing that they are more likely to fall for phishing attempts due to their lower awareness and preparedness.

Inconsistent adherence to security protocols is another consequence of subgrouping that can be exploited by phishers. Different subgroups may develop their own informal practices that deviate from the organization's official security policies. For example, a sales team might prioritize quick communication and therefore be more likely to bypass security checks such as verifying the authenticity of emails before responding. Phishers can take advantage of these informal practices by crafting phishing emails that align with the subgroup's specific behaviors and expectations, making the attacks more convincing and harder to detect.

Moreover, subgrouping can lead to a lack of cohesive response strategies to phishing attacks. When subgroups operate independently, they may not have a unified approach to handling phishing incidents. This can result in delayed or fragmented responses, allowing phishers to exploit the confusion and gain a foothold within the organization. For instance, if a phishing email targets multiple departments, the lack of a coordinated response can lead to inconsistent actions, with some employees reporting the email while others might inadvertently engage with it.

The distinct contexts experienced by subgroups can also be leveraged by phishers to craft highly targeted and convincing phishing attacks. By understanding the specific roles, responsibilities, and communication styles of different subgroups, attackers can tailor their phishing emails to appear more legitimate. For example, a phishing email targeting the finance department might mimic an invoice or payment request, while an email targeting the HR department might appear as a job application or employee benefits update. These tailored attacks are more likely to succeed because they align with the specific context and expectations of the targeted subgroup.

Furthermore, subgrouping can create a false sense of security within certain parts of the organization. Subgroups that have not experienced phishing attacks directly may become complacent, believing that they are not at risk. This complacency can lead to lax security practices and a higher likelihood of falling for phishing attempts. Phishers can exploit this false sense of security by targeting these complacent subgroups with well-crafted phishing emails that exploit their lack of vigilance.

Phishing attackers employ a variety of sophisticated tactics to target subgroups within organizations, exploiting the unique characteristics and vulnerabilities of these smaller units. One of the most common and effective methods is spear phishing. In spear phishing attacks, cybercriminals meticulously craft emails to appear as if they are coming from a trusted source, often someone within the target's own organization or social circle. By conducting thorough research on the target subgroup, attackers gather information from public sources like social media to make their phishing emails more convincing and personalized. This level of customization increases the likelihood that the recipient will trust the email and engage with its content.

Another tactic used by phishers is whaling, which specifically targets high-level executives or other high-profile individuals within an organization. These attacks are a form of spear phishing but are directed at "whales" or key decision-makers. Whaling emails often claim there is a legal issue or appear to be from a trusted authority, such as a law firm or government agency, to trick the victim into revealing sensitive corporate data. The high stakes and urgency often associated with these emails make them particularly effective.

Phishers also exploit the specific communication channels used by different subgroups within an organization. These channels might include team messaging apps, social media groups, or internal forums. By infiltrating these trusted communication platforms, attackers can launch phishing attacks from within, making their malicious messages appear more legitimate. For example, a scammer might join a professional networking group on a platform like LinkedIn and then send phishing messages to group members, pretending to be a fellow professional in need of assistance.

Impersonation is another common tactic. Attackers may pose as roles or functions that a particular subgroup would be familiar with such as IT support, HR, or customer service. This approach leverages the trust that subgroup members have in these internal roles, increasing the likelihood that they will follow the instructions in the phishing email. For instance, an email that appears to come from the IT department requesting a password reset might prompt immediate action from the recipient, who believes they are complying with a legitimate request.

Phishers also tailor their attacks to exploit the specific contexts and interests of the targeted subgroup. By understanding the subgroup's recent activities, projects, or personal situations, attackers can craft highly relevant phishing lures. For example, if a subgroup has recently attended a conference, a phishing email might reference that event and offer a link to download additional materials, which actually leads to a malicious site. This contextual relevance makes the phishing attempt more believable and harder to detect.

Moreover, attackers often exploit the inherent trust and familiarity within subgroups. For example, a phishing email might appear to come from a colleague or a known business partner, leveraging the existing trust relationships to deceive the recipient. This tactic is particularly effective in professional contexts where quick and unquestioning responses to emails are common.

To mitigate the negative effects of subgroup formation and inconsistent contexts, organizations can implement various strategies. Encouraging cross-functional teams, job rotations, and regular meetings or events that bring together employees from different subgroups can help bridge the gaps and foster a shared understanding of the organizational context. Developing and reinforcing a strong organizational culture and set of values can help align subgroups and create a sense of shared identity and purpose across the organization. Improving communication channels and promoting transparency in decision-making processes can help ensure that all subgroups have access to the same information and understand the broader organizational context. Effective leadership and role modeling from top management can help shape the desired organizational context and set the tone for how subgroups should interact and align with the broader organizational goals and values. Organizations should strive to create a fair and equitable environment by addressing power imbalances and status inconsistencies that may contribute to the formation of privilege.

By recognizing and addressing the factors that contribute to inconsistent contexts within subgroups, organizations can foster a more cohesive and aligned workforce, ultimately enhancing organizational effectiveness and performance.

2.5   The AI impact on user context

The integration of artificial intelligence (AI) into phishing has significantly altered the phishing landscape. Unlike traditional phishing, AI-powered phishing can analyze vast amounts of data, understand user behavior, and create highly personalized attacks. This enhanced capability not only increases the success rate of phishing attempts but also has profound psychological effects on users. AI-driven phishing attacks are designed to exploit human cognitive biases, manipulate emotions, and undermine the user's ability to discern legitimate from fraudulent communications.

2.5.1   *Advanced personalization and erosion of trust*

AI in phishing employs sophisticated data analysis to micro-target individuals, making deceptive messages highly relevant and personalized.

Micro-targeting allows an attacker to shape the user's context by crafting electronic communications that reference personal information obtained from social media or data breaches. An example of micro-targeting would be a phishing email that appears to come from a known contact, citing recent shared experiences or interests. This would give the user a contextual experience of communicating with a familiar entity.

This form of targeting is enabled by AI algorithms "scraping" and analyzing an individual's online behavior, identifying patterns and preferences. This behavioral analysis is used to design phishing attacks that align with the target's usual online activities. For instance, frequent online shoppers might receive phishing emails mimicking popular e-commerce platforms, complete with details reflecting their browsing history or wish list items.

Additionally, AI enables the psychological profiling of targets, allowing phishers to design messages that exploit specific psychological triggers such as urgency, fear, or curiosity. A common example of this is a phishing email falsely claiming suspicious activity in the recipient's bank account, prompting them to take immediate action.

Phishers also ensure their messages are contextually relevant, a tactic that becomes particularly effective during specific times, such as tax season, where phishing emails may mimic official communications from tax authorities [44].

The implications of micro-targeting in phishing are profound. These sophisticated attacks not only have a higher success rate but also present significant challenges in detection, both for individuals and traditional anti-phishing tools.

The advent of micro-targeting techniques has transformed phishing into a highly personalized and effective form of cybercrime, one in which the act itself can be automated, thus targeting thousands of users.

2.5.2   The role of contextual perception plays in AI-driven phishing

The user's contextual perception is pivotal in shaping how the user interprets and responds to a digital communication. As discussed previously, the idea of contextual perceptions refers to the user's understanding of their digital environment, influencing how they asses the authenticity of a communication.

In phishing, exploiting this perception involves creating scenarios that align with what the user expects in their digital interactions. The use of AI in phishing has significantly enhanced the ability of an attacker to manipulate this perception.

AI-enhanced phishing attacks excel at mimicking legitimate entities. By replicating the communication style, visual elements, and writing style of trusted organizations, these attacks skillfully exploit the user's trust. For example, an AI-generated email mimicking a bank's communication style can be highly deceptive, leading customers to believe in its authenticity [45].

The timing and contextual relevance of messages are also key factors in AI-driven phishing. AI algorithms are adept at determining the most opportune moments to launch attacks. This tactic exploits the increased likelihood of users expecting and trusting such communications [44].

Sophisticated social engineering tactics are a hallmark of AI-enabled phishing. AI can analyze social networks and communication patterns to impersonate authority figures or acquaintances effectively, thereby exploiting the user's trust in these individuals. This approach is particularly insidious as it leverages established social connections to facilitate deception.

2.5.3   Example of AI leveraging contextual variables in phishing

Hypothetical AI-driven phishing scenario integrating contextual features

Scenario overview

In this scenario, an AI system is designed to execute a phishing attack targeting employees of a specific corporation. The AI uses advanced data analysis to gather and synthesize contextual information about the company, its employees, and recent events that could be leveraged to enhance the credibility of the phishing attempt.

Data collection and analysis

1. **Corporate events and news**: The AI scours the internet for recent news articles, press releases, and social media posts related to the target corporation. It identifies an upcoming corporate merger and the introduction of a new IT security protocol as key events.
2. **Employee information gathering**: Through publicly available information and perhaps previous data breaches, the AI compiles a list of employees, focusing on those in departments most impacted by the merger and new IT policies such as the IT and HR departments.

Crafting the phishing message

Using the gathered data, the AI crafts an email that appears to be from the corporation's IT department, addressing the recipient by name and referencing the merger and new IT security protocol for added authenticity.

1. **Subject line and email content**: The subject line reads, "Immediate Action Required: Update Your Security Settings Following the Merger." The email content explains that as part of the integration process with the new company, all employees are required to update their security settings and provides a link to do so.
2. **Contextual integration**: The email meticulously integrates specific details about the merger, including the name of the company being merged with and the timeline, to build trust. It mentions the implementation of the new IT security protocol, using technical language and logos obtained from the company's official communications to enhance its legitimacy.
3. **Call to action**: The user is urged to click on a link to update their security settings. This link leads to a fake website that mimics the corporation's internal IT portal, where they are prompted to enter their login credentials.

The effectiveness of this phishing attempt lies in its use of specific, contextual information that the recipient would find relevant and urgent. By incorporating details about recent corporate events and changes, the email bypasses the recipient's initial skepticism, exploiting their likely familiarity and concern with these topics.

2.5.4 AI used to deepen cognitive and emotional vulnerabilities

The term cognitive bias refers to systematic patterns of deviation from norm or rationality in judgment. Cognitive biases play a crucial role in how individuals process information and make decisions.

AI-driven phishing exploits these biases to manipulate users into divulging sensitive information or executing harmful actions. The sophistication of AI allows for a nuanced understanding and exploitation of these psychological tendencies.

One common cognitive bias targeted by AI phishing is the familiarity bias. The familiarity bias involves individuals showing a preference for familiar or well-known entities. AI-driven phishing attacks often mimic the communication style, branding, and messaging of familiar entities relative to a given victim. This tactic leverages the user's natural inclination to trust familiar sources, increasing the likelihood of successful deception. For example, a phishing email that appears to be from a well-known service provider can exploit this bias, leading users to lower their guard and engage with the content [45].

A possible set of actions an AI could use to exploit the familiarity bias may involve the following steps:

1. **Data harvesting and analysis**: *The AI begins by gathering publicly available data on the target and their network. It focuses on close contact with the target, let's say a coworker named Alex. It analyzes past email exchanges, social media interactions, and any publicly shared documents to understand the nuances of Alex's communication style and topics of interest shared between Alex and the target.*

2. **Mimicry and message crafting**: *Using natural language processing and generation algorithms, the AI crafts a phishing email that closely mimics Alex's writing style. The email might reference a shared professional interest or an ongoing project, embedding familiarity cues to lower the target's guard. The message suggests downloading a document related to this shared interest, which is, in reality, malware or a link to a credential-harvesting site.*

3. **Deployment**: *The email is sent to the target from an address that closely resembles Alex's, with minor alterations that can be easily overlooked such as "alex.work@example.com" instead of "alex@example.com." The AI might even choose a time known for high email interaction between Alex and the target, increasing the chances of success.*

4. **Exploiting familiarity bias**: *The target, recognizing the familiar themes, communication style, and supposed sender, is much more likely to lower their defenses and engage with the email's content, leading to potential security breaches.*

Another form of cognitive bias involves the anchoring effect. Anchoring bias refers to action's users take when relying too heavily on the first piece of information (the "anchor") they encounter. In phishing, AI algorithms can craft messages that present misleading anchors, such as a fake security alert, which then influences subsequent decisions and judgments. Users, once anchored to the idea of a security threat, may be more likely to follow deceptive instructions without critically evaluating their authenticity [44].

Hypothetical AI-driven phishing scenario exploiting the anchoring effect

Scenario setup

In this scenario, an AI system targets individuals within a financial services firm. The attack begins with the AI collecting data on recent financial trends, specifically a significant stock market dip, and identifying employees through social media and professional networking sites who have shown an interest in investment or are part of the firm's investment division.

Phase 1: Establishing the anchor

1. ***Initial contact****: The employees receive an email that appears to come from the firm's Chief Investment Officer (CIO). This email discusses the recent market dip and emphasizes its potential impact on the company's investments. It mentions that the CIO will share important strategies in a follow-up message to mitigate any negative effects and capitalize on potential opportunities. This email sets the anchor by focusing the recipient's attention on the market dip and the promise of valuable insider information.*
2. ***Anchoring effect****: The mention of the market dip and the impending advice from the CIO creates a psychological anchor, making the recipients more likely to prioritize information related to this topic. Their judgment and actions are now influenced by the urgency and importance of the situation as presented in the initial email.*

Phase 2: Exploiting the anchor

1. ***Follow-up phishing email****: A few days later, the AI sends another email, supposedly from the CIO, with a subject line that references the promised strategies to navigate the market dip. The email contains a link to a document titled "Exclusive Market Recovery Strategies."*
2. ***Urgency and exclusivity****: The email stresses the urgency of acting quickly to implement these strategies and the exclusivity of the information, suggesting that it is only being shared with a select group of employees. This plays on the initial anchor, reinforcing the perceived importance of the information and the need for swift action.*
3. ***Deceptive link****: The link to the "strategies" document actually leads to a phishing site designed to mimic a login page for the firm's internal system. Employees are prompted to enter their credentials to access the document, which the attackers can then use to gain unauthorized access to sensitive company information.*

AI phishing attacks also leverage authority bias. An AI phishing attack leveraging AI authority bias involves situations in which individuals tend to obey authority figures. By impersonating figures of authority, such as company executives or government officials, AI-driven phishing attacks manipulate users into complying with requests or directives. This bias is particularly effective in corporate settings, where employees may not question instructions that seem to come from higher management.

Hypothetical AI-driven phishing scenario exploiting the authority bias

Scenario overview

In this scenario, an AI system targets the employees of a healthcare organization. The attack is designed to exploit the authority bias by impersonating the organization's Director of Information Security, a figure with recognized authority over cybersecurity practices within the organization.

Data collection and persona creation

1. **Data harvesting**: *The AI scours publicly available information, such as the organization's website, social media platforms, and professional networking sites, to gather detailed information about the organization's structure, its key personnel, and specifically, the Director of Information Security.*
2. **Persona crafting**: *Utilizing the information collected, the AI crafts a detailed persona of the Director, including their communication style, typical email format, signature, and even the type of language and terminology they use in official communications.*

Crafting the Phishing Message

1. **Email creation**: *The AI generates an email that closely mimics the Director's communication style. The email is addressed to a wide range of employees and discusses an urgent security protocol update due to an alleged increase in phishing attacks targeting the healthcare sector.*
2. **Urgency and authority**: *The email emphasizes the urgency of complying with the new protocols to protect patient data and the organization's reputation. It includes a call to action, urging employees to immediately click on a link to complete a mandatory cybersecurity training designed to combat the latest phishing threats.*
3. **Deceptive link**: *The link in the email redirects to a fake training website that mimics the organization's internal training platform. Employees are prompted to enter their login credentials to access the training material, which the attackers can then use to gain unauthorized access to the organization's systems.*

The effectiveness of this phishing attack relies on the perceived authority of the Director of Information Security. Employees, conditioned to follow security-related instructions from this figure, are more likely to comply without questioning the legitimacy of the request, especially when the communication style matches their expectations.

Scarcity and urgency biases are also exploited in AI-driven phishing. These biases induce a sense of urgency or fear of missing out, compelling users to act quickly without fully considering the consequences. AI algorithms can analyze the optimal timing for sending such phishing messages, like during busy periods or significant events, to exploit these biases more effectively.

Hypothetical AI-driven phishing scenario exploiting scarcity and urgency bias

Scenario overview

In this scenario, an AI targets customers of a popular online retail platform. The AI is designed to send out emails that mimic the platform's branding and communication style, announcing a limited-time offer supposedly available to a select group of "valued customers."

Data collection and email crafting

1. ***Target identification***: *The AI identifies potential targets through data obtained from previous breaches or publicly accessible information, focusing on individuals who have shown a preference for online shopping and have interacted with limited-time offers in the past.*
2. ***Email personalization***: *The phishing email is personalized for each recipient, using their name and referencing past purchases to enhance credibility. It announces an exclusive, one-time discount on a high-demand item or a special offer available only to recipients of the email.*

Exploiting scarcity and urgency

1. ***Limited availability***: *The email emphasizes that the offer is available to only a small number of customers, exploiting the scarcity bias. It suggests that the item or discount is highly sought after and that availability is extremely limited.*
2. ***Immediate action required***: *To exploit the urgency bias, the email states that the offer will expire within a few hours or by the end of the day, pressuring the recipient to act quickly. It includes a call to action, urging the recipient to click on a link to claim the offer before it's too late.*
3. ***Deceptive link***: *The link leads to a fake website that closely mimics the appearance and user interface of the legitimate retail platform. Visitors are prompted to log in with their customer account to access the special offer. This login page is designed to capture the user's credentials, which can then be used for unauthorized purchases or identity theft.*

The effectiveness of this phishing attack lies in its ability to create a compelling narrative that combines the scarcity of the offer with the need for immediate action, preying on natural human tendencies to react to such stimuli.

Key takeaways

- The human experience is informed by our respective context.
- Our context can shape how we direct our attention.
- Our context can shape how we prioritize our attention needs.
- Our experiential context may not align with our actual context.

- The social experience online is different than our physical social experiences.
- Because we relate to our online interactions differently than we do to our physical interactions, we may not perceive threats in the same manner online as we would in person.
- Cybersecurity training centered around phishing has to date not produced a significantly phishing resistant user population.
- AI-driven phishing attacks can dramatically impact a user's perceived context.
- AI-driven phishing attacks leverage personal information scraped from sources like social media, and timing to craft effectively deceptive phishing emails.
- AI-driven phishing attacks exploit several user biases increasing the likelihood a user will be exploited exponentially

Questions to reflect on

1. Interacting in cyberspace generates a wide variety of interpersonal experiences, experiences that differ from those we have in person. Taking this into account, how can organizations mitigate the exploitation of users by cybercriminals under these circumstances?
2. Can organizations influence the user's experience of their context?

Chapter 3
The technology comfort experience

If we understand fear to be an adaptive but phasic (transient) state elicited through confrontation with a threatening stimulus, coupled with the frequency in which users succumb to phishing attacks, an assumption could be made that users do not perceive the human–technology interaction as a threat, regardless of what appears on their screen. How does the design of user interfaces (UI) and anthropomorphizing of technology affect our phishing resistance? Does our very comfort experience with technology lower our defenses and make us vulnerable to attacks which would otherwise be readily detected?

Fear, by definition, is a primal emotion that arises in response to perceived threats, serving an adaptive purpose by preparing the organism to face or flee from danger. In the context of human–technology interaction, this fear response should theoretically extend to digital threats such as phishing attacks. Phishing attacks are deceptive practices by cybercriminals to trick individuals into revealing personal, sensitive information or installing malware. However, the observed high frequency of successful phishing attacks suggests that the typical user does not recognize or respond to these threats with the anticipated fear or caution. This discrepancy raises questions about how users perceive digital environments and their inherent threats.

The design of UI plays a pivotal role in shaping user experiences and perceptions in digital environments. Effective UI design can guide user attention, influence decisions, and enhance the usability of technology. However, in the context of cybersecurity, UI design can also be manipulated by cybercriminals to create convincing phishing sites that mimic legitimate ones. The sophistication of these designs often leaves users unable to distinguish between authentic and malicious sites, bypassing the fear response that a clearly dangerous situation would elicit.

Anthropomorphism, or the attribution of human characteristics to non-human entities, is increasingly prevalent in technology design. From virtual assistants to chatbots, technology is being designed to interact with users in more human-like ways. While this can enhance user engagement and satisfaction, it may also affect how users perceive and respond to threats within these interfaces. Anthropomorphizing technology can create a false sense of security, as users might subconsciously apply social trust norms to their interactions with technology.

As users become more comfortable and familiar with technology, they may lower their defenses and become less vigilant in identifying potential threats. This comfort level can lead to a form of complacency, where users overlook security warnings or fail to critically assess the legitimacy of the information presented to them.

The relationship between fear, UI design, anthropomorphism, and comfort with technology creates a complex landscape for cybersecurity. Understanding these factors is crucial in developing more effective strategies to combat phishing attacks. By examining how each of these elements influences user behavior and perception, we can begin to outline approaches for enhancing digital literacy, designing more secure UIs, and fostering a healthy skepticism among users without inducing undue fear. This comprehensive approach could significantly improve phishing resistance, ultimately leading to a safer digital environment for all users.

The following hypothetical case study illustrates how anthropomorphized features of technology can facilitate a successful phishing campaign:

Hypothetical case study: the friendly chatbot phishing campaign

In recent years, the integration of chatbot technology into customer service platforms has become increasingly common. These chatbots are often designed with anthropomorphic features such as names, personalities, and the ability to engage in seemingly human-like conversations. While these features enhance user experience and efficiency, they also open new avenues for cybercriminals to exploit user trust such as during a phishing campaign.

This hypothetical phishing campaign began with cybercriminals creating a fake website that closely mimicked the official site of a well-known financial institution. This fake site included a chatbot named "Eva." Eva was designed to appear as an intelligent assistant, capable of guiding users through various banking queries and tasks.

The criminals then disseminated emails to potential victims, claiming to be from the financial institution and advising them of a security concern with their account. These emails included a link directing users to the fake website for resolution. Upon arriving at the site, users were greeted by Eva, who helped with their security issue.

Eva was programmed to mimic the comforting and trustworthy interaction style of legitimate customer service chatbots. Users were asked to verify their identity by entering sensitive information such as login credentials, account numbers, and even social security numbers. Eva's human-like interactions, complete with empathetic responses and assurances of help, lowered the users' defenses and led them to trust the information they were providing would be handled with care and confidentiality.

Many users, lulled into a false sense of security by their interactions with Eva, complied with the requests. This led to a significant breach of personal and financial information, which the cybercriminals then exploited for unauthorized transactions, identity theft, and further phishing attempts within other contexts.

The case study highlights several critical lessons:

1. **The power of anthropomorphism in technology:** Users tend to attribute human-like trust and expectations to technology that exhibits human-like characteristics. This can cloud judgment and reduce vigilance against potential threats.
2. **The importance of awareness and education:** Raising awareness about the potential for such phishing attacks and educating users on how to recognize and respond to suspicious interactions is crucial. This includes verifying the authenticity of communication through multiple channels and being wary of providing sensitive information online.
3. **Enhanced security measures:** Institutions utilizing chatbots and other anthropomorphized technology must implement robust security measures, including two-factor authentication and regular security audits, to protect against phishing exploits.
4. **Continuous monitoring for phishing threats:** Organizations should employ continuous monitoring of their brand's presence online to quickly identify and take down fake sites and phishing campaigns.

This case study of the friendly chatbot phishing campaign underscores the sophisticated methods employed by cybercriminals to exploit human psychology, particularly using anthropomorphized technology.

3.1 My electronic devices "feel" like mine!

The goal of personalizing technology of making it appear, act, or seem human-like is a deliberate design goal. We customize and personalize our technology experience in such a way it becomes a function of our personal expression. It starts with the appearance of our backgrounds, icons, texts, our selection of emoji's, alert notifications and the sounds associated with them. Our phones, tablets and computers feel like "ours." It is deeply personal. Additionally, this level of customization is a deliberate design feature. Highly customizable technology sells well.

In addition to being highly customizable, the increasing prevalence of artificial intelligence introduces another layer of interaction that further complicates our relationship with technology. Applications like Amazon's Alexa or Apple's Siri not only manage our Internet-enabled home devices, but they can also tell us jokes, change their voices, select our music, ensure we catch the latest weather and arrive to our meetings on time. In many ways, our technology is assuming the role of humans, and in doing so, is being designed with human-like features.

However, while these highly desirable features add to the bottom line, they also afford the phisher an opportunity to exploit an unaware and comfortable user.

3.1.1 Human attentiveness and the deliberate design of technologies

The success of a technology is not only dependent upon its features but also how these features present to a user. In her book *100 Things Every Designer Needs to Know about People* [46], Susan Weinschenk, PhD states:

> Vision trumps all the senses. Half the brain's resources are dedicated to seeing and interpreting what we see. What our eyes physically perceive is only one part of the story. The images coming into our brains are changed and interpreted. It is really our brains that are 'seeing'.

In other words, the brain doesn't see per se, the brain interprets images and attempts to compile images into information that is meaningful. This basic principle is key in computer interface design.

Today, most people spend countless hours behind one screen or the other. The abundance of information coming across these screens would be unmanageable if we were to process every bit of information individually. However, typically users scan a screen and interact with it based upon past experiences and expectations, summarizing the information in lieu of attentive processing of individual data points [46].

To that end, over the years, users have come to expect certain images to mean certain things. Users see object images that indicate what actions they should take. These on-screen objects are often designed to mimic familiar three-dimensional objects such as buttons or keys. These familiar cues prompt a user to select these objects.

In addition to "filling in the blanks," our brains may omit a visual cue if it does not conform to expectations. This phenomenon is known as attention blindness or change blindness. The point is often illustrated in the now famous "Gorilla Video" (found at https://youtu.be/vJG698U2Mvo).

This video was used in a study conducted by Christopher Chabris and Daniel Simmons in 2010. Study participants were asked to watch the video while their eye movements were tracked using eye-tracking equipment. The video involved a basketball game which the study participants were instructed to watch and count the number of times the players in the white shirts passed the ball. In the middle of the game, a person in a gorilla suite walked across the court.

The eye-tracking equipment found that 100 percent of the participants looked at the gorilla. But only 50 percent were aware they had seen the gorilla. In other words, appearance of the gorilla on a basketball court did not conform to expectations, and therefore 50 percent of the study participants were experiencing attention blindness.

While people can be distracted by several things, they can also focus on specific things. This is referred to as selective attention. People can be intentionally selective in what they pay attention to, and they can also be unconsciously attentive to something [46].

Humans are also known to process information selectively. This may be intentional or unintentional. However, in general, humans can only apply selective attention to a matter for up to 10 minutes.

Additionally, actions performed regularly become habituated and can be relegated to subconscious processing. This manner of working through mental inputs serves to conserve mental resources and is deployed by all humans [47].

Another feature affecting attention involves the frequency of events. Our minds recognize and operate within the realm of pattern development. Event frequencies form mental patterns and help us quickly process a phenomenon based on its associated components relative to the pattern. Frequency of events is a data point within these mental models, and positions us to make assumptions about our environment and therefore decisions [46].

Lastly, humans can only sustain a level of focused attention for up to 10 minutes. If our sustained attention were required for a longer period of time, we would need to pause our activity and "change it up" to restart our 10-minute sustained attention window.

The brain's role in interpreting visual stimuli, as Weinschenk points out, underscores the profound impact of design on user experience and behavior. The fact that vision dominates our sensory intake means that the design of UI can powerfully influence how information is perceived and processed. In the realm of cybersecurity, this understanding becomes crucial when designing systems that are both user-friendly and secure against phishing attacks.

In addition to the fundamental role the brain plays in processing visual information, users bring a set of expectations to their interactions with technology, shaped by past experiences. These expectations influence how users interpret visual cues on a screen. Designers often use this to their advantage by incorporating familiar visual elements, such as icons and buttons, that suggest certain actions. This strategy enhances usability but can also be exploited by phishers who design fraudulent websites with familiar-looking elements to deceive users into thinking they are interacting with a legitimate site.

The phenomenon of attention blindness, as demonstrated by the "Gorilla Video," reveals how easily important visual information can be overlooked if it does not align with expectations. Phishers exploit this by crafting emails and websites that look legitimate enough briefly, knowing that many users will not notice subtle signs of fraud. By blending in with the expected digital environment, phishing attempts bypass the users' scrutiny.

Selective attention allows individuals to focus on relevant information while ignoring distractions. However, this focus can be manipulated by phishers who use urgent or alarming language to capture and steer users' attention toward specific actions, like clicking on a link or downloading an attachment, without giving them a reason to doubt the authenticity of the request.

The habitual nature of many online activities, such as checking email or logging into accounts, means that actions often become automatic. Phishers take advantage of this automatism by mimicking the look and feel of routine communications from trusted sources. Users, operating out of habit, may not critically assess these communications, leading to successful phishing attacks.

Humans are pattern-recognizing creatures, a trait that plays a significant role in how we process repeated events and make decisions. Phishing attempts often mimic the frequency or patterns of legitimate communications to blend in. By doing so, they increase the likelihood of being accepted as part of the user's expected digital environment, making it easier to deceive the user.

This deeper understanding of human perception and cognition highlights the importance of designing cybersecurity measures that account for these innate human tendencies. Educating users about these psychological vulnerabilities can help increase awareness and vigilance. Moreover, designing interfaces that not only meet user expectations for familiarity and usability but also incorporate elements that encourage critical assessment and verification can help mitigate the risk of phishing attacks.

The interplay between human perception, interface design, and cybersecurity forms a complex web that must be navigated with care. By acknowledging and addressing the psychological aspects of user interaction with technology, designers and cybersecurity professionals can work together to create safer digital environments. This holistic approach, which considers both the human and technological aspects of cybersecurity, is essential for developing effective defenses against the ever-evolving threat of phishing.

3.1.2   Human attention and the phish

If the nature of human attention is selective, habitual, and limited in scope and duration, attempting to discern a well-crafted, nefarious electronic communication that appears during routine daily activities would seem like a difficult task.

This difficulty might be mitigated if we were operating in an environment that triggers our natural defenses (the caution we would exercise in a dark alley). But we are often looking at the screens of our own electronic devices, which have been customized to our liking, complete with colors, sounds, and images. There is no dark alley experience when we are checking the emails on our iPhones.

If we understand fear to be an adaptive but phasic (transient) state elicited through confrontation with a threatening stimulus, nothing about our electronics signals to us that we are encountering a threatening stimulus.

If comforting design is enough to lower our guard, what will occur when we use anthropomorphized technologies? How do these variables affect a user's ability to focus and direct their attention?

The human ability to focus and pay attention during interactions with technology plays a crucial role in navigating the vast and complex landscape of the Internet. This capacity, however, is under constant assault from a barrage of notifications, multitasking demands, and the sheer volume of information available online. These factors contribute to a fragmented attention span, making users more susceptible to cybersecurity threats, notably phishing scams.

Human attention is a finite resource, divided into several types, including sustained, selective, and divided attention. Sustained attention allows us to focus on a single task over a prolonged period, whereas selective attention enables us to prioritize one stimulus over others. Divided attention, often referred to as multitasking, involves splitting focus between multiple tasks simultaneously. In the context of digital interaction, users frequently switch between these types of attention, attempting to navigate and process information from various sources.

The digital environment, with its constant notifications and endless streams of content, places considerable demand on our selective and divided attention. This demand can lead to a state of continuous partial attention, where the user is perpetually scanning the environment without fully engaging with any single task. This state is particularly conducive to overlooking subtle cues that might indicate a phishing attempt, such as slight discrepancies in a website's URL or unexpected requests for personal information.

A diminished focus can significantly impact a user's ability to identify and avoid phishing scams. Phishing attempts are designed to deceive, often mimicking legitimate communications from trusted entities. A user with fragmented attention is more likely to miss the telltale signs of such attempts, including typographical errors, unusual sender addresses, or incongruities in the email's layout and design.

Moreover, the psychology behind phishing exploits the human propensity for error under conditions of reduced attention. Cybercriminals rely on social engineering tactics, such as creating a sense of urgency or invoking fear, to prompt hasty actions from the target. For instance, an email claiming that an account will be closed unless immediate action is taken can provoke a quick, unconsidered response from someone already distracted by multiple other tasks.

The challenge of maintaining focus and attention in a digitally saturated environment is significant, with direct implications for cybersecurity. The nature of human attention, susceptible to distraction and fragmentation, creates vulnerabilities that phishing attempts exploit. Digitally saturated world, the constant barrage of information and stimuli from various devices and platforms can have a profound impact on our cognitive abilities, particularly our capacity for sustained focus and concentration. This phenomenon creates an environment ripe for exploitation by cybercriminals through phishing attacks.

The human brain is not wired to handle the overwhelming amount of digital input we encounter daily. Our attention spans are constantly divided as we juggle multiple tasks, switch between apps, and consume an endless stream of notifications, emails, and social media updates. This state of perpetual distraction and cognitive overload can lead to a phenomenon known as "digital burnout," where our mental resources become depleted, and our ability to concentrate and make sound judgments is compromised.

When our minds are in this fragmented state, we become more susceptible to the tactics employed by phishing attackers. Phishing emails and messages are designed to exploit our cognitive vulnerabilities by creating a sense of urgency, fear, or curiosity, prompting us to act impulsively without proper scrutiny. The constant bombardment of digital stimuli has conditioned us to react quickly to new information, making it easier for phishers to bypass our critical thinking faculties and lure us into their traps.

Moreover, the digital environment has fostered a culture of multitasking, where we constantly switch between different tasks and contexts. This cognitive switching comes at a cost, as our brains struggle to maintain focus and attention on any single task. In this state of divided attention, we become more susceptible to missing subtle cues or red flags that could indicate a phishing attempt, increasing the likelihood of falling victim to these attacks.

The impact of digital saturation on our ability to concentrate is further exacerbated by the addictive nature of many digital platforms and applications. These technologies are designed to capture and hold our attention through various psychological triggers such as intermittent rewards, social validation, and fear of missing out (FOMO). As we become increasingly dependent on these digital stimuli, our ability to disengage and focus on other tasks diminishes, leaving us vulnerable to the manipulative tactics employed by phishing attackers.

Furthermore, the constant exposure to digital media and the ease of sharing information online can shape our perceptions of reality and erode our critical thinking skills. We become accustomed to consuming information in bite-sized chunks, often without verifying its authenticity or considering alternative perspectives. This cognitive bias can make us more susceptible to believing the deceptive narratives presented in phishing attempts, as we may fail to scrutinize the information or question its legitimacy.

In this digitally saturated environment, where our attention spans are fragmented, our cognitive resources are depleted, and our critical thinking skills are compromised, phishing attacks thrive. Cybercriminals exploit our vulnerabilities, capitalizing on our distracted state and our tendency to react impulsively to digital stimuli. By understanding the psychological and cognitive impacts of digital saturation, we can better equip ourselves to resist these attacks and cultivate more mindful and focused digital habits.

The constant influx of information from various digital devices and platforms fragments our attention, making it easier for attackers to deceive and manipulate us. This exploitation of human distractions is a cornerstone of many social engineering attacks, particularly phishing.

The modern digital landscape is characterized by an overwhelming amount of stimuli, from emails and social media notifications to instant messages and news updates. This constant bombardment leads to cognitive overload, where our mental resources are stretched thin. Multitasking, often seen as a productivity booster, actually diminishes our ability to focus and increases the likelihood of errors. Each shift in attention results in a loss of cognitive energy and focus, making us more susceptible to phishing attacks. Cybercriminals are well aware of this and design their attacks to exploit these moments of distraction.

Cybercriminals use psychological tactics to exploit our natural tendencies and cognitive biases. For instance, they often create a sense of urgency or fear in their phishing messages, compelling recipients to act quickly without thorough scrutiny. This manipulation leverages the affect heuristic, where decisions are made based on emotions rather than rational thought. By inducing stress or panic, attackers can hijack our executive functioning, leading us to click on malicious links or provide sensitive information impulsively.

Decision fatigue is another critical factor that cybercriminals exploit. As we navigate through numerous choices and decisions daily, our mental energy depletes, making us more prone to errors. Attackers capitalize on this by sending phishing emails at times when people are likely to be tired or overwhelmed, such as late in the day or during busy periods. In these states, individuals are less likely to notice subtle cues that indicate a phishing attempt, increasing the chances of a successful attack.

Cybercriminals also exploit various cognitive biases to enhance the effectiveness of their attacks. For example, the authority bias makes individuals more likely to comply with requests from perceived authority figures. Phishers often impersonate executives or IT personnel to gain trust and prompt action. Similarly, the reciprocity bias can be exploited by offering a small favor or helpful information, making the target feel obligated to reciprocate by providing sensitive data.

With phishing being the most common form of social engineering attack that exploits human distractions, the efficacy of the attacks centers around sending emails that appear to come from reputable sources, tricking recipients into revealing sensitive information or downloading malware. Spear phishing, a more targeted form of phishing, uses personalized information gathered from social media and other sources to craft convincing messages. This personalization increases the likelihood of the target falling for the scam, as the message appears more legitimate.

Our extensive digital footprints provide cybercriminals with the information needed to craft convincing phishing messages. By analyzing social media profiles, attackers can gather personal details such as names, job titles, and interests, which they use to tailor their attacks. This level of customization makes it harder for individuals to distinguish between legitimate and malicious communications, especially when they are distracted or overwhelmed.

To mitigate the risk of falling victim to phishing attacks, it is crucial to adopt a mindful approach to digital interactions. Cybersecurity mindfulness involves being hyper-attentive to one's surroundings and digital activities, reducing the likelihood of making impulsive decisions. Training programs that emphasize the importance of focusing on one task at a time and recognizing the signs of phishing can significantly enhance an individual's ability to resist these attacks. Additionally, organizations should implement robust security awareness training that educates employees about cognitive biases and how to counteract them.

Given the relationship between attention and vulnerability to phishing, it is imperative to adopt strategies that enhance focus and cybersecurity awareness:

- **Digital hygiene practices:** Encouraging habits such as regularly checking for software updates, using strong, unique passwords, and employing multi-factor authentication can provide foundational security against phishing.
- **Focused browsing sessions:** Dedicate specific times for checking emails or browsing the internet, minimizing the number of open tabs, and disabling non-essential notifications to reduce distractions.
- **Education and training:** Regularly updated training sessions that simulate phishing attempts can increase awareness and improve the ability to identify potential threats.
- **Use of technological aids:** Tools such as ad blockers, email filters, and security software can reduce the cognitive load on users, allowing them to focus more effectively on potential threats.

3.2   A deeper look into anthropomorphized technology

The dictionary definition of anthropomorphism states anthropomorphism is an interpretation of what is not human or personal in terms of human or personal characteristics. As it relates to technology, anthropomorphized technology can be understood as a technology that is designed to have human-like characteristics such as a voice, face, or even a personality. Examples of anthropomorphized technologies include virtual assistants such as Siri and Alexa. However, what makes something anthropomorphized is not agreed upon by all.

A recent study conducted by Li and Suh resulted in participants dividing anthropomorphism into six categorical definitions [48].

- twenty-six percent of the respondents described it as a tendency to attribute human or humanlike characteristics to something non-human.
- twenty-three percent of the respondents described it as a user's perception of AI-enabled technology as human-like.
- eleven percent of the respondents described it as a technological stimuli that features human likenesses of AI-enabled technology (appearance, emotions, motions, etc.).
- six percent of the respondents described it as a process whereby individuals attribute human or human-like characteristics to technology.
- six percent of the respondents described it as conceptualized anthropomorphism which involves an inference made by users attributing human likenesses to technology.
- twenty-six percent of the respondents did not define anthropomorphism.

When combining these definitions, an overarching definition emerges and can be understood as:

Anthropomorphism is a tendency, perception, process, or inference of humanlike characteristics, human likenesses, or human appearance relative to a non-human or technology.

The genesis of anthropomorphized technology can be traced back to the human tendency to attribute human-like qualities, emotions, and intentions to non-human entities. This inclination stems from our social nature and the cognitive shortcuts our brains use to navigate the complex social environments we evolved in.

As technology has become increasingly integrated into daily life, the design of anthropomorphized technology has evolved to make these interactions more natural, engaging, and efficient. This approach to design leverages our innate social behaviors to foster more intuitive and meaningful interactions with technology.

Anthropomorphizing technology is a way to bridge the gap between humans and machines, making technological devices and applications more accessible and relatable to users. By designing technology that mimics human characteristics, creators aim to:

- **Enhance user engagement:** Anthropomorphized features such as conversational interfaces or character-like robots engage users on a more personal level, increasing the likelihood of continued use and loyalty to the technology or brand.

- **Improve usability:** Technologies designed with human-like interactions can reduce the learning curve for users. When devices behave in predictably human ways, users can apply their social and interpersonal understanding to interactions with technology.
- **Foster emotional connection:** By invoking empathy and emotional responses, anthropomorphized technologies can create a sense of companionship and trust. This is particularly evident in technologies like social robots and virtual assistants, where users often attribute personalities and emotions to these devices.

As such, the design of anthropomorphized technology offers significant financial benefits, driven by its impact on user engagement, satisfaction, and brand loyalty. Key financial advantages include:

- **Increased product adoption:** Products that users find relatable and easy to use are more likely to be adopted quickly, expanding the user base and accelerating market penetration.
- **Enhanced customer loyalty:** Emotional connections fostered by anthropomorphized design can translate into stronger brand loyalty. Satisfied users are more likely to become repeat customers and advocate for the brand within their social circles.
- **Premium pricing opportunities:** Products offering a more personalized and engaging experience can command higher prices. Consumers may be willing to pay a premium for products that not only meet their needs but also resonate with them on an emotional level.
- **Cross-selling and upselling potential:** Engaged and loyal customers are more receptive to additional offers from a brand they trust. Anthropomorphized technologies can facilitate deeper insights into user preferences, enabling more targeted and effective cross-selling and upselling strategies.
- **Competitive differentiation:** In crowded markets, the unique user experience provided by anthropomorphized technology can serve as a key differentiator. This uniqueness can attract media attention, generate buzz, and position the company as an innovator, further driving sales and market share.

While the financial and engagement benefits are compelling, designing anthropomorphized technology creates a tension in other areas.

In addition to the previously discussed cybersecurity challenges relative to anthropomorphized technologies, ethical considerations, such as privacy concerns and the potential for over-reliance on technology, must be addressed. Moreover, there is a delicate balance between making technology relatable and inadvertently creating unrealistic expectations of its capabilities, which could lead to user frustration and disengagement.

The genesis of anthropomorphized technology is deeply rooted in human psychology, aiming to make our interactions with technology more natural and engaging. By leveraging our propensity to attribute human qualities to non-human entities, designers can create products that are not only more intuitive and enjoyable to use but also capable of forging emotional connections with users.

The financial benefits of adopting this approach are significant, offering opportunities for increased adoption, loyalty, and revenue. However, it is crucial for designers and companies to navigate the ethical implications carefully, ensuring that anthropomorphized technology enhances human life without exploiting or diminishing the value of human interactions.

In many ways, anthropomorphizing technology was inevitable. Anthropomorphism has long been a part of human nature, as individuals have a tendency to ascribe human-like qualities to inanimate objects, animals, and even abstract concepts. This inclination stems from our innate desire to understand and relate to the world around us. In the context of technology, anthropomorphism can manifest in various forms such as voice assistants with human-like personalities, chatbots that mimic human conversation patterns, or UI designed to evoke emotional responses.

The study on voice intelligence highlights how anthropomorphic characteristics can enhance users' perceived ease of use, usefulness, and security, thereby increasing their willingness to adopt the technology. This increased trust and comfort with anthropomorphized technology could potentially make users more susceptible to phishing attacks if they are less vigilant about security protocols.

While not directly related to anthropomorphized technology, the case involving deepfake scammers demonstrates how advanced AI can create convincing forgeries of individuals, leading to significant financial losses. This example underscores the potential for human-like representations to deceive users into trusting and acting on fraudulent communications.

Similarly, the broader discussion on anthropomorphism in AI suggests that attributing human-like qualities to technology can lead to misplaced trust and reduced critical scrutiny. This psychological manipulation can be leveraged in phishing attacks, where users might be more inclined to follow instructions from a seemingly trustworthy source.

To illustrate how anthropomorphized technology could facilitate phishing attacks, consider the following hypothetical scenario:

Imagine a phishing attack where cybercriminals deploy chatbots designed to mimic human customer service representatives. These chatbots use natural language processing and anthropomorphic cues, such as personalized greetings and empathetic responses, to build rapport with users. By creating a sense of trust and familiarity, the chatbots could persuade users to disclose sensitive information or click on malicious links, believing they are interacting with a legitimate service representative.

In this scenario, the anthropomorphic qualities of the chatbots exploit the human tendency to trust and engage with entities that exhibit human-like traits, making it easier for cybercriminals to bypass users' defenses and gain access to sensitive information.

While the sources do not provide a direct case study of anthropomorphized technology facilitating phishing attacks, they underscore the potential risks associated with increased trust and reduced vigilance when interacting with human-like technology. Cybercriminals can exploit these psychological tendencies to craft more convincing and effective phishing schemes.

To mitigate these risks, it is crucial for individuals and organizations to remain vigilant and adopt a critical mindset when interacting with any technology, regardless of its anthropomorphic qualities. Cybersecurity awareness training programs should emphasize the importance of scrutinizing all communications, even those that appear to be from trusted sources, and encourage users to verify the authenticity of requests before taking any action. Additionally, organizations should implement robust security measures, such as multi-factor authentication, to add an extra layer of protection against phishing attacks, even in cases where users may have been initially deceived by anthropomorphized technology.

3.2.1   *Anthropomorphized technology and the dangers of phishing*

Customizing our technology to a point where we can relate to it in a deeply personal manner enhances our experience. However, with those experiences, we also enhance our "trust" relative to technology.

Recent studies involving chatbots suggest anthropomorphism positively influences trust and satisfaction with the chatbot. It also increases the chatbot's effectiveness [49].

Konya-Baumbach *et al.* [49] attributed these trust factor increases to an experience referred to as social presence. The authors state:

> Our work identifies social presence as the underlying psychological mechanism that drives the effectiveness of chatbot anthropomorphism.

Increasing our level of trust relative to our technologies, i.e. devices that we've already personalized, and made "comfortable," by including anthropomorphized features, such as a Chatbot, or a personal assistant, we've now stepped into a situation where resisting a cyberattack centered around deception becomes increasingly difficult.

Spear phishing campaigns use targeted methods to send communications we would normally expect. Such communications appear familiar to us, for example, an email from our bank. If the email includes an anthropomorphized character, such as an animated avatar, the recipient may be more likely to trust the message and click on a link or provide personal information.

Research has shown that people are more likely to trust information that is presented by an anthropomorphized character.

In a study conducted by Rhim *et al.* [50], participants were more likely to trust a message that was presented by an anthropomorphized chatbot than a message that was presented in a non-anthropomorphized manner.

Additionally, research has shown that people are more likely to disclose personal information to an anthropomorphized character than to a non-anthropomorphized character.

In a study conducted by Go and Sundar [51], participants were more likely to disclose personal information to an anthropomorphized chatbot than to a non-anthropomorphized chatbot. This finding suggests that people may be more susceptible to phishing attacks when using anthropomorphized technology.

With anthropomorphized technologies now exploding onto our technology landscape, a safe bet can be made that these technologies will be increasingly seen in phishing campaigns as well.

These newer technologies have created new opportunities for perpetrators to exploit anthropomorphized technologies on several levels. Anthropomorphized devices such as chatbots, voice assistants, and social robots have become integral to our daily lives. These technologies, designed to mimic human traits and behaviors, offer convenience and efficiency but also introduce new vulnerabilities that cybercriminals can exploit. By leveraging the human-like characteristics of these technologies, attackers can manipulate users' trust and familiarity, making them more susceptible to phishing attacks and other forms of cyber exploitation.

Anthropomorphism, the attribution of human traits to non-human entities, significantly impacts how users interact with technology. When users perceive technology as more human-like, they tend to develop a sense of trust and emotional connection. This psychological phenomenon can be exploited by cybercriminals to deceive users into revealing sensitive information or performing actions that compromise their security.

Voice assistants like Siri, Google Assistant, and Alexa are another form of anthropomorphized technology that can be exploited. Researchers have demonstrated how attackers can use inaudible commands, such as those embedded in ultrasound or obfuscated sounds, to control these devices remotely. These attacks can be executed without the user's knowledge, allowing cybercriminals to steal information, open malicious websites, or perform unauthorized actions. The trust users place in their voice assistants, combined with the devices' ability to process commands seamlessly, makes them vulnerable to such sophisticated attacks.

Social robots, designed to interact with humans in public and private spaces, can also be exploited for phishing and other malicious activities. These robots often exhibit human-like behaviors and emotions, which can create a strong bond with users. Cybercriminals can exploit this bond by programming social robots to engage in social engineering tactics such as impersonating trusted individuals or entities. For example, a social robot could be programmed to simulate a security alert, prompting users to provide sensitive information to resolve a fabricated issue.

A recent case involving deepfake technology highlights the potential for human-like representations to deceive users. In this instance, advanced AI was used to create convincing forgeries of individuals, leading to significant financial losses. While not directly involving anthropomorphized technology, this case underscores the risks associated with human-like representations and the ease with which they can be used to manipulate trust and extract sensitive information.

To mitigate the risks associated with anthropomorphized technology, it is crucial for individuals and organizations to remain vigilant and adopt a critical mindset when interacting with any technology, regardless of its human-like qualities. Cybersecurity awareness training programs should emphasize the importance of scrutinizing all communications, even those that appear to be from trusted sources, and encourage users to verify the authenticity of requests before taking any action.

Additionally, organizations should implement robust security measures, such as multi-factor authentication, to add an extra layer of protection against phishing attacks, even in cases where users may have been initially deceived by anthropomorphized technology.

3.3 Appeal versus alarm

The conditions under which we are using technologies are primarily designed to appeal to us. We can customize our technologies to a point that our devices and machines are experienced as something exclusively "ours." In addition, to the highly customizable features, we now see our technologies increasingly include anthropomorphized capabilities.

These converging experiences create a sense of social presence relative to our technologies, an experience of having a social interaction versus a machine-to-human interaction.

When adding the natural limitations of our attention to the equation, the ability to detect subtle, well-designed threats, such as a phishing email, becomes increasingly difficult.

However, it can be assumed that designing threatening, or foreboding technology, or even sanitized and sterile technology may involve undesirable effects relative to the bottom line. Additionally, introducing an experience of dread or foreboding into the computing experience raises several ethical questions. Do interface designers really want to inflict a traumatizing experience on users? Would productivity be reduced if technology were less appealing?

To each of these points, the intersection between user and technology forms the nexus at which the computing experience is staged. The moment the screen comes to life, our personal access codes entered and our private on-ramp to the virtualized world initialized, we transition mentally from our space into a psychological space, a space formulated by our own projections, experiences, and expectations. Conveying a need to add protections to this experience is counter intuitive and requires mental resources that are not infinite in nature. Maintaining vigilance relative to computer security is a difficult task. In addition, anthropomorphized devices such as chatbots, voice assistants, and social robots have become integral to our daily lives.

Anthropomorphism, the attribution of human traits to non-human entities, significantly impacts how users interact with technology. When users perceive technology as more human-like, they tend to develop a sense of trust and emotional connection. This psychological phenomenon can be exploited by cybercriminals to deceive users into revealing sensitive information or performing actions that compromise their security.

Chatbots, widely used in customer service and support roles, are prime targets for exploitation. Cybercriminals can create realistic fake chatbots or take over legitimate ones to conduct phishing attacks. These chatbots engage users in see-mingly genuine conversations, using natural language processing and anthro-pomorphic cues such as personalized greetings and empathetic responses. By building rapport and trust, these malicious chatbots can persuade users to disclose sensitive information, such as login credentials or financial details, or to click on malicious links.

Voice assistants like Siri, Google Assistant, and Alexa are another form of anthropomorphized technology that can be exploited. Researchers have demon-strated how attackers can use inaudible commands, such as those embedded in ultrasound or obfuscated sounds, to control these devices remotely. These attacks can be executed without the user's knowledge, allowing cybercriminals to steal information, open malicious websites, or perform unauthorized actions. The trust users place in their voice assistants, combined with the devices' ability to process commands seamlessly, makes them vulnerable to such sophisticated attacks.

Social robots, designed to interact with humans in public and private spaces, can also be exploited for phishing and other malicious activities. These robots often exhibit human-like behaviors and emotions, which can create a strong bond with users. Cybercriminals can exploit this bond by programming social robots to engage in social engineering tactics such as impersonating trusted individuals or entities. For example, a social robot could be programmed to simulate a security alert, prompting users to provide sensitive information to resolve a fabricated issue.

A recent case involving deepfake technology highlights the potential for human-like representations to deceive users. In this instance, advanced AI was used to create convincing forgeries of individuals, leading to significant financial losses. While not directly involving anthropomorphized technology, this case underscores the risks associated with human-like representations and the ease with which they can be used to manipulate trust and extract sensitive information.

To mitigate the risks associated with anthropomorphized technology, it is crucial for individuals and organizations to remain vigilant and adopt a critical mindset when interacting with any technology, regardless of its human-like quali-ties. Cybersecurity awareness training programs should emphasize the importance of scrutinizing all communications, even those that appear to be from trusted sources, and encourage users to verify the authenticity of requests before taking any action.

Additionally, organizations should implement robust security measures, such as multi-factor authentication, to add an extra layer of protection against phishing attacks, even in cases where users may have been initially deceived by anthro-pomorphized technology.

AI's ability to anthropomorphize technology, endowing non-human entities with human-like qualities, is perhaps the most concerning aspect of AI-driven phishing attacks. AI's evolution has led to systems capable of mimicking human conversation, learning from interactions, and even expressing emotions. Examples include conversational agents like Siri and Alexa, and AI in social robots.

As previously discussed, anthropomorphism in technology refers to the attri-bution of human characteristics, intentions, or emotions to non-human entities.

AI algorithms play a pivotal role in personalizing phishing emails to an unprecedented degree. As with shaping a user's context, by analyzing a user's online behavior, interests, and communication patterns, AI can tailor phishing messages to mimic human interactions, creating a sense of familiarity and trust.

Wright and Marett [44] have shown that personalized emails, which appear to understand the recipient's preferences and habits, are more likely to elicit a response than canned messages that appear to have been generated through automation. This level of personalization reaches a critical point when enabled by AI. Distinguishing the difference between a human-generated communication versus a communication generated by AI is nearly impossible for users to distinguish.

To that end, AI's capability to mimic human communication styles further enhances the effectiveness of phishing emails. AI-driven tools can replicate tone, language style, and specific phrases that are commonly used by individuals or within organizations. Such detailed replication in communication can make phishing emails appear as though they are from a known or trustworthy source. This increases the risk of recipients trusting and acting on the deceptive content of these emails.

Infusing phishing emails with emotionally charged language is a key feature of AI that employs anthropomorphism. Crafting messages that evoke fear, urgency, or empathy can manipulate recipients' emotions, a tactic proven to increase the success rate of phishing attacks.

Over 18 years ago, Jakobsson and Ratkiewicz [45] studied how emotional manipulation in phishing exploits human vulnerabilities, making recipients more likely to engage with the content without skepticism. Nothing since then has changed relative to user's ability to manage their emotional experience relative to electronic communications. This makes today's computing environment fertile ground for attackers leveraging AI-driven phishing attacks.

3.3.1   Ethical considerations

Anthropomorphizing AI raises several ethical questions, including the manipulation of emotional responses and the responsibility of AI developers in managing users' expectations.

The human tendency to anthropomorphize is deeply rooted in our psychology and evolutionary history. It is a cognitive bias that allows us to navigate our social world by applying familiar schemas to understand and predict the behavior of others, including non-human agents.

This inclination is amplified in the context of AI by design choices that encourage users to ascribe human-like qualities to these systems. The motivations for designers to anthropomorphize AI are multifaceted, including the desire to make technology more accessible, intuitive, and engaging for users.

Anthropomorphizing AI can significantly affect the nature of human–AI interactions. On one hand, it can enhance user engagement, trust, and satisfaction by making interactions more natural and less mechanical. On the other hand, it can lead to unrealistic expectations of AI's capabilities, misunderstanding of its limitations, and an overreliance on technology.

These outcomes can have both practical and ethical implications, influencing users' decision-making, privacy, and emotional well-being.

Designers and developers are tasked with creating systems that are not only efficient and user-friendly but also ethically sound and secure. The anthropomorphic features that make AI systems appear more relatable and trustworthy can also make them a tool for malicious exploitation. Phishing, a cybersecurity threat that relies heavily on social engineering, becomes particularly concerning in this context. By impersonating trusted entities, attackers can leverage the human-like qualities of AI to deceive users into divulging sensitive information.

One of the primary ethical considerations is the responsibility of designers to anticipate how the human-like attributes of AI can be misused. Designers must navigate the fine line between creating engaging, relatable AI interfaces and ensuring these interfaces do not mislead users about the nature of their interaction with machines. Transparency about the AI's non-human status, its capabilities, and its limitations becomes crucial in maintaining an ethical balance. Users should be made aware, in clear terms, that they are interacting with AI, potentially mitigating some risks associated with phishing attempts.

Furthermore, the design of anthropomorphized AI must prioritize user privacy and consent. The collection and use of personal data—practices often integral to the functionality of AI systems—must be handled with utmost care. Users must be informed about what data is being collected, how it is being used, and whom it is being shared with. This transparency not only upholds ethical standards but also builds trust with users, reducing the likelihood of successful phishing attacks by making users more wary of unsolicited requests for personal information.

The emotional and psychological engagement these AI systems foster can blur the lines between human and machine interaction, making users more susceptible to phishing schemes. Cybercriminals can exploit these emotional connections, crafting more convincing phishing messages that mimic the tone and personality of trusted AI assistants. Therefore, the security measures integrated into the design of anthropomorphized AI must be robust, capable of identifying and mitigating phishing threats without compromising the user experience.

To address these ethical and security concerns, a multidisciplinary approach is essential. Collaboration between AI developers, cybersecurity experts, and ethicists can lead to the development of guidelines and best practices for designing anthropomorphized AI. These guidelines should emphasize ethical considerations, such as transparency, privacy, and user consent, while also outlining specific security measures to counteract phishing and other cyber threats. Continuous evaluation and adaptation of these guidelines will be necessary as both AI technology and cybersecurity threats evolve.

The design of anthropomorphized AI presents a unique intersection of ethical and cybersecurity considerations. The very qualities that make these AI systems appealing and effective—human-like interaction and emotional engagement—also make them potential vectors for phishing attacks. Addressing these challenges requires a careful balance between innovation and responsibility, ensuring that the development of anthropomorphized AI is guided by ethical principles and robust security measures. As we continue to integrate AI more deeply into our digital lives, the importance of navigating these ethical and security considerations cannot be overstated. The future of anthropomorphized AI must be built on a foundation of ethical integrity and cybersecurity resilience, ensuring that these technologies serve the best interests of users while safeguarding against the ever-evolving landscape of cyber threats.

The following scenario, while hypothetical, demonstrates how anthropomorphized AI can exploit user bias and craft an effective phishing campaign:

Scenario: the AI-generated executive phish

In this scenario, cybercriminals use an advanced AI system capable of generating human-like text and voice simulations. This AI has been trained on vast datasets including social media posts, corporate communications, and publicly available voice recordings of company executives.

Development of the phish:

1. ***Data aggregation****: The AI begins by aggregating publicly available data on a target company, focusing on its organizational structure, ongoing projects, and the idiosyncratic communication styles of its executives.*
2. ***Simulation creation****: Leveraging this data, the AI crafts a series of emails and voice messages that closely mimic the tone, style, and usual content of communications from the company's CEO. The messages are tailored to address the recipients by name, referencing specific, real projects they are involved with, and incorporating recent corporate achievements or events to enhance believability.*
3. ***Execution****: The phishing attack is launched with emails and voice messages being sent to mid-level managers. The messages urge immediate action on a pressing matter, such as approving a payment or providing sensitive information, citing a fabricated, yet plausible, urgent business need. The requests direct the targets to a malicious website disguised as an internal corporate tool or prompt them to send information directly to a compromised email address.*

Anthropomorphizing effects and ethical implications

The effectiveness of this phishing attack hinges on the anthropomorphizing capabilities of the AI. By generating communications that appear genuinely human, complete with the nuances of individual communication styles and emotional cues, the AI exploits the human tendency to trust familiar voices and textual styles. This not only demonstrates the technical prowess of AI in mimicking human behaviors but also underscores several ethical concerns:

1. ***Deception enhanced by humanization****: The anthropomorphizing effect of the AI intensifies the deception, leading targets to believe they are interacting with a real, trusted human being. This raises ethical questions about the manipulation of human trust and the erosion of interpersonal trustworthiness in digital communications.*

2. ***Emotional manipulation****: By crafting messages that mimic the emotional tone of real executives, the AI manipulates the recipients' emotions, exploiting feelings of urgency, loyalty, or fear. This manipulation for malicious ends highlights the ethical dangers of using AI to engineer trust through anthropomorphization.*
3. ***Transparency and accountability****: The scenario reveals a lack of transparency and accountability in the use of AI for malicious purposes. It prompts a discussion on the need for mechanisms to trace AI-generated content and hold creators accountable, emphasizing the importance of ethical guidelines and regulations in AI development and deployment.*
4. ***Informed consent and autonomy****: Victims of this phishing attack are deceived into taking actions without informed consent, undermining their autonomy. This points to the broader ethical issue of consent in the digital age, where the line between human and machine-generated content is increasingly blurred.*

This hypothetical scenario of an AI-developed phishing attack showcases the complex interplay between anthropomorphizing AI and cybersecurity threats. It illustrates not only the technological capabilities of AI but also the profound ethical implications of using AI to mimic human attributes for deceptive purposes. Addressing these challenges requires a concerted effort from technologists, ethicists, policymakers, and the public to foster an environment where the benefits of AI can be realized without compromising ethical standards or human trust.

3.3.2   Where is cybersecurity in the development of anthropomorphized AI?

The rapid advancement of artificial intelligence (AI) and its integration into various technologies, particularly chatbots, has revolutionized how businesses operate and interact with customers. However, this swift progression has often outpaced the implementation of robust cybersecurity measures, leading to significant vulnerabilities. Understanding why cybersecurity has been omitted in the development of AI and chatbots requires an examination of several key factors, including the prioritization of innovation, the complexity of AI systems, the lack of standardized security protocols, and the insufficient awareness and expertise in AI security.

One of the primary reasons cybersecurity has been overlooked in the development of AI and chatbots is the intense focus on innovation and rapid deployment. In the highly competitive tech industry, companies are under constant pressure to release new features and products quickly to gain a market edge. This rush to innovate often results in a trade-off where functionality and user experience are prioritized over security. Developers may cut corners on security testing and vulnerability assessments to meet tight deadlines, leading to the deployment of AI systems and chatbots with inadequate security measures.

AI systems, including chatbots, are inherently complex, relying on vast amounts of data and sophisticated algorithms to function effectively. This complexity makes it challenging to identify and mitigate security vulnerabilities. Traditional security measures may not be sufficient to protect AI systems, which require specialized approaches to address unique threats such as adversarial attacks, data poisoning, and model inversion. The intricate nature of AI systems means that even minor oversights in security can lead to significant vulnerabilities that are difficult to detect and rectify.

The field of AI and chatbot development lacks standardized security protocols and best practices. Unlike more mature technologies, where industry standards and regulatory frameworks guide security practices, AI and chatbots operate in a relatively unregulated environment. This lack of standardization means that security measures can vary widely between different implementations, leaving some systems more vulnerable than others. The absence of a unified approach to AI security has resulted in inconsistent and often inadequate protection against cyber threats.

Many organizations lack the necessary awareness and expertise to implement effective cybersecurity measures for AI and chatbots. The rapid pace of technological advancement means that cybersecurity professionals must continuously update their knowledge and skills to keep up with emerging threats. However, there is a shortage of skilled cybersecurity professionals with expertise in AI security, leading to gaps in protection. This skills gap is exacerbated by the fact that many developers and engineers working on AI projects may not have a strong background in cybersecurity, resulting in the omission of critical security considerations during the development process.

The omission of cybersecurity in the development of chatbots has led to several specific vulnerabilities:

- Some chatbots operate without encryption, meaning that any information sent to the bot could potentially be accessed and read by unauthorized parties. This lack of encryption poses a significant risk to digital privacy, as sensitive information can be intercepted and exploited by cybercriminals.
- A chatbot is only as secure as its host network. If the network hosting the chatbot lacks robust security measures or suffers a cybersecurity attack, any information sent to the chatbot is also vulnerable. This dependency on the host network's security highlights the need for comprehensive security protocols that encompass both the chatbot and its underlying infrastructure.
- Chatbots can inadvertently provide back-door access for hackers, particularly if they do not follow established security protocols such as hypertext transfer protocol secure (HTTPS). This back-door access can be exploited to gain unauthorized entry into systems, leading to data breaches and other malicious activities.

To address the cybersecurity lag in AI and chatbot development, several steps must be taken:

- Prioritizing security in development. Security must be integrated into the development lifecycle of AI and chatbot systems from the outset. This includes conducting thorough security assessments, implementing secure coding practices, and performing regular vulnerability testing. Developers should prioritize security alongside functionality and user experience.
- Standardization and regulation. The development of standardized security protocols and regulatory frameworks for AI and chatbots is essential. Industry standards can provide guidelines for best practices, while regulatory

frameworks can enforce compliance and accountability. Collaboration between industry stakeholders, regulatory bodies, and cybersecurity experts is necessary to establish these standards.

- Education and training. Organizations must invest in education and training to raise awareness of AI security risks and build expertise in this area. This includes training developers, cybersecurity professionals, and end-users on best practices for securing AI and chatbot systems. Continuous learning and professional development are essential to keep pace with emerging threats.
- Advanced security measures. Advanced security measures, such as AI-specific threat detection and response, must be developed and implemented. This includes techniques for detecting and mitigating adversarial attacks, data poisoning, and model inversion. Continuous monitoring and incident response capabilities are also critical to detect and respond to security breaches in real time.

The omission of cybersecurity in the development of AI and chatbots presents significant risks that must be addressed to ensure the safe and secure deployment of these systems. By prioritizing security in development, establishing standardized protocols, investing in education and training, and implementing advanced security measures, organizations can mitigate these risks and protect their users and data from cyber threats. The integration of robust cybersecurity practices is essential to harness the full potential of AI and chatbot technologies while safeguarding against evolving threats.

3.3.3 Does it "pay" to include cybersecurity into the AI development life cycle?

The integration of cybersecurity into the AI development lifecycle is not only a technical necessity but also a strategic financial imperative. As AI technologies, particularly chatbots and other anthropomorphized systems, become increasingly prevalent across various industries, the financial incentives to embed robust cybersecurity measures from the outset are manifold. These incentives span cost savings, risk mitigation, regulatory compliance, and competitive advantage.

One of the most compelling financial incentives for integrating cybersecurity into AI development is the potential for significant cost savings through risk mitigation. Cybersecurity breaches can result in substantial financial losses due to data theft, system downtime, and damage to an organization's reputation. By proactively incorporating cybersecurity measures, organizations can prevent these costly incidents.

For instance, the cost of a data breach can be astronomical, encompassing direct financial losses, legal fees, regulatory fines, and the long-term impact on customer trust and brand reputation. According to the sources, the financial sector faces a myriad of cybersecurity risks, including data breaches and system intrusions, which can have devastating financial consequences. By embedding cybersecurity into the AI development lifecycle, organizations can significantly reduce the likelihood of such breaches, thereby avoiding these substantial costs.

Regulatory compliance is another critical financial incentive for integrating cybersecurity into AI development. Governments and regulatory bodies are increasingly focusing on the ethical and secure use of AI technologies. Compliance with these regulations is not optional; failure to adhere can result in hefty fines and legal repercussions. For example, the U.S. Department of the Treasury emphasizes the importance of applying appropriate risk management principles to AI development to prevent data poisoning, data leakage, and data integrity attacks. Financial institutions, in particular, must navigate a complex regulatory landscape that mandates stringent cybersecurity measures. By ensuring that AI systems are developed with robust security protocols, organizations can avoid the financial penalties associated with non-compliance and maintain their operational licenses.

Incorporating cybersecurity into AI development also provides a competitive advantage. As consumers and businesses become more aware of cybersecurity risks, they are increasingly likely to choose products and services that prioritize security. Organizations that can demonstrate a commitment to cybersecurity are better positioned to attract and retain customers, thereby driving revenue growth. The sources highlight that AI-driven tools can enhance cybersecurity by automating threat detection and response, leading to improved security outcomes and increased customer trust. By integrating cybersecurity into AI development, organizations can differentiate themselves in the market, offering secure and reliable AI solutions that meet the growing demand for data protection.

AI technologies, when combined with robust cybersecurity measures, can enhance operational efficiency and reduce costs associated with manual security processes. AI-driven cybersecurity tools can automate tedious tasks such as monitoring logs and analyzing network traffic, freeing up human resources to focus on more complex security issues. This automation not only improves the efficiency of security operations but also reduces the likelihood of human error, which can be a significant source of security vulnerabilities.

Investing in cybersecurity during the AI development phase can contribute to long-term financial stability. Organizations that suffer from frequent security breaches may face long-term financial instability due to ongoing remediation costs, loss of customer trust, and potential legal liabilities. By integrating cybersecurity from the beginning, organizations can build resilient AI systems that withstand evolving cyber threats, ensuring long-term financial health and stability.

Finally, a strong focus on cybersecurity can encourage innovation and attract investment. Investors are more likely to fund companies that demonstrate a commitment to security, as this reduces the risk of financial losses due to cyber incidents. Additionally, a secure AI development environment fosters innovation by allowing developers to experiment and create new solutions without the constant threat of security breaches.

The financial incentives to integrate cybersecurity into the AI development lifecycle are substantial and multifaceted. From cost savings and risk mitigation to regulatory compliance, competitive advantage, operational efficiency, long-term financial stability, and encouraging innovation, the benefits of embedding robust cybersecurity measures into AI development are clear. As AI technologies continue to evolve and become more integral to business operations, the importance of prioritizing cybersecurity in their development cannot be overstated. Organizations that recognize and act on these financial incentives will be better positioned to thrive in an increasingly digital and interconnected world.

3.3.4   The talent gap

The rapid advancement of AI has brought about transformative changes across various industries, including cybersecurity. However, the integration of AI into cybersecurity practices has been hampered by a significant shortage of skilled professionals who possess both cybersecurity expertise and a deep understanding of AI. This talent gap poses a critical challenge in effectively addressing cybersecurity within the AI development lifecycle, leading to vulnerabilities that can be exploited by cybercriminals. To mitigate these risks, it is essential to understand the root causes of this talent shortage and implement strategies to bridge the gap.

The cybersecurity industry has long grappled with a talent shortage, with an estimated 3.5 million unfilled positions globally. This shortage is exacerbated by the rapid evolution of cyber threats and the increasing complexity of cybersecurity tasks. The demand for cybersecurity professionals has outpaced the supply, leading to a significant skills gap that affects organizations of all sizes. This gap is particularly pronounced in the field of AI, where the need for specialized knowledge and expertise is critical.

AI systems, including those used in cybersecurity, are inherently complex and require a deep understanding of both AI technologies and cybersecurity principles. AI algorithms must be trained on vast datasets, and their effectiveness depends on the quality and diversity of the data used. Additionally, AI systems must be continuously monitored and updated to adapt to evolving threats. This complexity necessitates a workforce that is proficient in both AI and cybersecurity, a combination that is currently in short supply.

The lack of cybersecurity talent who understands AI creates several challenges in addressing cybersecurity within the AI development lifecycle:

- Inadequate security measures. Without skilled professionals to design and implement robust security measures, AI systems are vulnerable to various cyber threats. For example, adversarial attacks, where malicious actors manipulate input data to deceive AI algorithms, can compromise the integrity of AI systems. These attacks require specialized knowledge to detect and mitigate, which is lacking due to the talent shortage.
- Bias and data privacy concerns. AI algorithms can inherit biases present in their training data, leading to discriminatory outcomes and oversight of particular threats. Additionally, the use of vast datasets raises significant data privacy concerns. Addressing these issues requires expertise in both AI and cybersecurity to ensure that AI systems are fair, unbiased, and compliant with data protection regulations.
- Inability to keep up with evolving threats. The cybersecurity landscape is constantly evolving, with new threats emerging regularly. AI systems must be continuously updated to adapt to these changes. However, the lack of skilled professionals means that organizations struggle to keep their AI systems up-to-date, leaving them vulnerable to new and sophisticated attacks.

To address the cybersecurity talent shortage and improve the integration of cybersecurity into the AI development lifecycle, several strategies can be implemented.

Organizations and educational institutions must invest in comprehensive education and training programs that focus on both AI and cybersecurity. This includes developing specialized curricula that cover the intersection of these fields and providing hands-on training opportunities. Encouraging continuous learning and professional development is essential to keep pace with the rapidly evolving threat landscape.

Promoting diversity and inclusion within the cybersecurity workforce can help address the talent shortage. By attracting individuals from diverse backgrounds, organizations can tap into a broader talent pool and benefit from varied perspectives and problem-solving approaches. Efforts should be made to encourage underrepresented groups, such as women and minorities, to pursue careers in cybersecurity and AI.

AI can be used to augment human capabilities and alleviate the burden on cybersecurity professionals. For example, AI-driven tools can automate routine tasks such as threat detection and incident response, allowing cybersecurity experts to focus on more complex and strategic activities. This human–machine partnership can enhance the overall effectiveness of cybersecurity efforts.

Collaboration between industry and academia is crucial to address the talent shortage. By working together, these sectors can develop targeted training programs, conduct research on emerging threats, and create opportunities for students to gain practical experience. Industry partnerships can also provide funding and resources to support educational initiatives and research projects.

Organizations should implement internal training and upskilling programs to develop the skills of their existing workforce. This includes offering certifications, workshops, and on-the-job training to help employees gain expertise in AI and cybersecurity. By investing in their current employees, organizations can build a more capable and resilient cybersecurity team.

The lack of cybersecurity talent who understands AI is a significant challenge that hinders the effective integration of cybersecurity into the AI development lifecycle. Addressing this talent shortage requires a multifaceted approach that includes investing in education and training, promoting diversity and inclusion, leveraging AI to augment human capabilities, fostering collaboration between industry and academia, and implementing internal training programs. By taking these steps, organizations can build a skilled workforce capable of developing and maintaining secure AI systems, ultimately enhancing their cybersecurity posture and protecting against evolving threats.

However, the current cybersecurity workforce often lacks the specific skills necessary to effectively develop and secure AI systems. This skills gap poses a significant challenge, as the complexity and sophistication of AI technologies require a unique blend of expertise that is not commonly found in traditional cybersecurity roles.

One of the most critical skills lacking in the current cybersecurity workforce is a deep understanding of machine learning algorithms and their applications.

Machine learning, a subset of AI, involves training algorithms on large datasets to recognize patterns and make predictions. Cybersecurity professionals need to be proficient in various machine learning techniques, such as supervised and unsupervised learning, neural networks, and reinforcement learning. This knowledge is essential for developing AI-driven security tools that can identify and respond to emerging threats in real time. However, many cybersecurity experts have not received formal training in these advanced AI methodologies, limiting their ability to leverage machine learning effectively.

Another essential skill that is often missing is data science proficiency. AI systems rely heavily on data to function correctly, and the quality of the data used for training these systems is paramount. Cybersecurity professionals must be adept at data collection, preprocessing, and analysis to ensure that AI models are trained on accurate and representative datasets. This includes skills in data cleaning, feature engineering, and statistical analysis. Additionally, understanding how to handle and protect sensitive data is crucial, as AI systems often process large volumes of personal and confidential information. The lack of data science expertise in the cybersecurity workforce can lead to poorly trained AI models that are prone to errors and biases, ultimately compromising their effectiveness.

Furthermore, expertise in adversarial machine learning is a critical yet underrepresented skill in the cybersecurity field. Adversarial machine learning involves understanding how AI models can be manipulated or deceived by malicious actors. Cybercriminals can exploit vulnerabilities in AI systems by introducing adversarial examples—subtly altered inputs designed to mislead the model. Cybersecurity professionals need to be familiar with techniques for detecting and defending against such attacks, including robust model training, anomaly detection, and the development of adversarial defenses. Without this specialized knowledge, AI systems remain vulnerable to sophisticated attacks that can undermine their reliability and security.

Ethical considerations and bias mitigation are also areas where the current cybersecurity workforce often falls short. AI systems can inadvertently perpetuate biases present in their training data, leading to unfair or discriminatory outcomes. Cybersecurity professionals must be equipped to identify and address these biases to ensure that AI-driven security tools are fair and unbiased. This requires an understanding of ethical AI principles, as well as techniques for bias detection and mitigation. The ability to conduct thorough ethical reviews and implement fairness measures is essential for building trustworthy AI systems that uphold the principles of equity and justice.

Additionally, proficiency in AI-specific regulatory and compliance requirements is crucial for cybersecurity professionals working with AI technologies. As governments and regulatory bodies increasingly focus on the ethical and secure use of AI, understanding the legal landscape is essential. Cybersecurity experts need to be familiar with regulations such as the General Data Protection Regulation (GDPR) and the California Consumer Privacy Act (CCPA), as well as emerging AI-specific guidelines. This knowledge ensures that AI systems are developed and deployed in compliance with legal standards, reducing the risk of regulatory penalties and enhancing overall security.

Finally, interdisciplinary collaboration skills are vital for the successful integration of AI into cybersecurity. AI development often requires collaboration between data scientists, machine learning engineers, and cybersecurity professionals. Effective communication and teamwork are essential to bridge the gap between these disciplines and ensure that AI systems are both technically sound and secure. Cybersecurity professionals must be able to work closely with AI experts to understand the intricacies of AI models and provide valuable input on security considerations. This collaborative approach fosters a holistic understanding of AI security and promotes the development of robust and resilient systems. In conclusion, the current cybersecurity workforce lacks several critical skills necessary for effective AI development, including machine learning expertise, data science proficiency, adversarial machine learning knowledge, ethical and bias mitigation capabilities, regulatory and compliance understanding, and interdisciplinary collaboration skills. Addressing this skills gap requires targeted education and training initiatives, as well as a commitment to fostering a multidisciplinary approach to AI security.

Key takeaways

- Phishing uses both psychological and technical deceptions to compel users into engaging activities that work against their interests.
- However, users often do not recognize attacks.
- Our technologies are designed to appeal to us, not illicit alarm.
- Our technologies can be highly personalized creating an experience of "mine." Something that is "mine" is not often perceived as dangerous.
- Our experience with our technologies creates a psychological space.
- Within this psychological space, we experience our own projections and thoughts.
- When coupled with our personalized technologies, the notion of "danger" is potentially too far removed from our experience to take seriously.
- AI-driven phishing attacks greatly enhance the anthropomorphism of an electronic communications.
- AI-driven phishing attacks appear genuinely human and easily deceive an unsuspecting attacker.
- The inclusion of AI in today's phishing attacker's toolbox has dramatically increased an attacker's likely success rate.
- AI has been quite readily adapted to anthropomorphizing technologies and has thus created considerable cybersecurity risks, particularly those centered around social engineering.
- Cybersecurity is lagging behind the development of AI.
- The lag in cybersecurity protections relative to AI centers around lack of an AI-knowledgeable cybersecurity workforce.

Chapter 4

Failing to regulate the ascriptions of threats/ vulnerability knowledge to technology

The information a person ingests to assess a situation consists of senses information, learned behaviors, training, education, and experiences. When interacting online, the input users receive is often limited to text, with some sounds and possibly visual displays. The entire array of information humans normally use to determine if they should or should not expose themselves to a situation is not available in this context. When considering what humans need to assess the severity of a threat, the traditional workspace (consisting of a computer, telephone, and possibly various other electronic devices) does not create a setting in which threat information can be processed adequately. To that end, the question arises: what does a threat aware workplace look like?

Creating a threat-aware working environment, particularly in the digital realm where phishing and similar online threats are rampant, requires a multifaceted approach that goes beyond traditional cybersecurity awareness campaigns.

Given that typical awareness methods like posters and emails often fail to make a lasting impact, and the fact that employees can become desensitized to constant warnings, a more integrated, continuous, and interactive strategy is needed. These are some of the traditional steps organizations take to raise the awareness of a cyber risk:

- **Regular, interactive cybersecurity training:** Engage employees with regular, hands-on training sessions that simulate real-world phishing attacks and other common cybersecurity threats. Interactive elements, such as quizzes and role-playing games, can help reinforce learning.
- **Phishing simulations:** Conduct controlled phishing campaigns within your organization to test employees' reactions to suspicious emails. Follow up these simulations with feedback sessions to discuss what happened, why, and how to recognize similar threats in the future.
- **Advanced threat protection tools:** Implement advanced threat protection solutions that can identify and neutralize phishing attempts and other threats before they reach end-users. Educate employees about how these tools work as part of their cybersecurity training.
- **User behavior analytics (UBA):** Utilize UBA tools to monitor for unusual activity that could indicate a security threat. By understanding normal user

behavior, these tools can alert on deviations that may signify an attack such as phishing.

- **Promote a security-first mindset:** Encourage a company culture where cybersecurity is everyone's responsibility. This includes making security practices a natural part of everyday work life and encouraging employees to speak up if they notice anything amiss.
- **Empower employees as cybersecurity assets:** Instead of relying solely on IT departments to handle security, empower all employees to be proactive in identifying and reporting potential threats. Recognition programs for those who demonstrate strong cybersecurity awareness can incentivize participation.
- **Regular updates and feedback:** Provide regular updates about new threats and share stories of successful threat identification within the organization. Encourage open communication about the challenges employees face in recognizing threats.
- **Cybersecurity surveys:** Use surveys to understand employee attitudes toward cybersecurity, identify gaps in knowledge, and tailor training programs accordingly.
- **Adaptive learning programs:** Recognize that different employees have varying levels of knowledge and experience with cybersecurity. Use adaptive learning programs that adjust the difficulty and focus of training based on individual progress and performance.
- **Skeptical but not paranoid:** Educate employees on the importance of skepticism when dealing with emails, links, and attachments, especially from unknown sources. They should verify the authenticity of requests through direct communication when possible and report suspicious activities.

A threat-aware workplace is one that evolves continuously, adapting to new threats and leveraging the latest in technology and training methods. By fostering an environment where cybersecurity is part of the fabric of daily operations, organizations can significantly reduce their vulnerability to online threats like phishing. The following case study illustrates how employees did not perceive their technology to be a vector through which threats could target them.

Case: medical payment cyber fraud

A typical practice among small, rural hospitals is to contract for services and automate the invoicing and billing process. This was the case with a small, rural hospital that contracted for Emergency Department (ED) services with an emergency medical group. The group sent invoices to the hospital and was paid monthly through electronic funds transfers from the hospital.

In June 2019, the hospital received an email invoice from the ED group with instructions to send funds to a new account. The hospital sent $200,500 to this new account on July 10. The payment was returned with a notice that the account had been frozen. On July 16, the hospital received another invoice with another new account, to which the hospital sent the returned $200,500. In August, the hospital received another invoice with instructions to send funds to yet another new account. The hospital sent $206,500 on August 13.

Ultimately, it was discovered that these emails directing the hospital to send funds to new accounts were fraudulent. The ED group never sent the email requests. The cyber criminals in all collected $407,000 from this phishing campaign [52].

In reviewing the medical payment fraud case, one could certainly identify several points at which hospital staff may have identified the potential fraud. However, if we analyze how the human–computer interaction (HCI), design, context, and personal dispositions play into this scenario, it can become understandable how such a phishing campaign could lead to success for the perpetrators.

4.1 Human–computer interaction

HCI is described as those activities involved with humans interacting with computers to produce something or achieve some end. The efficacy of an HCI is often spoken of in terms of useability. Usability is characterized by ease of use, resultant productivity, efficiency, retainability, learnability, and user satisfaction (noticeably missing: security).

HCI evolved as a discipline in the early 1980s. It evolved out of the field of Human Factors, and more specifically ergonomics. Ergonomics is understood as both a science and an engineering discipline. While HCI is centered around the interactions of humans with computers, HCI researchers and engineers draw upon multiple disciplines, to include cognitive psychology, behavioral sciences, experimental psychology, sociology, anthropology, linguistics and of course computer science.

Technology features we consider today as integral components of our computing experience evolved from the field of HCI. Some of these innovations include the mouse, track pads, and tablets.

In researching HCI whether we investigate the interaction between a user and a tablet, or that of a high-performance computing environment and a user, the interaction invariably consists of some form of input, processing, and output. Regardless of the form or type of interaction, the beginning of this interaction centers around the input device, which is the interface between the human and the computing device. Therefore, interface design is a central focal point of research in HCI.

Since the success of an interface relative to the HCI metrics involves measures of the human experience, HCI interface research leans heavily into theories of human cognition. While in the early days of HCI research, HCI applied theories of cognitivism which included classical conceptions (i.e. cognition was understood as mental processes independent of the body), today most HCI research incorporates theories of cognition involving embodied or situated perspectives. In other words, HCI research now seeks to understand how the physical experience of an interface informs cognitive processes when users interact with an interface.

This distinction between cognition being strictly mental processes independent of the body versus cognition informed by a physical experience deriving inputs from an environmental context is potentially key to understanding threat perceptions, and more importantly rating threats relative to the likelihood of occurrence, severity, and impact. To wit, a closer examination of the theory of predictive coding may hold some clues.

4.2   Predictive coding and electronic threat experiences

Traditional theories relative to perception posit that we ingest the elements of an experience, process said data points, and formulate a reaction. However, a new theory is gaining ground and suggests that our brains are essentially predictive machines, capturing various data points and formulating a model of a given situation.

The predictive coding theory suggests our brains accomplish predictive coding using hierarchical generative models. For example, the initial data points would include low-level features, such as colors, shapes, sounds, and emerge as a cohesive model when higher cognition, such as memory, and language are applied to this data. In combination, these data points then form the mental model with the brain providing inferences to the model where actual data may be missing. This process allows humans to "make sense" of the world as opposed to seeing and experiencing individual, discreet data points individually. As Lange and Haefner [53] stated:

> The brain has learned a probabilistic generative model of the world where data come in through the senses analogously to raw pixels from a camera, and percepts correspond to structures in the world that the brain infers as the causes of the data.

When applying predictive coding theory to threat perceptions, we can see how senses information, past experiences, learned information, and personal extrapolations can formulate a model of a threat in our minds and compel us to adapt a cautious posture (Figure 4.1). For example, as we contemplate traversing a dark alley to take a short cut to our destination, our senses information tell us the alley is dark, it may smell of old liquor, you may hear empty cans blowing across the ground, you may hear dripping water. This sensory information sets the stage for you to formulate a mental model in your mind that this alley could have hidden threats with little to no help should you encounter something dangerous. While you have no direct evidence such a threat exists in this alley, you have sufficient data points to extrapolate the danger, and therefore you decide to take the longer route to your destination. If we, as human beings, formulate threats using this modeling approach, how do the design principles of HCI influence our threat modeling process?

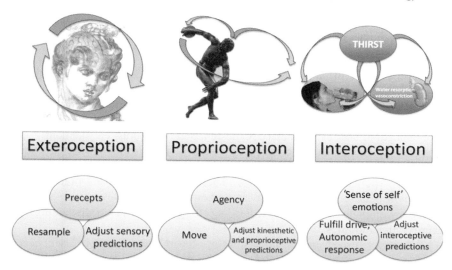

Figure 4.1   Predictive coding, E.Q. Knight, Harvard Medical School, and K.J. Friston, University College, London, September 2014

If we accept the premise of predictive coding, accepting that the brain is constantly generating and updating a mental model of the environment to predict sensory inputs, then it is also conceivable that a "whole body experience" is associated with our perception of threats. Predictive coding relies heavily on both physical and cognitive inputs to make these predictions accurate.

In the context of HCI, the interaction is predominantly visual and auditory, with tactile feedback being limited and other sensory inputs like smell and taste being virtually nonexistent. This sensory limitation can fundamentally alter the way threats are perceived and assessed. There are limitations of HCI in threat perceptions due to the following factors:

1. **Lack of sensory input:** In a digital environment, users are deprived of the full range of sensory inputs that inform risk assessment in the physical world. For example, the subtle cues that might indicate danger in a physical environment—such as an unusual sound or a sudden change in temperature—are not present in digital interactions. This limitation can dull the users' instinctual reactions to potential threats.
2. **Overreliance on visual and textual cues:** HCI primarily engages users through visual and textual content, making them overly dependent on these forms of communication to identify threats. Phishing emails and fraudulent websites often exploit this by mimicking legitimate sources closely, tricking users whose threat detection is tuned to these limited cues.
3. **Emotional and physical detachment:** Physical experiences can evoke strong emotional responses that influence decision-making and risk assessment. In a digital context, the detachment from physical presence can lead to

underestimating the severity of threats. The emotional response to a threatening email may be significantly less intense than to a physical threat, leading to complacency.

To counteract these limitations, incorporating elements that can simulate or evoke the missing sensory and emotional responses is crucial. To help users develop a more nuanced mental model of a digital environment that enhances their ability to detect threats, it is essential to focus on several key strategies that integrate education, user experience (UX) design, and continuous engagement.

First, understanding the concept of mental models is crucial. Mental models are internal representations of how users perceive and understand systems based on their past experiences and knowledge. These models help users predict how a system will behave and guide their interactions with it. However, mental models can be incomplete or inaccurate, leading to potential misunderstandings and security vulnerabilities.

To improve users' mental models, we must start with education. Users need to be informed about the fundamental principles of digital environments and cybersecurity. This education should cover how data flows through networks, common types of cyber threats (such as phishing, malware, and social engineering), and basic security practices. Educational materials should be designed to be engaging and accessible, using a mix of text, visuals, and interactive elements to cater to different learning styles.

Incorporating UX design principles is another critical aspect. UX design can bridge the gap between users' existing mental models and the actual system. Designers should strive to create interfaces that are intuitive and align with users' expectations. For instance, using familiar metaphors and consistent navigation patterns can help users form accurate mental models. Additionally, providing clear and concise explanations for security features and alerts can guide users in understanding the importance of these elements and how to respond to them.

One effective method to uncover and refine users' mental models is through user research techniques such as task analysis, surveys, and think-aloud sessions. Task analysis helps identify common user goals and the steps they take to achieve them, revealing potential gaps in their understanding. Surveys and questionnaires can gather insights into users' current knowledge and behaviors, while think-aloud sessions allow designers to observe how users interact with the system and verbalize their thought processes. These insights can inform the design of more user-friendly and secure interfaces.

Continuous engagement with users is also vital. Security awareness should not be a one-time effort but an ongoing process. Regular updates, reminders, and training sessions can reinforce good security practices and keep users informed about new threats. Gamification elements, such as quizzes and challenges, can make learning about cybersecurity more engaging and memorable.

Moreover, leveraging advanced technologies like artificial intelligence (AI) and machine learning (ML) can enhance threat detection and user education. AI-driven systems can analyze user behavior and provide personalized feedback and recommendations to improve security practices. For example, if a user frequently falls for phishing attempts, the system can offer targeted training modules to help them recognize and avoid such threats in the future.

In summary, developing a more nuanced mental model of a digital environment for users involves a multifaceted approach that combines education, user-centered design, continuous engagement, and advanced technologies. By providing users with the knowledge and tools they need to understand and navigate digital environments securely, we can empower them to detect and respond to threats more effectively.

This can be achieved through:

1. **Enhanced simulation of physical experiences:** Virtual reality (VR) and augmented reality (AR) technologies have the potential to bridge some of the sensory gaps by providing more immersive experiences. These technologies can simulate physical cues and evoke stronger emotional responses to digital threats.
2. **Interactive training:** Utilizing interactive, scenario-based training can help mimic real-world experiences more closely. By engaging users in simulations that closely resemble potential threats, they can learn to recognize and react to these threats more effectively.
3. **Adapting predictive coding for digital environments:** Encouraging users to develop a more nuanced mental model of digital environments can improve threat detection. This involves educating users about the common tactics used by cybercriminals and using predictive coding theory to understand how these threats might evolve.

The idea of mental models and understanding how predictive coding plays into a user's vulnerabilities to phishing may beg the questions of how perpetrators know to exploit such specific psychological principles.

Phishers, despite not being trained psychologists, have become adept at exploiting user mental models through a deep understanding of human behavior and psychological principles. This expertise is not necessarily derived from formal education in psychology but rather from practical experience and observation of how people interact with digital environments. The psychology that explains this phenomenon is rooted in social engineering, cognitive biases, and emotional manipulation.

Phishers leverage social engineering techniques, which are strategies designed to manipulate individuals into divulging confidential information. These techniques are based on the understanding that humans have predictable responses to certain stimuli. For instance, phishers know that people are more likely to comply with requests from perceived authority figures. By impersonating a trusted entity, such as a bank or a company executive, phishers can exploit this trust to elicit sensitive information. This manipulation taps into the psychological principle of authority bias, where individuals tend to follow orders from those they perceive as authoritative.

Another critical aspect of how phishers exploit user mental models is through the use of cognitive biases. Cognitive biases are mental shortcuts that people use to make decisions quickly. These biases, while useful in everyday life, can be exploited by phishers to bypass rational thinking. For example, confirmation bias leads individuals to favor information that confirms their preexisting beliefs. Phishers craft messages that align with these beliefs, making it more likely for the victim to accept the message as genuine without scrutinizing it further.

Emotional manipulation is another powerful tool in the phisher's arsenal. By triggering strong emotional responses such as fear, urgency, or curiosity, phishers can override the victim's logical thinking processes. This tactic is often referred to as "amygdala hijack," where the emotional part of the brain (the amygdala) takes over, leading to impulsive actions. For instance, an email that threatens account suspension unless immediate action is taken can create a sense of panic, prompting the recipient to act without verifying the legitimacy of the request.

The psychology behind these manipulative tactics is well-documented. Daniel Kahneman's model of two systems of thinking—System 1 (fast, intuitive, and emotional) and System 2 (slow, deliberate, and logical)—explains why phishing is so effective. Phishers aim to keep users in System 1 mode, where decisions are made quickly and based on heuristics, rather than allowing them to engage System 2, which would involve more careful and rational consideration.

Phishers also exploit the human tendency for social connection. By impersonating friends or colleagues, they can create a sense of familiarity and trust, making it more likely for individuals to share confidential information or click on malicious links. This exploitation of social bonds is a form of social engineering that takes advantage of our innate desire for social interaction and trust in our social networks.

Phishers exploit user mental models by leveraging social engineering, cognitive biases, and emotional manipulation. They understand that humans have predictable responses to certain stimuli and craft their attacks to trigger these responses. This knowledge, while not necessarily derived from formal psychological training, is based on a keen observation of human behavior and the application of psychological principles to deceive and manipulate victims. By understanding these tactics, we can better protect ourselves from falling prey to phishing attacks.

While HCI presents challenges in perceiving and assessing online threats, understanding these limitations through the lens of predictive coding offers a pathway to mitigate them. By enriching digital interactions with more immersive and realistic simulations and training, individuals can develop a more accurate mental model for predicting and responding to cyber threats. Adapting our digital environments and educational approaches to align more closely with human cognitive and sensory processing capabilities will be key to enhancing cybersecurity awareness and actions in the digital age.

4.3   HCI interface design principles

As a science and an engineering discipline, HCI interface design is driven by a set of principles which emerged from research and implementation. Foundational elements of this work were conducted in the 1980s and produced interface design principles that remain in effect today.

While there are several different sets of HCI interface principles, these can be summed up into the following guidelines:

1. **Uniformity across technologies:** Uniformity aids in creating a seamless UX across different platforms and technologies. However, this uniformity can be exploited by malicious actors who create phishing sites or malicious links that mimic the look and feel of legitimate sites, leading users to trust them based on familiar visual cues.
2. **Consistency:** Consistency in interface design allows users to leverage knowledge from one context in another, reducing learning curves. Cybercriminals exploit this by ensuring that malicious emails and websites mimic the consistent patterns of legitimate ones, tricking users' brains into recognizing them as safe.
3. **Simplicity and ease of use:** While simplicity enhances accessibility and usability, it also means that users are less likely to question the processes that seem straightforward, even if they're malicious. Phishing attempts often use simple and direct prompts to action, exploiting our preference for easy-to-use interfaces.
4. **Intuitiveness:** Intuitive design helps users navigate new platforms with minimal conscious thought, relying on ingrained habits and understandings. Unfortunately, this intuition can lead users to click on malicious links that appear legitimate at a glance, bypassing critical evaluation.
5. **Informative feedback:** Users depend on feedback from their actions to understand system states. Malicious sites can provide false feedback that mimics legitimate processes, confirming erroneous actions as correct, further deceiving the user.
6. **Easy error correction:** The ease of correcting errors in well-designed interfaces encourages experimentation and exploration. In a malicious context, however, it may allow users to overlook warning signs, assuming they can easily backtrack from any mistake.

Adherence to these HCI design guidelines can be found in every conceivable computer/human interface today. Without even thinking about it, we know light blue, highlighted, and underlined text signifies a hyperlink which can be clicked on, regardless of which website we may find ourselves traversing. Certain sounds indicate we may have made an erroneous selection, buttons invite us to select them, and arrows direct us to move to the next page.

We are accustomed to these features, and we can create a mental model and predict how the outcome should appear to us when we select an onscreen feature. Is it any wonder that we fall victim to a phishing attack by clicking on a link that reads www.myhometown-bank.com (nefarious site) versus www.myhometown–bank.com (correct site)? Even if we consider ourselves attentive and aware, the very mechanism our brains use to predict outcomes would offer us a model suggesting that the nefarious site is a legitimate site and nothing dangerous should be expected from this selection.

To wit, our predisposition to create mental models of our computing experience significantly influences our vulnerability to phishing attacks simply because our mental models are the cognitive frameworks that help us understand and predict how systems work based on our past experiences and knowledge. These models guide our interactions with digital environments, but they can also lead to misconceptions and oversights that cybercriminals exploit.

One of the primary ways mental models contribute to phishing vulnerability is through the simplification of complex systems. Users often rely on heuristic shortcuts to navigate the digital world, which can lead to overconfidence in their ability to detect threats. For instance, a user might develop a mental model that equates familiar visual cues, such as logos or email formats, with legitimacy. Phishers exploit this by mimicking these cues to create convincing fake messages. As a result, users may not scrutinize these messages closely, leading to successful phishing attempts.

Research has shown that users' mental models of cybersecurity are often metaphorical, likening digital threats to physical dangers or diseases. While these metaphors can help users grasp abstract concepts, they may not always translate into effective security behaviors. For example, thinking of a phishing attack as a "virus" might lead users to believe that antivirus software alone can protect them, overlooking the need for vigilance and critical evaluation of suspicious communications. This metaphorical thinking can create a false sense of security, making users more susceptible to phishing.

Moreover, the study "Thwarting Instant Messaging Phishing Attacks: The Role of Self-Efficacy and the Mediating Effect of Attitude towards Online Sharing of Personal Information" highlights an intriguing paradox: higher self-efficacy, or confidence in one's ability to learn and apply anti-phishing knowledge, can actually increase susceptibility to phishing. This counterintuitive finding suggests that overconfidence may lead users to process information quickly and superficially, rather than with the deliberate caution required to identify phishing attempts. Users who believe they are well-equipped to handle threats might not take the necessary precautions, such as verifying the authenticity of a message or link, thereby increasing their risk of falling victim to phishing.

Additionally, the study "Do Different Mental Models Influence Cybersecurity Behavior?" suggests that the effectiveness of security warnings is more critical than the mental models users employ. Users often bypass security warnings because they do not fully understand or trust them. This behavior indicates that improving the clarity and reliability of security communications could be more effective than attempting to reshape users' mental models. For instance, instead of presenting abstract warnings, systems could redirect users to safe sites and provide clear explanations for the redirection, thereby reinforcing safe behaviors without relying on users' interpretations.

Our mental models of computing experiences shape how we perceive and respond to potential threats. While these models help us navigate digital environments, they can also create vulnerabilities that phishers exploit. Overconfidence, reliance on familiar cues, and metaphorical thinking can all contribute to phishing susceptibility. Therefore, enhancing the effectiveness of security communications and fostering a deeper understanding of digital threats are crucial steps in mitigating these vulnerabilities. By addressing the limitations of our mental models and improving the design of security systems, we can better protect users from phishing attacks.

To counterbalance these inherent vulnerabilities, a multifaceted approach is necessary:

1. **Enhanced user education:** Regular training that goes beyond traditional cybersecurity awareness, focusing on the nuances of HCI design principles and how they can be exploited, can prepare users to be more skeptical and discerning.
2. **Advanced security features:** Incorporating advanced security features that can detect and alert users to potential phishing attempts in real time, such as analyzing the legitimacy of links and the security of sites, can provide an additional layer of defense.
3. **Designing for skepticism:** Design interfaces that encourage critical evaluation, such as subtle cues for external links or potentially unsafe actions, can help users pause and reconsider before proceeding.
4. **Regularly updating security protocols:** As cybercriminals evolve their strategies, so too must the security protocols that protect against them. Regular updates and patches can close vulnerabilities that might be exploited.
5. **Customized feedback for security:** Tailoring feedback mechanisms to specifically highlight security concerns, such as warning users more forcefully about the risks of proceeding with a suspicious action, can elevate their vigilance.

While the principles of HCI design have undoubtedly made technology more accessible and intuitive, they also present challenges in the context of cybersecurity. By understanding how these principles can be exploited and implementing strategies to mitigate these risks, it's possible to create a safer online environment that respects both usability and security. This balanced approach requires continuous effort from designers, cybersecurity professionals, and users alike, evolving in response to new threats and technologies.

4.4   AI's influence on HCI

The realm of HCI has been profoundly transformed by the advent and integration of AI. This transformation has redefined how humans interact with digital systems, making these interactions more intuitive, efficient, and personalized. AI's influence in HCI is evident in various facets, from voice and gesture recognition to predictive text and personalized content recommendations.

One of the most notable influences of AI in HCI is the enhancement of UX through personalization. AI algorithms can analyze user data to tailor digital experiences to individual preferences and behaviors. This personalization extends to content recommendation systems on platforms like Netflix and Spotify, where

AI curates content based on past interactions. Such personalization not only improves user satisfaction but also increases engagement with digital platforms.

Natural language processing, a subset of AI, has revolutionized HCI by enabling more natural and effective communication with computers. Voice assistants like Alexa and Siri exemplify this revolution. They understand and respond to spoken language, making digital interaction more accessible and convenient, especially for users with disabilities or those not comfortable with traditional computing interfaces.

Gesture recognition and AR are other areas where AI significantly impacts HCI. AI-powered gesture recognition allows users to interact with digital systems through natural movements, enhancing the intuitiveness of interfaces. In AR applications, AI enhances the user's perception of the real world by superimposing digital information, creating immersive experiences.

Predictive analytics, powered by AI, enables systems to anticipate user needs and provide relevant information or actions. This aspect of HCI, often referred to as anticipatory design, reduces user decision-making and streamlines interactions, making them more efficient.

As AI continues to shape HCI, ethical considerations, particularly regarding user privacy and data security, have become increasingly important. The future of HCI will likely focus on balancing the benefits of AI integration with the protection of user rights and data integrity.

AI's influence on HCI is a testament to its potential in enhancing human interaction with digital systems. As AI technologies evolve, they will continue to redefine the boundaries of HCI, making digital interactions more personalized, intuitive, and efficient. While these advances benefit the overall UX, it also increases the likelihood users will fall prey to a phishing attack leveraging HCI advances.

The following scenario depicts a situation where a phishing email is crafted by an AI to exploit the user using HCI design features:

Hypothetical AI-developed phishing example: the "Trusted Bank Update"

Scenario overview

An AI system is programmed to craft a phishing campaign targeting customers of a well-known bank. The AI is trained on a dataset containing genuine customer service emails from various banks, learning their tone, style, and typical content. Additionally, it's fed information about common user interfaces (UI) and interaction patterns in banking apps and websites, as well as common security practices.

Phishing email creation

1. ***Subject line and greeting****: The AI generates a subject line that mimics those used by the bank for important notifications, such as "Action Required: Verify Your Account Information." The email greeting personalizes the message using the recipient's name, a detail easily scraped from public databases or social media.*
2. ***Content and context****: The body of the email explains that the bank has updated its security systems and requires all customers to verify their account*

information to continue using online services. It highlights the importance of security and the bank's commitment to protecting customer data, appealing to users' trust in the institution.

3. **HCI exploitation**: *The email includes a button labeled "Verify Now," designed to mimic the look and feel of legitimate buttons in the bank's official communications and website. The color scheme, font, and layout are consistent with the bank's brand, exploiting users' familiarity and trust in the visual design. This is where the HCI vulnerability is exploited; users are conditioned to expect certain design elements and are more likely to click on something that looks familiar and trustworthy.*

4. **Urgency and consequence**: *The message creates a sense of urgency by stating that failure to verify the account within 24 hours will result in temporary account suspension. This exploits the human tendency to avoid loss, prompting quick action without thorough verification.*

The phishing link

1. **Deceptive URL**: *The "Verify Now" button links to a phishing website that closely replicates the bank's login page. The URL may use a lookalike domain name, exploiting common HCI oversights where users fail to notice small discrepancies in website addresses.*

2. **Data harvesting mechanism**: *On the phishing site, users are prompted to enter their login credentials, and possibly additional sensitive information like social security numbers or security questions. This data is collected by the attackers for future fraudulent activities.*

This scenario showcases how AI can be leveraged to create highly sophisticated phishing attacks that exploit human vulnerabilities related to HCI design. The combination of personalized content, visual design mimicry, and exploitation of urgency and trust can make these phishing attempts particularly dangerous. Users need to be vigilant, double-check URLs, and verify the authenticity of such communications through official channels before responding.

Interestingly, as discussed in previous chapters, in spite of introducing these vulnerabilities cybersecurity is not well integrated into the AI development life cycle, to include the use of AI in HCI design.

However, AI is increasingly integrated into HCI design, transforming how users interact with digital systems. This integration is driven by AI's ability to learn from user behavior, adapt to user needs, and provide more personalized and intuitive experiences. The convergence of AI and HCI is reshaping various domains, from UIs to complex system interactions, and is characterized by several key developments.

One significant application of AI in HCI design is the creation of intelligent UIs. These interfaces leverage AI to understand and predict user behavior, enabling more responsive and adaptive interactions. For instance, AI-powered chatbots are now commonplace in customer service, providing conversational interfaces that mimic human interaction. These chatbots can understand natural language, process user queries, and offer personalized responses, enhancing the UX by making interactions more seamless and efficient. An example is the AI chatbot used by H&M, which assists customers in their online shopping by understanding their preferences and providing tailored product recommendations and styling advice. AI also plays a crucial role in enhancing accessibility in HCI design. By incorporating AI, designers can create interfaces that cater to users with diverse abilities. For example, the Apple Watch includes a range of accessibility features powered by AI, such as haptic feedback for users with hearing impairments and voice-over functionality for users with visual impairments. These features ensure that the device is usable by a broader audience, demonstrating how AI can make technology more inclusive.

In the realm of VR and AR, AI is used to create more immersive and interactive experiences. AI algorithms can process and interpret vast amounts of data in real time, enabling dynamic and responsive virtual environments. For instance, Audi's virtual showroom allows customers to explore and interact with cars in a virtual space, providing a more engaging and personalized experience than traditional product displays. This use of AI in VR and AR not only enhances user engagement but also opens new possibilities for how products and services are marketed and experienced.

Moreover, AI is instrumental in the development of smart cities, where it is used to optimize urban infrastructure and services. By analyzing data from sensors and Internet of Things (IoT) devices, AI can predict traffic congestion, adjust traffic signals dynamically, and optimize energy consumption in buildings. These applications demonstrate how AI can improve the efficiency and sustainability of urban environments, making them more livable and responsive to the needs of their inhabitants.

In healthcare, AI's integration into HCI design is revolutionizing patient care and medical diagnostics. AI algorithms can analyze vast datasets of genetic information and patient records to identify patterns and predict health outcomes with high accuracy. When combined with digital engineering, such as 3D printing, AI enables the creation of customized medical implants and prosthetics tailored to individual patients' anatomical features. This fusion of AI and HCI not only improves the precision and effectiveness of medical treatments but also enhances the overall patient experience by providing more personalized care. The synergy between AI and HCI is also evident in the field of education, where AI-driven personalized learning systems are transforming traditional pedagogical approaches. AI can provide intelligent tutoring systems that adapt to individual learners' needs, offering personalized feedback and adaptive learning paths. This application of AI in HCI design helps create more effective and engaging educational experiences, supporting learners in achieving their goals more efficiently.

Overall, AI's integration into HCI design is driving significant advancements across various domains. By leveraging AI's capabilities to learn, adapt, and personalize, HCI designers can create more intuitive, responsive, and inclusive interfaces. This convergence of AI and HCI not only enhances UXs but also opens new possibilities for innovation and efficiency in how we interact with digital systems.

4.4.1   If AI is being used to design HCI, where is the cybersecurity?

The integration of AI into HCI design is a significant trend that enhances UXs by making systems more intuitive, adaptive, and personalized. However, the question arises whether cybersecurity is being integrated into HCI design alongside AI, and if not, why this might be the case.

AI's role in HCI design primarily focuses on improving user interaction through intelligent interfaces, personalized experiences, and adaptive systems. These advancements are driven by AI's ability to process large amounts of data, learn from user behavior, and make real-time adjustments. For instance, AI-powered chatbots, virtual assistants, and recommendation systems are all examples of how AI enhances HCI by making interactions more seamless and efficient.

However, the integration of cybersecurity into HCI design is not always as straightforward. While AI can significantly enhance cybersecurity by detecting and responding to threats more effectively, the incorporation of these security measures into the UI and experience design often lags behind. This discrepancy can be attributed to several factors.

First, the primary focus of HCI design is often on usability and UX. Designers aim to create interfaces that are easy to use, intuitive, and engaging. Security measures, on the other hand, can sometimes be perceived as adding complexity or friction to the UX. For example, multi-factor authentication (MFA) and frequent security prompts, while essential for security, can be seen as cumbersome by users. This tension between usability and security can lead to a situation where security features are not fully integrated into the HCI design or are implemented in ways that are not user-friendly.

Second, the rapid pace of AI development in HCI can outstrip the implementation of corresponding security measures. As AI technologies evolve and are quickly adopted to enhance user interactions, the development and integration of robust cybersecurity measures may not keep pace. This can result in AI-enhanced systems that are highly effective in terms of UX but potentially vulnerable to security threats.

Moreover, the complexity of integrating AI and cybersecurity into HCI design requires interdisciplinary collaboration between AI developers, cybersecurity experts, and HCI designers. This collaboration can be challenging to achieve in practice, as it involves aligning different priorities, expertise, and workflows. For instance, AI developers may focus on optimizing algorithms for performance and accuracy, while cybersecurity experts prioritize threat detection and mitigation, and HCI designers concentrate on UX. Coordinating these efforts to create a cohesive and secure UI can be a complex and resource-intensive process.

Despite these challenges, there are promising developments in integrating AI-driven cybersecurity into HCI design. AI can enhance cybersecurity by providing real-time threat detection, anomaly detection, and automated responses to security incidents. For example, AI algorithms can analyze user behavior to identify unusual patterns that may indicate a security breach, and then prompt users with security alerts or automatically take protective actions. These capabilities can be integrated into the UI in ways that are minimally intrusive and maintain a positive UX.

While AI is being integrated into HCI design to enhance user interactions, the integration of cybersecurity measures is not always as seamless. This is due to the inherent tension between usability and security, the rapid pace of AI development, and the complexity of interdisciplinary collaboration. However, as awareness of cybersecurity's importance grows and technologies continue to evolve, there is potential for more effective integration of AI-driven security measures into HCI design, ultimately creating systems that are both user-friendly and secure.

4.5   The intersection of predictive coding, HCI and AI influenced HCI

Building upon the intricate relationship between predictive coding, AI-influenced HCI, and their implications for user behavior, especially in the context of phishing attacks, it's crucial to delve into how these psychological and technological dynamics interplay to affect user vigilance and susceptibility to online threats.

When predictive coding is applied to HCI, especially in interfaces that are AI-enhanced to adapt and respond to user behavior, the system becomes a dynamic participant in the user's cognitive process. This adaptive interactivity can shape user expectations, making certain patterns of interaction seem more "natural" or expected than others.

1. **Tailored experiences and trust:** AI-driven personalization in HCI tailor's UXs so effectively that users grow to expect certain types of content and interaction patterns, building a sense of trust and comfort with the system. Phishing attempts exploit this trust by mimicking these personalized interaction patterns, making malicious content appear as if it's part of the expected personalized content stream.
2. **Mimicry and predictive patterns:** Phishing attacks often leverage design mimicry, utilizing visual and textual elements that users have learned to associate with trustworthy sources. This mimicry fits seamlessly into the predictive patterns established through regular, legitimate interactions, further blurring the lines between safe and malicious content.
3. **Exploitation of urgency:** Predictive coding also primes users to react more promptly to perceived urgency, a tactic frequently exploited in phishing attempts. Because AI-driven interfaces can highlight content that requires immediate attention, users may become conditioned to respond quickly to such cues, behavior phishers mimic by creating a false sense of urgency.

The convergence of predictive coding, AI, and HCI has significantly transformed the digital landscape, enhancing UXs and system efficiencies. However, this integration has transformed an adversary's landscape as well, particularly in the context of phishing attacks.

Understanding how these elements interact to increase user susceptibility to phishing requires a deep dive into the mechanisms of each and their combined effects.

When considering that predictive coding is a theory from cognitive neuroscience that explains how the brain processes information, we learn this theory posits that the brain continuously generates and updates a model of the environment to predict sensory input. When there is a mismatch between the prediction and the actual input, the brain adjusts its model to reduce the prediction error. This process allows for efficient information processing by focusing on unexpected or novel stimuli.

In the context of HCI, predictive coding can be applied to understand how users interact with digital interfaces. Users develop mental models based on their experiences and expectations, which guide their interactions with technology. When these expectations are met, interactions are smooth and efficient. However, when there is a discrepancy, users must adjust their mental models, which can lead to confusion and errors.

AI, particularly ML and deep learning, has revolutionized HCI by enabling systems to learn from user behavior and adapt in real time. AI algorithms can analyze vast amounts of data to identify patterns and make predictions, enhancing the personalization and responsiveness of digital interfaces. For example, AI can power recommendation systems that suggest content based on user preferences, or chatbots that provide instant customer support. These advancements make interactions more intuitive and satisfying, but they also create opportunities for exploitation.

Phishers have become adept at leveraging AI to craft sophisticated and convincing phishing attacks. AI-powered tools can scrape data from various sources, such as social media profiles and public records, to create highly personalized phishing messages. These messages can mimic the language and style of legitimate communications, making them difficult to distinguish from genuine ones. For instance, large language models (LLMs) like GPT-3 can generate text that is grammatically correct and contextually relevant, eliminating traditional red flags such as poor grammar and spelling mistakes. This level of sophistication increases the likelihood that users will fall for phishing attempts, as the messages align closely with their expectations and mental models.

The integration of AI into HCI further complicates the detection of phishing attacks. AI-driven interfaces are designed to be user-friendly and seamless, often automating tasks and reducing the cognitive load on users. While this enhances the UX, it can also lead to complacency. Users may become overly reliant on automated systems and less vigilant about potential threats. For example, an AI-powered email filter that successfully blocks most spam and phishing emails can create a false sense of security. Users may assume that any email that reaches their inbox is safe, making them more susceptible to sophisticated phishing attempts that bypass the filter. Predictive coding also plays a role in this dynamic. As users interact with AI-enhanced systems, their mental models evolve to incorporate the reliability and efficiency of these systems. When a phishing email arrives that closely matches their expectations of legitimate communication, the prediction error is minimal, and the user is less likely to scrutinize the message. This is particularly true for spear-phishing attacks, where the phisher has gathered detailed information about the target to craft a highly personalized message. The alignment between the user's mental model and the phishing message reduces the likelihood of detection.

Moreover, the continuous adaptation of AI systems can be exploited by phishers. AI algorithms learn from user interactions and adjust their behavior to improve performance. This adaptability can be used to refine phishing techniques. For example, if a phishing attempt is detected and reported, the AI can analyze the failure and adjust future attempts to avoid similar detection. This iterative process makes phishing campaigns more effective over time, as the AI learns to exploit specific vulnerabilities in user behavior and system defenses. The interplay between predictive coding, AI, and HCI also affects the design of security measures. Traditional security approaches often rely on explicit user actions such as entering passwords or responding to security prompts. However, AI-driven interfaces aim to minimize user effort and streamline interactions, which can conflict with these security measures. For instance, biometric authentication methods like facial recognition or fingerprint scanning are designed to be quick and unobtrusive, but they can also be spoofed using advanced techniques. The challenge is to design security measures that are both effective and seamlessly integrated into the UX, without creating additional cognitive load or friction.

To mitigate these vulnerabilities, it is essential to incorporate cybersecurity considerations into the design of AI-enhanced HCI systems. This involves not only implementing robust technical defenses but also fostering a culture of security awareness among users. Continuous education and training are crucial to help users recognize and respond to phishing attempts. Interactive and engaging training modules, powered by AI, can provide personalized learning experiences that address individual weaknesses and reinforce good security practices.

The integration of predictive coding, AI, and HCI has transformed digital interactions, making them more intuitive and efficient. However, this convergence also introduces new vulnerabilities, particularly in the context of phishing attacks. Phishers leverage AI to craft sophisticated and convincing messages that exploit users' mental models and expectations. The seamless and automated nature of AI-driven interfaces can lead to complacency and reduced vigilance, increasing susceptibility to phishing. To address these challenges, it is essential to incorporate cybersecurity considerations into HCI design and foster a culture of security awareness among users. By understanding the interplay between predictive coding, AI, and HCI, we can develop more effective strategies to protect users from phishing attacks and other cyber threats. Given that the interplay between predictive coding and AI-influenced HCI can significantly impact user behavior, particularly in the context of security:

1. **Reduced vigilance:** As users become accustomed to personalized, predictive interfaces, their vigilance toward anomalous or malicious content may wane.

The system's adaptation to user expectations can create a false sense of security, making it harder to discern when those expectations are being manipulated.

2. **Overreliance on familiar cues:** Users may over-rely on familiar cues (like design elements or urgency signals) to gauge legitimacy, neglecting deeper verification processes. This overreliance can be particularly problematic when those cues are precisely what phishers replicate.

3. **Challenges in authenticity verification:** The seamless integration of personalized and predictive content makes verifying the authenticity of communications more challenging. Users may struggle to distinguish between genuine and phishing attempts, especially when the latter are designed to exploit these predictive models.

To counter these challenges, several strategies can be employed:

1. **Enhanced education on AI and predictive coding:** Educating users about how AI influences their online experiences and the role of predictive coding in shaping their expectations can foster a more critical approach to online content.

2. **Advanced security protocols:** Implementing advanced security protocols that can detect AI-driven mimicry and alert users to potential phishing attempts can provide a crucial safety net.

3. **Promoting critical engagement:** Encouraging users to engage critically with content, especially when cues of urgency are present, can help counteract the exploitation of predictive coding by phishers.

4. **Verification mechanisms:** Providing users with easy-to-use tools for verifying the authenticity of suspicious communications can empower them to take control of their security.

The dynamic between predictive coding, AI-influenced HCI, and user behavior offers both opportunities and vulnerabilities in the context of cybersecurity. By understanding how these elements interact, users and developers can better prepare for and defend against the sophisticated phishing attempts that exploit these very human and technological nuances.

4.6   Future predictions for HCI

Predictions for UI design for 2024+ take us deeper into the personalized experience of the user. The integration of the user's total experience is now reflected in their technology and design interfaces. Whether the interface involves wearable technologies, mobile technologies, home automation, entertainment, or work, the design emphasis is on an even deeper personalized experience of our technologies. Some of these trends translate into:

- Enhanced cross-platform UXs
- Scrollytelling (scrolling + storytelling)
- Data storytelling

- Minimalism
- Buttonless interfaces
- Design process using generative AI
- Mixed reality (MR) and AR
- Emotional design

To wit, the emphasis of UI/UX design will be centered around a deeper, more personalized experience. The landscape of UI/UX design is moving toward a future where technology is not just a tool but an extension of the individual's personal space and preferences. This shift impacts various aspects of technology interaction, from wearable devices to home automation systems, affecting how designers approach UI/UX to create intuitive and engaging experiences. Below, we delve into each trend, exploring its implications and how it shapes our interaction with technology.

4.6.1   Enhanced cross-platform user experiences

The seamless integration of UXs across different devices and platforms is becoming increasingly important. Users expect to transition smoothly from their smartphones to their laptops, wearables, or home automation systems without a hitch in usability or experience. This trend emphasizes the need for a cohesive design language and interoperability standards that ensure a unified experience, regardless of the device or platform.

To achieve the experience of a seamless transition between computing platforms, a multifaceted approach is required to achieve true seamlessness in user interaction across the spectrum of digital devices and platforms. This seamless integration is not just about aesthetic consistency or the superficial transfer of features from one device to another; it's about creating a unified, coherent user journey that transcends individual devices, harnessing the unique capabilities of each to enrich the overall experience. The core elements of this capability include:

1. **Cohesive design language:** A unified design language is paramount. It ensures that visual elements, interactive behaviors, and usability norms are consistent across platforms, making the transition between devices feel intuitive. This language must be adaptable, capable of expressing the same principles in varied contexts—from the compact screen of a smartwatch to the expansive display of a desktop.
2. **Interoperability standards:** Interoperability is the technical backbone that supports seamless cross-platform experiences. Standards and protocols must be in place to allow devices and applications to communicate and work together effortlessly. This includes data syncing, activity continuation features, and shared authentication systems to ensure that users can pick up exactly where they left off, regardless of the device they switch to.
3. **Context-aware design:** Understanding and adapting to the user's context— such as their location, the time of day, or the specific device they are using— allows for a more personalized and relevant UX. For instance, a health app

might prioritize displaying upcoming workouts on a smartphone while focusing on real-time heart rate and activity data on a wearable device.

4. **Adaptive UI:** Interfaces must dynamically adjust not just in appearance but in functionality, optimizing for the strengths and limitations of each device. On mobile devices, this might mean simplifying navigation and prioritizing touch interactions, whereas on desktops, more complex functionalities can be offered with an emphasis on keyboard and mouse input.

5. **Unified accounts and data syncing:** A single user account should provide access across all platforms, with cloud-based data syncing ensuring that users' preferences, history, and data are consistent and up to date everywhere. This synchronicity eliminates the friction of device switching, making the experience truly seamless.

Achieving enhanced cross-platform UXs is not without its challenges. Designers and developers must navigate the fine line between consistency and device-specific optimization, ensuring that the experience is not just uniform but also maximally beneficial on each device. Privacy and security considerations also come to the fore, as seamless experiences often rely on the sharing and syncing of personal data across devices and platforms.

Looking ahead, emerging technologies such as AI, ML, and the IoT promise to further enhance cross-platform experiences. AI can help predict user needs and adapt interfaces preemptively, while IoT enables a broader range of devices to integrate into the user's digital ecosystem. As these technologies evolve, the potential for truly intelligent, anticipative cross-platform experiences becomes increasingly tangible.

Enhanced cross-platform UXs represent the convergence of design, technology, and user-centric thinking, aiming to create a digital environment where transitioning between devices and platforms is not just possible but effortless and enriching. As we move forward, the focus will increasingly shift toward not just enabling these transitions but making them a meaningful part of the user's digital narrative, enhancing productivity, engagement, and satisfaction across the board.

4.6.2 Scrollytelling (scrolling + storytelling)

Scrollytelling combines the act of scrolling with storytelling, creating an interactive narrative that unfolds as the user scrolls through a webpage or digital environment. This approach leverages dynamic animations, interactive elements, and multimedia content to engage users more deeply, making information consumption more immersive and memorable.

4.6.3 Data storytelling

With the exponential increase in data generation, data storytelling emerges as a crucial trend in UI design. It involves presenting data in a compelling, easy-to-understand format, using visualizations, interactive charts, and narratives that make complex information accessible. This trend underscores the importance of data literacy in design, enabling users to glean insights and make informed decisions based on the data presented to them.

Scrollytelling is an innovative storytelling technique that transforms the passive act of scrolling through digital content into an engaging narrative journey. By integrating multimedia elements, animations, and interactivity into the scrolling experience, scrollytelling captivates users, guiding them through stories in a way that is both intuitive and immersive. The following is a deeper exploration of its components, advantages, and how it is shaping the future of digital storytelling:

1. **Dynamic animations:** As users scroll, animations unfold in a synchronized manner, illustrating key points of the story. These animations can range from simple motion graphics to complex sequences that react to the user's scroll speed, enhancing the narrative's impact.
2. **Interactive elements:** Interactivity is a cornerstone of scrollytelling, inviting users to engage with the content actively. This might include clicking, dragging, or other actions that reveal additional information or alter the course of the narrative, providing a sense of agency.
3. **Multimedia integration:** Scrollytelling often incorporates various media types—text, images, videos, and audio—into a cohesive story. This multimodal approach caters to different learning styles and preferences, enriching the user's experience.
4. **Layered information:** Information in scrollytelling is presented in layers, with the base layer offering a straightforward narrative and additional layers providing deeper insights. This allows users to explore topics at their own pace and depth of interest.
5. **Responsive design:** Ensuring that scrollytelling experiences are responsive and accessible across devices is crucial. This adaptability ensures that the narrative's integrity and engagement are maintained, regardless of screen size or device type.

The advantages of scrollytelling include:

1. **Enhanced engagement:** By making the story interactive and visually compelling, scrollytelling captures users' attention more effectively than traditional, static content.
2. **Improved retention:** The immersive nature of scrollytelling aids in memory retention, as users are more likely to remember information presented in a narrative and interactive format.
3. **Accessibility:** Scrollytelling can make complex information more accessible and understandable, breaking down barriers to comprehension through visualization and interaction.
4. **Emotional impact:** The combination of visuals, text, and interactivity can evoke emotions more powerfully, making the narrative more impactful and memorable.

Scrollytelling has found its place in journalism, education, marketing, and storytelling, offering a novel way to present reports, explain concepts, promote products, or tell stories. Its ability to convey complex narratives in an engaging and accessible manner has made it a popular choice for data journalism, educational content, and brand storytelling.

Looking ahead, advancements in web technologies, AI, and AR/VR are set to take scrollytelling even further. AI could personalize narratives based on user interests or behaviors, while AR/VR integration might offer even more immersive scrollytelling experiences, bringing stories to life in three-dimensional spaces.

Scrollytelling represents a significant evolution in digital storytelling, offering a rich, immersive, and interactive way to engage with content. As technology advances and creators continue to innovate, the future of scrollytelling promises even more captivating and immersive narrative experiences, transforming the simple act of scrolling into a journey of discovery and engagement.

As scrollytelling adds yet another dimension of a personalized experience to our interactions with technologies, cybersecurity professionals should consider these aspects in assessing risks, and create risk scenarios, test them and investigation mitigations. The issue is particularly acute when considering the technological needs required to facilitate easy transitions between technologies.

4.6.4 Minimalism

The minimalist design trend continues to prevail, focusing on simplicity and the removal of superfluous elements. This approach values clean layouts, ample white space, and a limited color palette to enhance usability and reduce cognitive load. Minimalism in UI design not only appeals aesthetically but also facilitates quicker decision-making and easier navigation for the user.

4.6.5 Buttonless interfaces

The move toward buttonless interfaces reflects the industry's exploration of gesture-based and voice-controlled interactions. This trend is partly driven by the increasing capabilities of AI and ML algorithms that can interpret complex gestures and voice commands, offering a more natural and intuitive way of interacting with technology.

4.6.6 Design process using generative AI

Generative AI is revolutionizing the design process by enabling the creation of personalized content, layouts, and visuals at scale. By leveraging data on user preferences and behaviors, generative AI can assist designers in crafting interfaces that are highly tailored to individual users or segments, pushing the boundaries of personalized experiences.

4.6.7 Mixed reality and augmented reality

MR and AR are blurring the lines between the digital and physical worlds, offering new ways to interact with technology. These technologies allow for immersive experiences that can enhance learning, entertainment, and work, integrating digital content into the user's physical environment in a contextually relevant manner.

These technologies have the potential to revolutionize a wide range of fields, from education and training to entertainment and workplace productivity, by overlaying digital information onto the physical world in ways that are both contextually relevant and spatially aware. However, as these technologies become more pervasive, they also introduce new vectors for cybersecurity threats, particularly phishing, that merit careful consideration.

MR and AR technologies leverage advanced computing power, sophisticated sensors, and innovative display systems to create immersive experiences that can significantly enhance the way we learn, work, and play. For example:

- **Education and training:** MR and AR can transform educational content into interactive 3D models and simulations, making complex concepts easier to understand and remember.
- **Entertainment:** These technologies offer new forms of entertainment, turning users' surroundings into interactive gaming environments or immersive storytelling experiences.
- **Workplace productivity:** In professional settings, MR and AR can provide workers with real-time data overlays, hands-free access to information, and spatial computing capabilities that improve efficiency and accuracy.

As MR and AR technologies integrate more closely with our daily lives, the boundary between the digital and physical worlds becomes increasingly porous, creating new opportunities for cyber threats. The immersive and convincing nature of MR and AR experiences can be exploited by malicious actors in several ways:

- **Phishing in immersive environments:** Phishing attempts could become more sophisticated within MR and AR environments. For instance, malicious entities could create highly realistic simulations or overlays that trick users into revealing personal information or credentials, leveraging the immersive nature of these technologies to fabricate scenarios or entities that users are more likely to trust.
- **Contextual and spatially aware phishing:** The context-aware capabilities of MR and AR could enable a new breed of phishing attacks that are tailored to the user's physical environment or activities. For example, receiving a seemingly legitimate notification overlay about a security breach in your smart home system while you are using an AR interface could prompt immediate and less scrutinized responses.
- **Exploiting trust in digital overlays:** Users may develop a high level of trust in the digital overlays provided by MR and AR applications, making them less critical of potential threats. This trust can be exploited to present malicious links or requests for information as part of what appears to be a legitimate application or experience.
- Addressing the cybersecurity challenges posed by MR and AR requires a multifaceted approach.
- **Enhanced security protocols:** Developing security protocols specifically designed for MR and AR environments, including authentication methods that can effectively differentiate between legitimate and malicious content.
- **User education and awareness:** Just as with traditional digital environments, educating users about the potential risks and signs of phishing within MR and

AR experiences is crucial. Awareness campaigns should emphasize the unique aspects of security in immersive environments.

• **Advanced threat detection systems:** Leveraging AI and ML to detect unusual patterns or malicious content within MR and AR applications, providing an additional layer of protection against sophisticated attacks.

The advent of MR and AR technologies offers exciting possibilities for enriching our digital interactions and enhancing our perception of the world. However, the immersive and convincing nature of these experiences also opens new avenues for cyber threats, particularly phishing attacks that exploit the blurred lines between digital and physical realities. By recognizing these risks and implementing comprehensive strategies to mitigate them, we can safeguard the integrity of these transformative technologies and ensure that they remain powerful tools for positive change, rather than becoming conduits for cybercrime.

4.6.8   Emotional design

Emotional design focuses on creating products that elicit positive emotions and connections with users. By understanding and incorporating emotional intelligence into UI/UX design, technologies can provide more satisfying and meaningful experiences, fostering a deeper bond between the user and the technology.

As UI/UX design trends evolve toward more personalized and immersive experiences, the double-edged sword of these advancements becomes apparent. While they offer unparalleled convenience and engagement, they also present new challenges in safeguarding against nefarious activities like phishing. The trust and comfort built into these designs, while beneficial in creating a seamless UX, could potentially be exploited by malicious actors. As such, it is imperative for designers and users alike to remain vigilant, integrating robust security measures and awareness into the fabric of these advancing technologies to protect against potential threats.

This design approach, rooted in the principles of emotional intelligence, significantly enhances user engagement by designing experiences that are more intuitive, satisfying, and resonant on a personal level. Emotional design leverages colors, shapes, animations, and interactive elements in a way that aligns with the psychological and emotional responses of the user, thereby crafting experiences that are memorable and meaningful.

As the UI/UX design landscape evolves to prioritize these emotionally resonant experiences, it inadvertently introduces complexities into the cybersecurity domain, particularly concerning phishing threats. The emotional bonds and trust built through thoughtful design can, paradoxically, become vulnerabilities that malicious actors exploit.

1. **Trust exploitation:** Emotional design cultivates a sense of trust and safety among users. Phishers can mimic these emotionally engaging designs to create fraudulent sites or messages that feel familiar and trustworthy, coaxing users into lowering their guard and sharing sensitive information.

2. **Urgency and fear appeals:** Emotional design can also manipulate emotions such as urgency or fear, common tactics in phishing scams. Users accustomed

to responding emotionally to designed cues might not critically evaluate urgent requests for action or information, especially if they mimic the look and feel of legitimate communications.

3. **Personalized phishing attacks:** The same data driving personalized and emotionally resonant designs can be used to craft highly targeted phishing attacks. These "spear-phishing" attacks use personal information to create messages or websites that are not only convincing but also emotionally manipulative, increasing the likelihood of deception.

4. **Overconfidence in familiar interfaces:** Users might develop a false sense of security when interacting with well-designed, emotionally engaging interfaces, assuming that a familiar and positive UX equates to a secure one. This over-confidence can make them more susceptible to sophisticated phishing attempts that replicate the look and feel of legitimate platforms.

To counteract the potential for emotional design to be exploited in phishing attacks, a multifaceted approach is necessary, combining user education, advanced security measures, and ethical design practices:

1. **Educating users:** Empower users with the knowledge to recognize phishing attempts, emphasizing that visually appealing or emotionally engaging content can still be malicious. Regular training and updates about the latest phishing tactics are essential.

2. **Leveraging technology:** Implement advanced security technologies such as phishing detection algorithms, two-factor authentication, and secure browsing tools. These can provide an additional layer of protection against phishing attempts, even if the user is deceived by the design.

3. **Transparent design:** While striving for emotional engagement, designers should also prioritize transparency and clarity in conveying information about security and privacy. Clear indicators of website authenticity, secure connections (e.g., HTTPS), and privacy policies can help users navigate digital spaces more safely.

4. **Ethical design practices:** Designers should adhere to ethical guidelines that prevent the manipulation of user emotions in ways that could compromise security. This includes avoiding designs that overly rely on urgency or fear to prompt user actions.

The evolution of emotional design in UI/UX presents both opportunities for enhanced user engagement and challenges in maintaining cybersecurity, particularly against phishing. By understanding the nuanced ways in which emotional design can influence user behavior, designers and cybersecurity professionals can collaborate to create experiences that are not only emotionally resonant but also secure, protecting users from the ever-evolving threats in the digital landscape.

4.7   HCI trends: closing thoughts

In summarizing the exploration of HCI and its latest trends in UI/UX design, it's clear that the drive toward more personalized, immersive, and emotionally engaging digital experiences presents a double-edged sword for technology vendors and their users alike.

On one side, these advancements in UI/UX design significantly contribute to the bottom line of technology vendors by enhancing user satisfaction, engagement, and loyalty. Personalized experiences, emotional design, and immersive technologies like MR and AR not only differentiate products in a crowded market but also foster a deeper connection between users and technology, leading to increased usage and, ultimately, higher revenue streams.

However, these same innovations that propel technology vendors forward can also open the door to sophisticated cybersecurity threats, particularly phishing attacks. The very elements that make modern UI/UX designs so appealing—seamless integration across platforms, emotionally resonant interactions, and personalized content—can be exploited by malicious actors to create more convincing and effective phishing campaigns. These campaigns leverage the trust and comfort established through thoughtful design to deceive users, often with devastating consequences.

The challenges presented by these HCI trends underscore the need for a balanced approach that marries innovative design with robust cybersecurity measures:

1. **Design with security in mind:** Integrating security considerations from the outset of the design process can help mitigate risks without compromising on UX. This involves adopting secure design principles that anticipate potential threats and design accordingly.
2. **User education and awareness:** As UI/UX designs become more sophisticated, so too should the efforts to educate users about the potential risks. By understanding the tactics used by phishers, including how they may exploit design elements, users can become more vigilant in their interactions with digital content.
3. **Leverage advanced security technologies:** Employing cutting-edge security technologies that can detect and neutralize phishing attempts before they reach the user is critical. These technologies can act as a safeguard, protecting users even when design elements are exploited for malicious purposes.
4. **Ethical use of design:** Finally, technology vendors must commit to the ethical use of design principles, ensuring that efforts to engage users emotionally and personally do not inadvertently compromise their security. This includes being transparent about how user data is used and ensuring that users are informed about the security measures in place to protect them.

While the latest trends in HCI and UI/UX design offer tremendous market opportunities, their security will depend on how well vendors address the cybersecurity risks. Addressing these will require a concerted effort from designers, developers, security professionals, and users themselves. By fostering an environment where innovative design and stringent security measures coexist, technology vendors can not only capitalize on the benefits of these advancements but also protect their users from the evolving threats in the digital world.

Key takeaways

- HCI are designed to facilitate useability through: consistency, uniformity, simplicity, and ease of use.
- By conforming to HCI design principles, fraudulent emails can appear strikingly authentic.
- Even if small anomalies existed between an authentic email and a fraudulent email, unless a user is specifically trained and looking for fraud, the human brain is wired to project probable appearances into our awareness, and thus bypassing the anomalies.
- In the context of a routine, and busy day a user who is accustomed to seeing numerous emails coming from known sources, the chances of a user detecting a fraudulent email are questionable.
- With an increase in AI use, phishing attacks leveraging HCI features can be expected.
- As AI, HCI, and predictive modeling continue to integrate and leverage features, phishers gain more opportunities to exploit users.
- Integrating cybersecurity into the HCI development process, particularly when it involves the use of AI, is a consideration technologists must take into consideration.

Chapter 5

Assessing the phishing risk

In general, organizational leaders require actionable risk information to be available relative to their organization. For obvious reasons, actional risk information allows organizations to formulate mitigation strategies, contingency plans, and risk tolerance decisions. A statement of risk that includes a probability statement that a particular event will occur with a quantifiable expected loss is something an organization can work with.

If an expected loss is upward of $5M, and the likelihood the event will occur that year is at 95 percent, then a mitigation strategy costing $400,000 for that year deserves serious consideration. While receiving information along these lines relative to risk appears reasonable, common cybersecurity frameworks do not measure or articulate cybersecurity risks in this manner. In short, quantitative risk calculations are not used in most cybersecurity risk management frameworks, and this includes frameworks that attempt to measure phishing risks. Therefore, before we examine possible methods for evaluating phishing risks, the next section investigates cyber risk management frameworks writ large.

5.1   A sampling of cyber risk frameworks

Cybersecurity risk frameworks consist of standards, guidelines, implementation methodologies, templates, and auditing tools such as implementation assessment checklists.

One of the most used frameworks in the U.S. Federal Government is the Risk Management Framework (RMF) developed by the National Institute of Standards and Technology (NIST). The RMF consists of seven steps that include:

- Preparation—activities involving preparing the organization for managing security and privacy risks.
- Categorization—categorizing organizational information relative to high, moderate, and low impacts to the organization's mission.
- Selection—selecting security controls that mitigate threats to the organizations information and applied to the respective categories selected in the Categorization phase.
- Implementation—implementing the selected security controls.
- Assessment—assessing the implemented security controls relative to the categorization needs.

The NIST RMF is supported by numerous publications that detail the recommended process for implementing the complete framework.

A framework developed by the Computer Emergency Readiness Team (CERT) at Carnegie Mellon University is the Operationally Critical Threat, Asset, and Vulnerability Evaluation (OCTAVE). OCTAVE defines an evaluation methodology that enables organizations to identify information assets that are important to their respective missions. It allows organization to map threats to those assets as well as vulnerabilities associated with those assets. This critical asset to vulnerabilities to threats mapping allows organizations to understand which assets are at risk, and design strategies to reduce their overall risk exposure (Table 5.1).

Table 5.1   Example of a cyber risk heat map

		Impact				
	# Findings	No Threat	Minimal	Moderate	High	Critical
Likelihood of occurrence		1	2	3	4	5
	Very Likely	4				2	2
	Likely	2	1				1
	Moderate	5	2	1		1	1
	Rare	1	1				
	Unlikely	1	1				

The Control Objectives for Information and Related Technology (COBIT) is a framework developed by ISACA. This framework is specifically business focused and defines a set of processes for managers of information technology. Processes are characterized by inputs, outputs, key activities, objectives, performance measures, and a basic maturity model. The COBIT 2019 framework includes implementation resources, guidance, and insights as well as training options.

The MITRE Corporation developed the Threat Assessment and Remediation Analysis (TARA) framework. This framework is used to identify and assess cyber vulnerabilities and deploy countermeasures relative to these vulnerabilities. TARA is component of MITRE's security engineering practices.

While these cyber risk management frameworks do not represent the sum total of frameworks available, they do represent the general approach found in most frameworks: identify critical information assets, identify vulnerabilities relative to critical information assets, identify threats that could exploit these vulnerabilities, and define a means to mitigate said threat.

A common feature within many risk management frameworks is risk scoring systems that often include charts referred to as "heat maps." Risk scoring systems rate vulnerabilities relative to their risk and produce a qualitative characterization of the risks.

One of the most frequently used vulnerability scoring system is the Common Vulnerability Scoring System (CVSS). This system is owned and managed by

FIRST.Org.Inc, a U.S.-based non-profit. The CVSS is in essence a glossary of vulnerabilities that have been evaluated against a threat level relative to that vulnerability. The CVE score is then used by a consumer of this information to prioritize the mitigation of the vulnerability.

This type of scoring system is used to evaluate discreet technology components which are then "rolled up" into an overall impact statement and expressed in terms of a qualitative measure. These heat maps are generated to provide organizations with a visual depiction of their organizational cyber risk relative to these measures.

As an outcome, decision-makers are provided with a qualitative statement expressing something to the extent that a given set of assets have vulnerabilities which if exploited could produce high impacts on business operations and the likelihood of such an event occurring is very likely.

While this statement in and of itself could raise considerable alarm among decision-makers (and it usually does), one must ask the question whether this set of information is sufficient to make an informed decision about a corrective course of action relative to a potential cybersecurity risk?

Currently, there is no research evidence this approach reduces cyber risks [54]. While it can be argued that mitigating vulnerabilities does involve good cybersecurity practices, whether doing so in this manner provides organizations with actional information is the primary question.

In their book *How to Measure Anything in Cybersecurity Risk*, Hubbard and Seiersen make a compelling case for assessing cybersecurity risks using quantitative methodologies. While measuring cyber risks quantitatively would appear like an intuitive fit, interestingly this is not how measuring cyber risks evolved within the cybersecurity disciplines [54].

For the most part, in the early days of cybersecurity somewhere along the line perceptions arouse that cybersecurity risk could not be measured. In addition, a belief evolved that cybersecurity lacked sufficient data to formulate meaningful, quantitative statements of risk and in particular certainty was missing. For whatever reasons, early adopters of cybersecurity risk management frameworks did not investigate the use of Bayesian frameworks, or other statistical modeling approaches that could be used in situations where data was either missing or continuously evolving.

However, what did evolve was an approach expressing cybersecurity risks by way of qualitative measures that were mapped to quantitative measures and from that amalgamation an overall statement of risk was formulated. To say this approach is mathematically problematic is putting the issue mildly.

5.2   Variables involving human factors in most cyber risk frameworks

The manner in which human factors are measured in cybersecurity evolved possibly out of systems engineering maturity models, one of the originals being the Systems Engineering Capability Maturity Model (SE-CMM) designed by Carnie Melon in 1986. The maturity of an organizations engineering capability was measured relative to the level of formalization of its engineering process. In other words, if an organization had engineering processes that were documented by policy, guidelines, and procedures, implemented relative to these documents, and measured relative to their efficacy, then the organization's engineering processes were considered mature and would presumably produce quality engineering products when these steps were documented, were repeatable, and regularly evaluated.

The Systems Security Engineering Capability Maturity Model (SSE-CMM) followed in 1995 and adopted a similar pattern. These capability maturity models aligned organizational processes with "good outcomes."

As both the SE-CMM and the SSE-CMM were adopted by the U.S. Federal Government, extrapolations were made in U.S. Federal cybersecurity working groups that cyber risks involving humans could be managed and reduced by way of formalized processes. It was presumed that cybersecurity risks relative to human behaviors would be reduced if organizations had cybersecurity policies for acceptable use of information technology, guidelines, procedures, and training.

This evolution can be found within the NIST RMF Controls. Organizations found to be missing a user training program, for example, or user policies will receive a higher risk rating than those that have these processes in place. However, while cybersecurity professionals can agree that policies, procedures, guidelines, and training are sound management practices, how and by how much cybersecurity risks are reduced by implementing these measures is not known.

The reliance on formalized processes and policies to manage human factors in cybersecurity has been a common approach, but it has limitations in effectively addressing the complex and dynamic nature of human behavior. While having documented policies, guidelines, and training programs is essential for establishing a baseline of cybersecurity awareness and expectations within an organization, these measures alone do not guarantee a significant reduction in cyber risks related to human factors.

One of the challenges in quantifying the impact of these measures on reducing cybersecurity risks is the difficulty in isolating and measuring the specific effects of human behavior on overall cybersecurity posture. Human factors encompass a wide range of variables, including individual knowledge, attitudes, beliefs, and motivations, as well as organizational culture, communication, and leadership. These factors interact in complex ways and can be influenced by external factors such as social engineering tactics, evolving threat landscapes, and technological changes.

Moreover, the effectiveness of policies, procedures, and training programs in reducing human-related cyber risks depends on several factors such as the quality and relevance of the content, the frequency and format of delivery, and the level of employee engagement and adoption. Simply having these measures in place does not ensure that employees will consistently follow best practices or make sound cybersecurity decisions in their day-to-day activities.

To better address human factors in cybersecurity risk management, organizations need to adopt a more holistic and adaptive approach that goes beyond relying solely on formalized processes. This may involve:

1. Continuous monitoring and assessment of employee cybersecurity behaviors and attitudes through surveys, simulations, and incident reporting.

2. Tailoring training and awareness programs to the specific needs and roles of different employee groups, using engaging and interactive formats to promote active learning and retention.

3. Fostering a culture of cybersecurity responsibility and accountability at all levels of the organization, with clear communication and support from leadership.

4. Implementing technical controls and safeguards that minimize the potential impact of human errors or negligence such as multi-factor authentication (MFA), data encryption, and network segmentation.

5. Regularly reviewing and updating policies, procedures, and training content to keep pace with evolving threats, technologies, and regulatory requirements.

While having formalized processes and policies is a necessary foundation for managing human factors in cybersecurity, organizations must recognize the limitations of these measures and adopt a more comprehensive and adaptive approach to effectively reduce cyber risks related to human behavior. This requires ongoing monitoring, assessment, and improvement of cybersecurity practices, as well as a strong commitment to fostering a culture of cybersecurity awareness and responsibility throughout the organization.

5.3   Current trends in phishing risk management

Current trends in phishing risk management focus primarily on training, and more specifically, on training tools that allow organizations to launch an internal phishing campaign and document when users succumb to the training phish. When a user "falls" for the phish, organizations have adopted several actions. These actions can include mandatory cybersecurity training, to letters of reprimand, to disciplinary action and dismissals.

Since 2010, a variety of tools have emerged as leaders in this market and include:

- KnowBe4
- Cofense
- SANS Institute
- PhishLabs
- Broadcom
- Sophos
- Infosec
- Proofpoint
- Rapid7

In addition to commercial tools, open-source tools are available such as Gophish.

While simulated phishing campaigns have gained in popularity, the efficacy of such campaigns stands in question. In particular, a long-term (15 months) study involving over 14,000 participants was conducted by Lain *et al.* evaluating phishing in organizations [55]. In this study, research question 3 included an inquiry into the efficacy of simulated phishing campaigns that included an embedded training component. This study found that not only did simulated phishing campaigns with embedded training not improve the phishing resiliency of the user population, but in fact, possibly made the user population more susceptible to phishing campaigns.

While simulated phishing campaigns can provide organizations with interesting data points (such as click rate, re-click rates, report rates), these tools do not provide insight into the reasons behind individuals clicking on a phish [56]. Without understanding why an individual clicked on a phish, or under what circumstances they clicked (their context), it is difficult to assess an organization's actual phishing risk.

In addition to not understanding the reason why users clicked, simulated phishing campaigns offer a finite set of phishing examples (even if the phish datasets are updated frequently) and may not represent the types of threats a given organization could face. In short, the users could be trained to avoid a certain type of phish, while remaining unaware of other types of phish.

Lastly, several large organizations have reported that simulated phishing campaigns have had unexpected and undesirable results. Of these results, one of the most notable is a loss of productivity. The loss of productivity occurred when employees became skeptical of most emails sent to them and either reported them as a phish or discarded them all together. In each case, these actions slowed down corporate communications and caused uncertainty and concern among the workforces.

In all, whether an organization is attempting to assess cybersecurity risks using frameworks, such as the NIST RMF, or is attempting to understand its phishing risks using simulated phishing campaigns, the information derived from these methods is at best anecdotal, and in a worst case, misleading. In our next section, we'll exam how we can potentially improve assessing an organization's cybersecurity risks, and how actionable information can be obtained from various datasets.

In addition to the trends and challenges in phishing risk management, there are several emerging approaches that aim to address the limitations of traditional simulated phishing campaigns and provide more effective and comprehensive solutions.

One such approach is the use of machine learning and artificial intelligence (AI) to analyze user behavior and identify potential phishing risks in real-time. By monitoring factors such as email content, sender reputation, and user interactions, AI-powered systems can detect and flag suspicious emails more accurately than traditional rule-based filters. These systems can also adapt to evolving phishing tactics and provide personalized recommendations to users based on their individual risk profiles.

Another trend is the integration of phishing awareness training into the broader context of an organization's security culture and governance framework. Rather than treating phishing as an isolated issue, this approach recognizes the importance of fostering a holistic security mindset among employees and embedding phishing awareness into everyday work processes and decision making. This may involve incorporating phishing scenarios into regular team discussions, encouraging open communication about security concerns, and rewarding positive security behaviors.

Some organizations are also exploring the use of gamification and interactive learning experiences to make phishing awareness training more engaging and effective. By using game-like elements such as points, leaderboards, and challenges, these approaches aim to increase user motivation and retention of key concepts. Interactive simulations and role-playing exercises can also help users develop practical skills and experience in recognizing and responding to phishing attempts in a safe and controlled environment.

Finally, there is a growing recognition of the need for more collaborative and information-sharing approaches to phishing risk management. By participating in industry forums, threat intelligence networks, and cross-organizational initiatives, organizations can stay up-to-date on the latest phishing trends, share best practices, and leverage collective knowledge to improve their defenses. This may involve exchanging anonymized data on phishing incidents, collaborating on research and analysis, and developing shared standards and frameworks for phishing risk assessment and mitigation.

While these emerging trends and approaches show promise in addressing the limitations of traditional phishing risk management, it is important to recognize that there is no single silver bullet solution. Effective phishing risk management requires a multifaceted and adaptive approach that combines technical controls, user awareness and education, organizational culture and governance, and ongoing monitoring and improvement. By continuously evaluating and refining their phishing risk management strategies based on the latest research, best practices, and contextual factors, organizations can develop more robust and resilient defenses against this persistent and evolving threat.

5.4   Assessing phishing risks that involve AI

The advent of AI has not only revolutionized the landscape of technology but also introduced new vectors for cyber threats, notably in the domain of phishing attacks. Traditionally, phishing has relied on social engineering tactics to deceive individuals into divulging sensitive information, such as login credentials or financial details, by masquerading as a trustworthy entity in electronic communications. However, with the integration of AI, these deceptive practices have evolved, becoming increasingly sophisticated and challenging to detect.

This section aims to dissect the risks associated with AI-driven phishing attacks, contrasting them with conventional phishing methodologies, and evaluating the probability of success in mitigating these risks.

As risk management often involves considerable financial resources, this section will explore possible trajectory of phishing attacks as AI technologies continue to evolve.

5.4.1   *Risk mitigation requires an understanding of IA-driven phishing*

AI-driven phishing attacks present a formidable challenge for organizations seeking to protect their assets and personnel from cyber threats. By leveraging the power of machine learning and data analytics, attackers can create highly targeted and persuasive phishing campaigns that are far more difficult to detect and defend against than traditional phishing attempts.

One of the key ways in which AI enhances phishing attacks is through the automated analysis of vast amounts of publicly available data on potential targets. This may include information gleaned from social media profiles, professional networking sites, online forums, and other sources. By mining this data, AI algorithms can build detailed profiles of individuals, including their interests, affiliations, communication patterns, and potential vulnerabilities.

Armed with this information, attackers can then craft phishing messages that are meticulously tailored to each target. For example, an AI-generated phishing email might reference a recent conference the recipient attended, mention a mutual acquaintance, or use industry-specific jargon to establish credibility. These personalized touches can make the message appear far more authentic and trustworthy, increasing the likelihood that the target will take the desired action.

AI can also be used to optimize the timing and delivery of phishing messages. By analyzing data on when targets are most likely to be online and responsive, attackers can ensure that their messages arrive at the most opportune moments. This may involve sending phishing emails during busy work periods when targets are more likely to be distracted and less vigilant, or targeting individuals when they are away from their usual work environment and may be accessing email from less secure devices.

Another way in which AI can enhance phishing attacks is using natural language processing (NLP) and text generation techniques. By training machine learning models on large datasets of legitimate emails and online communications, attackers can create phishing messages that are virtually indistinguishable from genuine correspondence. This may involve mimicking the writing style, tone, and formatting of trusted sources, or even generating fake email threads that appear to be part of an ongoing conversation.

Phishing attacks, in their conventional form, typically involve the mass distribution of fraudulent emails designed to trick recipients into taking action that compromises their security. These actions may include clicking on a malicious link, downloading an infected attachment, or providing confidential information through a deceitful website. While effective to a degree, these traditional phishing attempts often rely on a "numbers game." This involves sending the same message to a vast audience with the hope that a small percentage will take the bait.

To mitigate the risks posed by AI-driven phishing attacks, organizations need to adopt a multilayered approach that combines technological solutions with user education and awareness training. This may involve:

1.   Advanced email filtering and threat detection systems that use machine learning algorithms to identify and block suspicious messages based on a range of indicators such as sender reputation, content analysis, and behavioral patterns.

2. Conducting regular phishing simulations and awareness training programs that educate users on the latest tactics and techniques used by attackers, and provide practical guidance on how to recognize and report potential phishing attempts.

3. Encouraging a culture of security awareness and vigilance, where employees are encouraged to question the legitimacy of unsolicited messages and are empowered to report suspicious activity without fear of reprisal.

4. Monitoring and analyzing user behavior and email traffic to detect anomalies and potential indicators of compromise, and having incident response plans in place to quickly contain and mitigate the impact of successful phishing attacks.

5. Collaborating with industry partners, threat intelligence providers, and security researchers to stay informed about the latest developments in AI-driven phishing attacks and to share best practices and defensive strategies.

By taking a proactive and adaptive approach to phishing risk management, organizations can better protect themselves against the evolving threat of AI-driven attacks. However, it is important to recognize that no single solution can provide complete protection, and that ongoing vigilance, education, and collaboration will be essential to staying ahead of this rapidly evolving threat landscape.

5.4.2 The AI technology used in phishing attacks

Understanding the AI technologies used in phishing attacks is the first step in finding a means to reduce the cybersecurity risk AI-driven phishing attacks pose. These technologies include:

Machine Learning Algorithms for Personalized Emails: AI-driven phishing employs machine learning to analyze vast amounts of data on potential targets, identifying patterns and preferences that can be used to tailor phishing emails. This personalization increases the likelihood that recipients will perceive these emails as credible and act on them.

Natural Language Processing for Convincing Messages: NLP enables cyber-criminals to generate convincing and grammatically correct text that mimics the style and tone of legitimate communications from trusted entities. This capability makes it harder for individuals to distinguish between genuine and phishing emails based solely on content quality.

Data Mining for Targeted Phishing Campaigns: By analyzing publicly available data or previously breached databases, attackers can identify potential victims and gather personal information. This information is then used to craft phishing messages that are highly relevant and engaging to the target, significantly increasing the chances of a successful attack.

The sophistication and adaptability of AI-driven phishing attacks present a significant challenge to cybersecurity defenses, necessitating a reevaluation of traditional detection and prevention strategies. As we delve deeper into the risks posed by these advanced phishing techniques, it becomes clear that a multifaceted approach is required to effectively mitigate the threat.

In addition to the AI technologies mentioned, there are other advanced techniques that cybercriminals employ to enhance the effectiveness of their phishing attacks. One such method is the use of deep learning algorithms for image and video manipulation. Deep learning, a subset of machine learning, enables attackers to create highly realistic fake images and videos that can be used to deceive targets. For instance, an attacker might use deep learning to generate a fake profile picture or video of a company executive, which can then be used in a phishing email to establish trust and credibility with the recipient.

Another emerging trend in AI-driven phishing is the use of reinforcement learning for optimizing attack strategies. Reinforcement learning algorithms can learn from the outcomes of previous phishing attempts and adapt their approach to maximize the chances of success. This means that the phishing attacks can continuously evolve and become more effective over time, making it even more challenging for organizations to defend against them.

The use of AI in phishing attacks also extends to the realm of voice phishing or "vishing." With advancements in speech synthesis and voice cloning technologies, attackers can now generate highly realistic voice messages that mimic the voice of a trusted individual such as a company executive or a customer service representative. These voice messages can be used in combination with other phishing techniques to add an extra layer of authenticity and persuasion to the attack.

To combat the growing threat of AI-driven phishing, organizations need to adopt a proactive and adaptive approach to cybersecurity. This involves staying informed about the latest AI technologies and techniques used by attackers and continuously updating their defense strategies accordingly. Organizations should also invest in advanced threat intelligence solutions that can help them identify and respond to AI-driven phishing attacks in real time.

Furthermore, fostering a culture of cybersecurity awareness and vigilance among employees is crucial. Regular training and education programs should be conducted to help employees recognize and report suspicious emails, even if they appear to be highly personalized and convincing. Encouraging a healthy level of skepticism and critical thinking when dealing with unsolicited emails can go a long way in reducing the risk of falling victim to AI-driven phishing attacks.

Collaborating with industry partners, sharing threat intelligence, and participating in cybersecurity research and development initiatives can also help organizations stay ahead of the curve in the fight against AI-driven phishing. By working together and pooling resources, organizations can develop more effective and resilient defense strategies that can withstand the ever-evolving threat landscape.

5.4.3   Assessing the risks of AI-driven phishing attacks

The escalation of phishing attacks utilizing AI presents a complex array of risks that diverge significantly from traditional phishing vectors. These risks are not only attributed to the enhanced sophistication and believability of AI-generated phishing content but also to the broader implications for individual and organizational security.

5.4.3.1   Factors contributing to the increased risk

Improved personalization and believability: AI-driven phishing attacks leverage detailed personal information to craft messages that resonate on a personal level with the target. This deep personalization increases the difficulty for recipients to discern phishing attempts from legitimate communications, thereby elevating the risk of successful deception.

Increased scalability of attacks: With AI, cybercriminals can automate the creation and distribution of phishing messages on an unprecedented scale. This automation allows for targeting vast numbers of individuals with personalized messages, expanding the potential reach and impact of phishing campaigns without the need for proportional increases in resources or effort.

Difficulty in detection: Traditional phishing detection methods, such as analyzing email content for known phishing indicators or relying on user reports, are less effective against AI-driven attacks. The sophistication and variability of AI-generated content mean that it can often bypass standard detection mechanisms, posing a significant challenge to existing cybersecurity defenses.

5.4.3.2   An example of AI-driven phishing attempts

Spear phishing campaigns: There have been documented instances where AI was used to mimic the writing style of a high-ranking official within an organization to request sensitive information or initiate fraudulent financial transactions. These campaigns utilized machine learning to analyze previous communications and craft highly convincing emails.

In a notable case, attackers used AI-generated audio deepfakes, simulating a CEO's voice, to instruct a subordinate to transfer funds to a fraudulent account. The realism of the deepfake audio significantly contributed to the success of the phishing attempt, highlighting the potential for AI to enhance the effectiveness of social engineering tactics.

To effectively assess an organization's risk of AI-driven phishing attacks, a comprehensive and multifaceted approach is necessary. This assessment should consider not only the technical aspects of the organization's cybersecurity defenses but also the human factors that contribute to vulnerability.

One key component of risk assessment is evaluating the organization's current level of awareness and preparedness regarding AI-driven phishing threats. This involves assessing the knowledge and skills of employees at all levels, from frontline staff to executives, in recognizing and responding to sophisticated phishing attempts. Regular training and simulation exercises can help identify gaps in awareness and provide opportunities for improvement.

Another crucial aspect is analyzing the organization's existing cybersecurity infrastructure and its ability to detect and mitigate AI-driven phishing attacks. This may include evaluating the effectiveness of email filtering systems, intrusion detection and prevention mechanisms, and incident response procedures. It is essential to determine whether these systems are capable of identifying and blocking AI-generated phishing content and if they can adapt to the evolving tactics employed by attackers.

Conducting a thorough audit of the organization's data security practices is also vital in assessing the risk of AI-driven phishing. This involves examining how personal and sensitive information is collected, stored, and accessed within the organization. Identifying potential vulnerabilities in data handling processes can help mitigate the risk of attackers obtaining valuable information that could be used to craft highly targeted phishing campaigns.

Engaging in threat intelligence sharing and collaboration with industry partners can provide valuable insights into the latest AI-driven phishing techniques and trends. By participating in information-sharing forums and working with cybersecurity experts, organizations can stay updated on emerging threats and best practices for defense.

Regularly monitoring and analyzing the organization's email traffic and user behavior can also help detect potential AI-driven phishing attempts. Anomalies such as unusual email patterns, suspicious attachments, or atypical user actions should be investigated promptly. Implementing user behavior analytics tools can assist in identifying deviations from normal patterns that may indicate a phishing attack.

Furthermore, conducting periodic risk assessments and penetration testing can provide a more comprehensive understanding of the organization's vulnerability to AI-driven phishing. These assessments should simulate realistic AI-driven phishing scenarios to evaluate the effectiveness of the organization's defenses and identify areas for improvement.

In addition to technical measures, fostering a culture of cybersecurity awareness and responsibility within the organization is crucial. Encouraging employees to adopt a cautious and skeptical approach when handling emails, even those that appear highly personalized, can help reduce the risk of falling victim to AI-driven phishing. Providing clear reporting channels and incentivizing employees to report suspicious activities can further strengthen the organization's defenses.

Finally, developing and regularly updating incident response plans specific to AI-driven phishing attacks is essential. These plans should outline clear procedures for detecting, containing, and mitigating the impact of successful phishing attempts. Regular drills and simulations can help ensure that the organization is prepared to respond effectively in the event of an actual attack.

By adopting a comprehensive and proactive approach to risk assessment, organizations can better understand their vulnerability to AI-driven phishing attacks and implement appropriate measures to mitigate the threat. Continuous monitoring, adaptation, and improvement of cybersecurity strategies are necessary to keep pace with the ever-evolving landscape of AI-driven phishing.

The following outline represents some common practices organizations could use to assess their risk to AI-driven phishing attacks:

Risk assessment plan: AI-driven phishing attacks

1. Objective
   To assess the organization's current vulnerability to AI-driven phishing attacks and identify areas for improvement in order to mitigate the risk of successful phishing attempts.

2. Scope
   This risk assessment covers all employees, departments, and systems within the organization that may be targeted by AI-driven phishing attacks.

3. Risk assessment team
   Assemble a cross-functional team consisting of representatives from IT, cybersecurity, human resources, legal, and relevant business units.

4. Data collection and analysis
   Conduct a comprehensive inventory of the organization's data assets, including sensitive information that could be targeted by phishing attacks.
   (a) Analyze the organization's email traffic and user behavior patterns to identify potential vulnerabilities and anomalies.
   (b) Review existing cybersecurity policies, procedures, and technical controls related to email security and phishing prevention.

5. Employee awareness and training assessment
   (a) Assess the current level of cybersecurity awareness among employees through surveys, quizzes, and interviews.
   (b) Evaluate the effectiveness of existing phishing awareness training programs and identify gaps in knowledge and skills.
   (c) Conduct simulated AI-driven phishing exercises to test employees' ability to recognize and report suspicious emails.

6. Technical controls assessment
   (a) Evaluate the effectiveness of the organization's email filtering and anti-phishing technologies in detecting and blocking AI-generated phishing content.
   (b) Assess the adequacy of the organization's intrusion detection and prevention systems in identifying and mitigating AI-driven phishing attempts.
   (c) Review the security configurations of email clients, web browsers, and other relevant software applications.

7. Incident response and reporting assessment
   (a) Review the organization's incident response plan specific to AI-driven phishing attacks and assess its adequacy in detecting, containing, and mitigating the impact of successful phishing attempts.
   (b) Evaluate the effectiveness of the organization's reporting mechanisms for suspicious emails and phishing incidents.

8. Third-party risk assessment
   (a) Identify and assess the risks associated with third-party vendors, partners, and service providers that have access to the organization's data or systems.
   (b) Review the cybersecurity practices and controls of these third parties, particularly in relation to AI-driven phishing prevention.

9. Risk identification and prioritization
   (a) Based on the data collected and assessments conducted, identify and prioritize the key risks related to AI-driven phishing attacks.
   (b) Consider factors such as the likelihood of occurrence, potential impact on the organization, and the effectiveness of existing controls.

10. Risk treatment and mitigation plan
     (a) Develop a comprehensive risk treatment and mitigation plan based on the identified risks and priorities.
     (b) Outline specific actions, such as implementing advanced email filtering technologies, enhancing employee training programs, and strengthening incident response capabilities.
     (c) Assign responsibilities and timelines for each action item.
11. Continuous monitoring and improvement
     (a) Establish a process for continuous monitoring and assessment of the organization's vulnerability to AI-driven phishing attacks.
     (b) Regularly review and update the risk assessment plan based on changes in the threat landscape, new technologies, and organizational factors.
     (c) Encourage ongoing employee feedback and reporting of phishing incidents to facilitate continuous improvement of the organization's defenses.
12. Reporting and communication
     (a) Prepare a comprehensive report summarizing the findings of the risk assessment, including identified risks, prioritized actions, and progress on mitigation efforts.
     (b) Communicate the results and action plans to key stakeholders, including senior management, board members, and relevant departments.
     (c) Foster an open and transparent communication channel to ensure that all employees are aware of the risks and their roles in preventing AI-driven phishing attacks.

5.4.4   Strategies for mitigating AI-driven phishing risks

Addressing the risks posed by AI-driven phishing attacks requires a multifaceted approach that encompasses both technological solutions and human-centric strategies.

5.4.4.1   Technological solutions

Deploying an AI and machine learning-based detection systems that can adapt to and recognize the evolving tactics of AI-driven phishing attempts is a crucial tool in detecting AI-driven phishing attacks. These systems can analyze patterns and anomalies in communications that may indicate a phishing attempt, even when traditional indicators are absent.

In addition to AI and machine learning-based detection systems, organizations should consider advanced authentication methods. These include implementing MFA and behavioral biometrics which can significantly reduce the success rate of phishing attacks by adding layers of security that are not easily bypassed through social engineering.

MFA requires users to provide two or more verification factors to gain access to a digital resource, significantly reducing the likelihood that an attacker can gain access through stolen credentials alone. MFA can include something the user knows (a password), something the user has (a security token or a mobile phone), and something the user is (biometric verification).

Pattern recognition and anomaly detection tools utilize machine learning algorithms to analyze email content, metadata, and sending patterns. These tools learn to distinguish between legitimate communications and potential phishing attempts. These tools learn by continuously updating their models based on new data, which allows them to adapt to new phishing strategies that may not have been seen before.

NLP techniques allow a tool to analyze the textual content of emails and websites for signs of phishing. This includes evaluating the tone, style, and urgency of messages (which are common manipulation tactics in phishing). NLP can help in identifying subtle cues that might not be apparent to human readers.

Evaluating the sender's reputation and conducting a behavioral analysis of an email's sender can also be an effective way to identify phishing attempts. This involves analyzing past sending patterns, the frequency of communications, and the typical content shared by the sender. Deviations from these patterns can trigger alerts that a message may be malicious.

While technological solutions are crucial in mitigating AI-driven phishing risks, it is equally important to address the human factor. After all, even the most advanced security systems can be compromised if employees are not adequately prepared to identify and respond to sophisticated phishing attempts. Therefore, a comprehensive strategy should include a strong focus on employee education, awareness, and empowerment.

One key human-centric strategy is the implementation of regular and engaging cybersecurity awareness training programs. These programs should go beyond the traditional "one-size-fits-all" approach and instead be tailored to the specific roles, responsibilities, and risk profiles of different employee groups. For example, employees in finance or HR departments, who often handle sensitive information, may require more intensive training on recognizing and reporting targeted spear-phishing attempts.

To make the training more effective, organizations should consider using a variety of interactive and immersive learning techniques. These can include gamified modules, simulated phishing exercises, and real-life case studies that help employees understand the tactics used by attackers and the potential consequences of falling victim to a phishing attempt. By making the training engaging and relatable, organizations can foster a culture of cybersecurity awareness and encourage employees to take an active role in protecting the company's assets.

Another human-centric strategy is to establish clear and easy-to-follow reporting procedures for suspicious emails and potential phishing incidents. Employees should feel confident and empowered to report any concerns without fear of reprisal or criticism. To facilitate this, organizations can set up dedicated reporting channels, such as a central email address or a ticketing system, and ensure that all reports are promptly investigated and acted upon.

In addition to reporting, organizations should also encourage open communication and knowledge sharing among employees. This can be achieved through regular cybersecurity updates, newsletters, and forums where employees can discuss the latest phishing trends, share their experiences, and learn from each other. By fostering a sense of community and collaboration, organizations can create a more resilient and adaptable workforce that is better equipped to deal with the evolving threat of AI-driven phishing.

Another effective human-centric strategy is to implement a system of positive reinforcement and rewards for employees who demonstrate good cybersecurity practices. This can include recognizing and celebrating employees who successfully identify and report phishing attempts, or who go above and beyond in promoting cybersecurity awareness within their teams. By highlighting and rewarding positive behaviors, organizations can create a culture of cybersecurity excellence and encourage employees to take a proactive approach to risk mitigation.

Finally, it is important for organizations to regularly assess and update their human-centric strategies based on the changing threat landscape and the evolving needs of their employees. This can involve conducting periodic surveys and focus groups to gather feedback on the effectiveness of training programs and reporting procedures, as well as monitoring key metrics such as the number of reported phishing incidents and the time taken to resolve them.

By combining robust technological solutions with comprehensive human-centric strategies, organizations can create a multilayered defense against AI-driven phishing attacks. This holistic approach recognizes that cybersecurity is not just a technical challenge, but also a human one, and that effective risk mitigation requires the active participation and commitment of every employee. By empowering and educating their workforce, organizations can build a strong and resilient cybersecurity culture that can adapt to the ever-evolving threat of AI-driven phishing.

The following outline provides a starting point in developing a mitigation strategy for AI-driven phishing risks:

(I)   Introduction
    (a)   Background on the growing threat of AI-driven phishing attacks
    (b)   Importance of a comprehensive mitigation strategy

(II)   Technological solutions
    (a)   AI and machine learning-based detection systems
       (i)   Adaptive learning models to identify evolving phishing tactics
       (ii)   Integration with existing email security infrastructure
    (b)   Advanced authentication methods
       (i)   MFA
       (ii)   Combination of factors: knowledge, possession, and inherence
       (iii) Reduced risk of unauthorized access through stolen credentials
    (c)   Behavioral biometrics
       (i)   Analysis of user behavior patterns for anomaly detection
       (ii)   Continuous authentication and risk assessment

     (d)   Pattern recognition and anomaly detection tools
- (i) Machine learning algorithms for analyzing email content and metadata
- (ii) Continuous model updates based on new data and emerging threats

     (e)   NLP techniques
- (i) Analysis of textual content for signs of manipulation and deception
- (ii) Identification of subtle cues and anomalies in communication patterns

     (f)   Sender reputation and behavioral analysis
- (i) Evaluation of past sending patterns and communication frequency
- (ii) Alerting on deviations from established norms

(III)   Human-centric strategies

     (a)   Tailored cybersecurity awareness training
- (i) Role-specific training based on risk profiles and responsibilities
- (ii) Interactive and immersive learning techniques (e.g., gamification, simulations)
- (iii) Regular updates and refresher courses to address evolving threats

     (b)   Clear and accessible reporting procedures
- (i) Dedicated reporting channels for suspicious emails and incidents
- (ii) Prompt investigation and response to all reported concerns
- (iii) Non-punitive environment encouraging employees to report potential threats

(IV)   Open communication and knowledge sharing

     (a)   Regular cybersecurity updates and newsletters

     (b)   Forums and discussion groups for employees to share experiences and insights

     (c)   Fostering a sense of community and collaboration in the face of evolving threats

(V)   Positive reinforcement and rewards

     (a)   Recognition and celebration of employees demonstrating good cybersecurity practices

     (b)   Incentives for proactive reporting and prevention of phishing incidents

     (c)   Promoting a culture of cybersecurity excellence and shared responsibility

(VI)   Continuous assessment and improvement

     (a)   Periodic surveys and focus groups to gather employee feedback

     (b)   Monitoring of key metrics (e.g., reported incidents, resolution time)

     (c)   Regular updates to training programs and policies based on evolving needs and threats

(VII)   Integration and governance
    (a)   Alignment with overall cybersecurity strategy
        (i)   Integration of AI-driven phishing mitigation into broader security frameworks
        (ii)   Consistency with existing policies, procedures, and standards
    (b)   Roles and responsibilities
        (i)   Clear assignment of ownership and accountability for mitigation efforts
        (ii)   Collaboration between IT, security, HR, and other relevant departments
    (c)   Metrics and reporting
        (i)   Establishment of key performance indicators (KPIs) for mitigation effectiveness
        (ii)   Regular reporting to senior management and stakeholders

(VIII)   Conclusion
    (a)   Importance of a multilayered, adaptive approach to mitigating AI-driven phishing risks
    (b)   Continuous improvement and collaboration as keys to long-term success

5.4.5   Implementation challenges

While these technological solutions offer substantial benefits in the fight against AI-driven phishing attacks, their implementation is not without challenges.

One of the primary concerns is the balance between detecting actual phishing attempts and minimizing false positives. High rates of false positives can lead to legitimate communications being blocked or marked as suspicious, potentially disrupting business operations.

Another challenge involves cyber criminal's ability to adapt to security controls. Cybercriminals are continually evolving their strategies to bypass detection mechanisms. Technological solutions must, therefore, be adaptable and continuously updated to counter new threats effectively as well.

Technological solutions can be resource intensive. Deploying and maintaining advanced AI-based detection systems and authentication methods can cost time and money. Often these solutions require a significant investment in technology and skilled personnel.

Addressing these challenges requires a strategic approach, focusing on optimizing the balance between security effectiveness and operational efficiency.

5.4.5.1   Enhancing accuracy and reducing false positives

Continuous learning and model tuning: AI and machine learning-based detection systems must be designed for continuous learning, allowing them to adapt to new phishing tactics as they emerge. Regularly updating the models with new data and outcomes helps refine their accuracy, reducing the likelihood of false positives without compromising the ability to detect genuine threats.

To ensure the effectiveness of AI and machine learning-based detection systems, it is crucial to establish a robust feedback loop that allows for the incorporation of real-world data and outcomes. This involves not only regularly updating the models with new data but also implementing mechanisms for capturing and analyzing false positives and false negatives.

By continuously monitoring the performance of the detection systems and making data-driven adjustments, organizations can optimize their accuracy and adaptability over time. This process should involve collaboration between security teams, data scientists, and end-users to ensure that the models are aligned with the organization's specific needs and evolving threat landscape.

User feedback integration: Incorporating feedback mechanisms where end-users can report suspected phishing attempts or false positives can significantly improve the accuracy of detection systems. This real-world input can be used to tune the models, enhancing their relevance and effectiveness in identifying phishing emails.

Layered detection approaches: Employing a multilayered approach to detection, which combines different techniques such as sender reputation analysis, content inspection, and anomaly detection, can provide a more robust defense against phishing. This redundancy ensures that even if one method fails to identify a threat, others may succeed, thereby minimizing the risk of false negatives and reducing false positives.

Collaborative threat intelligence sharing: Participating in threat intelligence sharing networks can help organizations stay ahead of emerging phishing tactics. By sharing information about new threats and attack vectors with other organizations and security professionals, companies can proactively update their defense mechanisms to counteract novel phishing strategies.

Participating in threat intelligence sharing networks is a valuable strategy for staying informed about emerging phishing tactics and enhancing organizational defenses. However, it is important to establish clear guidelines and protocols for sharing and receiving threat intelligence to ensure the security and confidentiality of sensitive information.

Organizations should carefully evaluate the reputation and trustworthiness of the sharing networks they engage with and implement secure communication channels for exchanging threat data. Additionally, it is crucial to have dedicated personnel or teams responsible for analyzing and actionizing the shared intelligence, ensuring that it is effectively integrated into the organization's security practices and technologies.

Investing in research and development: Allocating resources to the research and development of new security technologies and approaches is essential for staying ahead of cybercriminals. This includes exploring advancements in AI, machine learning, and behavioral biometrics to develop more sophisticated and resilient detection systems.

5.4.5.2   Managing resource intensity

Cost-benefit analysis: Organizations should conduct thorough cost-benefit analyses to understand the financial implications of implementing advanced technological solutions versus the potential costs of phishing attacks. This analysis can help prioritize investments in technologies that offer the highest return on investment in terms of enhanced security.

Cloud-based security solutions: Leveraging cloud-based security services can help reduce the resource intensity of deploying advanced security measures. These services often provide scalable, up-to-date security solutions that can be more cost-effective and easier to manage than in-house systems.

Training and skills development: Investing in training for IT staff to effectively manage and operate advanced security systems is crucial. Skilled personnel can ensure that technologies are utilized to their full potential, maximizing their effectiveness in detecting and preventing phishing attacks.

Investing in training and skills development for IT staff is essential for the effective implementation and management of advanced security systems. However, it is equally important to extend this training to employees across the organization to foster a culture of cybersecurity awareness and resilience. This can involve developing targeted training programs for different departments and roles, focusing on the specific phishing risks and best practices relevant to their work. Regular phishing simulations and exercises can help employees recognize and respond to AI-driven phishing attempts in a safe and controlled environment.

Additionally, providing opportunities for continuous learning and professional development can help employees stay updated on the latest threats and defensive strategies, empowering them to be active participants in the organization's cyber-security efforts.

5.4.5.3   Human-centric solutions

In addition to technology solutions, organizations should consider implementing human-centric risk reduction approaches. This involves adding the specific risks AI poses to users by way of AI-driven phishing attacks to an organizations education and training awareness programs.

Regular training sessions and awareness campaigns can equip individuals with the knowledge to identify and respond to phishing attempts, including those powered by AI. Emphasizing AI's ability to send highly personalized phishing emails is a critical piece of information that users must assimilate in order to become phishing resistant.

Users should also be encouraged to take precautions such as verifying the sender's details before responding to requests for sensitive information and avoiding clicking on links or downloading attachments from unknown sources. Developing cautionary behaviors and making these habitual can help users avoid falling victim to AI-driven phishing attacks.

By combining advanced technological defenses with informed and vigilant human behavior, it is possible to reduce the likelihood and impact of AI-driven phishing attacks. However, as AI technology continues to evolve, so too will the strategies and tools needed to combat its malicious use in phishing campaigns.

While educating users about the risks of AI-driven phishing attacks is crucial, it is also important to recognize that humans are inherently vulnerable to manipulation and deception. Therefore, organizations should strive to create a supportive and non-punitive environment that encourages employees to report potential phishing incidents without fear of repercussions. This can involve establishing clear and accessible reporting channels, providing prompt and helpful feedback to employees who report suspicious activities, and fostering a culture of open communication and collaboration around cybersecurity issues. By creating a psychologically safe environment, organizations can encourage employees to be more proactive and vigilant in identifying and responding to AI-driven phishing attempts.

5.4.6   Future trends in AI-driven phishing defense

As the cybersecurity landscape continues to evolve, particularly with the increasing sophistication of AI-driven phishing attacks, anticipating and preparing for future trends in defense mechanisms becomes paramount. This forward-looking perspective is crucial for staying one step ahead of cybercriminals who are constantly innovating.

The following section discusses potential trends in AI and machine learning that could affect the cybersecurity landscape in the near future.

5.4.6.1   Advancements in AI and machine learning

Future defense systems will likely leverage predictive analytics to identify potential phishing attacks before they occur. By analyzing patterns and trends in data, AI can predict likely targets and methods of attackers, enabling proactive defense measures.

AI systems with autonomous response capabilities could automatically take action against detected phishing threats, such as isolating suspicious emails or blocking malicious links, thereby reducing the window of opportunity for attackers.

As AI's ability to understand and interpret human language improves, we can expect more sophisticated NLU models that can better discern the subtleties of phishing content, including the context and intent behind messages.

5.4.6.2   Blockchain technology integration

Another potential future trend in AI-driven phishing defenses is the integration of blockchain technology. Blockchain's decentralized and immutable nature could be leveraged to create secure, tamper-proof databases of known phishing threats and attackers. By sharing this information across a distributed network, organizations could quickly and reliably identify, and block potential phishing attempts based on historical data. Additionally, blockchain-based identity management systems could help verify the authenticity of email senders, making it more difficult for attackers to impersonate legitimate sources.

5.4.6.3   Collaborative AI defense networks

As AI-driven phishing attacks become more complex and targeted, the need for collaborative defense efforts will likely increase. Future trends may involve the development of AI-powered defense networks that enable organizations to share threat intelligence and collectively train their defense models. By pooling data and resources, these networks could create more robust and adaptive defense systems that can keep pace with evolving phishing tactics. Collaborative AI could also enable real-time threat detection and response across multiple organizations, providing a more comprehensive and coordinated defense against phishing campaigns.

5.4.6.4   Explainable AI and interpretability

As AI and machine learning models become more complex, there is a growing need for explainable AI (XAI) and interpretability in phishing defense systems. XAI aims to make AI decision-making processes more transparent and understandable to human users. In the context of phishing defense, XAI could help security analysts better understand why a particular email or link was flagged as suspicious, enabling them to make more informed decisions and refine defense strategies. Interpretable AI models could also help organizations comply with regulatory requirements and maintain user trust by providing clear explanations for security actions.

5.4.6.5   Continuous adaptive learning

Future AI-driven phishing defenses will likely emphasize continuous adaptive learning to keep pace with the rapidly evolving threat landscape. Instead of relying on static rule-based systems, these defenses will employ dynamic machine learning models that can automatically update and adapt based on new data and feedback. By continuously learning from real-world phishing incidents and user interactions, these adaptive systems can improve their accuracy and effectiveness over time. This approach will be crucial for staying ahead of attackers who are constantly refining their tactics and exploiting new vulnerabilities.

5.4.6.6   Integration with zero trust architectures

As organizations increasingly adopt zero trust security models, future AI-driven phishing defenses will need to integrate seamlessly with these architectures. Zero trust assumes that no user, device, or network should be inherently trusted, requiring continuous authentication and authorization. AI can play a key role in enforcing zero trust principles by constantly monitoring user behavior, device health, and network activity for signs of phishing compromise. By leveraging AI-powered risk assessment and adaptive authentication, organizations can create a more granular and dynamic approach to phishing defense that aligns with the core tenets of zero trust.

As the AI-driven phishing landscape continues to evolve, it is crucial for organizations to stay informed about emerging defense technologies and strategies. By anticipating and preparing for these future trends, organizations can proactively strengthen their phishing defenses and maintain a resilient security posture in the face of increasingly sophisticated attacks. However, it is important to recognize that technology alone is not a silver bullet; effective phishing defense requires a holistic approach that combines advanced AI and machine learning capabilities with robust security processes, user education, and collaborative threat intelligence sharing.

5.4.7   Defenses against future threats: integration of human intelligence

The future of phishing defense will likely see a greater integration of human intelligence with AI systems. These collaborative systems can combine the adaptability and intuition of humans with the scalability and analytical power of AI, leading to more effective identification and mitigation of phishing attacks.

Leveraging the collective knowledge and experience of a global community of cybersecurity experts and enthusiasts through crowdsourcing can enhance the detection of phishing threats. This approach can bring diverse perspectives to the challenge, uncovering vulnerabilities and tactics that AI alone might miss.

The development of international cybersecurity frameworks and standards can facilitate more effective collaboration and information sharing between organizations and nations. These frameworks can help establish best practices for defending against AI-driven phishing attacks and coordinating responses to large-scale incidents.

In the face of increasingly sophisticated AI-driven phishing attacks, the necessity for a unified global approach to cybersecurity has never been more apparent. A global cybersecurity collaborative framework aims to bridge the gaps between nations, industries, and organizations, fostering an environment of shared knowledge, resources, and strategies to combat cyber threats collectively.

A cornerstone of global collaboration involves establishing a set of universally accepted standards and protocols which is crucial for ensuring consistent and effective cybersecurity practices worldwide. These standards would cover aspects such as threat intelligence sharing, incident response, and the ethical use of AI in cybersecurity.

A core element of such a framework should include a means to share threat intelligence across borders in real-time. This would enable countries and organizations to quickly disseminate information about emerging threats, such as new AI-driven phishing tactics, and coordinate their responses.

Collaborative R&D initiatives can accelerate the development of innovative cybersecurity technologies and strategies. By pooling resources and expertise, participants can tackle complex challenges more efficiently and push the boundaries of current defense capabilities.

To address the global cybersecurity skills gap, the framework should include comprehensive education and training programs. These programs would aim to raise awareness of cyber threats and best practices, as well as to cultivate the next generation of cybersecurity professionals.

Partnerships between the public and private sectors, as well as across different industries, are crucial for developing comprehensive defense strategies on a global scale. These partnerships can enable the sharing of threat intelligence, resources, and technological innovations, creating a unified front against phishing attacks.

By facilitating the sharing of threat intelligence and best practices, a global framework can improve the collective ability to detect and respond to phishing and other cyber threats. A unified approach helps to reduce fragmentation in how different countries and organizations approach cybersecurity, leading to more cohesive and effective defense strategies.

5.4.7.1   Challenges and strategies for implementation

Political and regulatory hurdles: Differences in national policies, privacy laws, and regulatory environments can complicate collaboration efforts. Diplomatic engagement and the development of flexible, adaptable frameworks that respect local regulations are essential for overcoming these obstacles.

Trust and confidentiality concerns: Establishing trust between participants is critical, especially when it involves sharing sensitive information. Implementing strict confidentiality protocols and secure communication channels can help mitigate these concerns.

Resource allocation: The success of a global framework depends on the willingness of participants to contribute resources, including funding, expertise, and data. Creating mechanisms for equitable contribution and benefit sharing can encourage broader participation.

5.4.7.2   Case studies on international cybersecurity cooperation

International cooperation in cybersecurity has yielded to date significant victories against cyber threats. It has demonstrated the power of collective action.

The following case studies highlight successful collaborations between nations, organizations, and sectors, offering insights into the strategies employed and the lessons learned. By examining these examples, we can understand the potential impact of a global cybersecurity collaborative framework and identify best practices for future initiatives.

Case study 1: the takedown of the GameOver Zeus botnet

Background: GameOver Zeus was a sophisticated botnet used for banking fraud and distributing ransomware. It was estimated to have infected up to 1 million computers worldwide, causing significant financial losses.

International collaboration: The operation to dismantle GameOver Zeus involved law enforcement agencies from over ten countries, including the United States, United Kingdom, and Canada, along with private sector partners.

Strategies employed: The collaborative effort included seizing servers, disrupting the botnet's command and control infrastructure, and providing tools for victims to remove the malware.

Outcome: The operation successfully disrupted the botnet, showcasing the effectiveness of international law enforcement and private sector collaboration in combating cyber threats.

Case study 2: the coordination against WannaCry ransomware

Background: In 2017, the WannaCry ransomware attack affected hundreds of thousands of computers across 150 countries, crippling critical infrastructure and businesses.

International collaboration: The response to WannaCry involved a coordinated effort among cybersecurity entities worldwide, including national cybersecurity agencies, private cybersecurity firms, and international organizations.

Strategies employed: Sharing of threat intelligence in real-time, development and distribution of decryption tools, and collective efforts to patch vulnerable systems were key components of the response.

Outcome: The rapid sharing of information and collaborative action significantly mitigated the impact of the attack and helped prevent further spread of the ransomware.

Case study 3: the formation of the Cybersecurity Tech Accord
Background: Recognizing the need for a more secure and stable online environment, the Cybersecurity Tech Accord was established in 2018 as a commitment among global technology companies to protect against cyber threats.

International collaboration: Over 100 companies from around the world have joined the Accord, pledging to defend customers from cyberattacks and not to assist governments in launching cyberattacks.

Strategies employed: The Accord facilitates collaboration on cybersecurity initiatives, sharing best practices, and advancing the development of security technologies through joint R&D efforts.

Outcome: The Cybersecurity Tech Accord has enhanced cooperation among technology companies, contributing to a safer online community and fostering innovation in cybersecurity measures.

These case studies demonstrate the effectiveness of international cooperation in addressing complex cybersecurity challenges. Key success factors include:

The lessons learned from these collaborations underscore the importance of developing a comprehensive global cybersecurity framework that leverages the strengths and capabilities of all stakeholders to combat emerging cyber threats more effectively.

Key takeaways

- Commonly used cybersecurity risk frameworks articulate human factors security risks in terms of missing policies, procedures, guidelines, and training.
- Commonly used cybersecurity risk frameworks express risk in terms of qualitative measures that do not quantify potential impact or losses.
- Phishing risks are often assessed by way of simulated phishing campaigns designed to identify users who have the propensity to succumb to phishing attacks.
- Simulated phishing campaigns, however, do not quantify phishing risks in a manner that would enable an organization to model potential future risks.

- Simulated phishing campaigns have unexpected outcomes such as making a user population more susceptible to phishing attacks and reducing productivity.
- AI-driven phishing attacks will require organizations to rethink their strategies, technologies, and human-centric mitigations. In addition, organizations must include considerations for future mitigations as cyber criminals using AI have the opportunity to quickly adapt to mitigation strategies more rapidly than before.

Chapter 6
The organization is at risk: start where you are

While several enabling factors can contribute to an organization's overall phishing risk, a recent long-term study conducted by Lain *et al.* found that regardless of the factors, over time an organization's risk to phishing attacks increases [55]. In other words, phishing is a function of an organization's exposure (also known as its attack surface) and the myriad of vulnerabilities humans interacting with computers introduce to a computing environment.

6.1   No silver bullet

While the phishing problem can be broken down into very simple truths:

> *Humans, who are using internet facing computing devices to work and socialize, are susceptible to deceptive tactics and are therefore being used as entry points for cyber-attacks, solving the problem is far from simple or straight forward.*

In Chapters 2–4, we discussed the various complexities involved with the human experience while interacting with technology. In Chapter 5, we discussed various frameworks that, regardless of their effectiveness, are being used to address cyberattacks. Even though we seemingly know a great deal about the user experience and risk mitigation frameworks, this information, nonetheless, does not make our organizations more phishing resistant. Unfortunately, to completely avoid phishing attacks, it would take staying off the internet entirely.

Since the internet is an inherent and vital component of most organizations today, to achieve some semblance of phishing resistance, organizations need to deploy a multilayered approach to phishing, while simultaneously understanding, anticipating, and planning for the inevitable attack. In the following subsections, we will examine a multilayered approach to managing the cybersecurity threats involved with phishing.

6.2   Establish the scope of your exposure and quantify the risk

Phishing is a cybersecurity risk and can be addressed as such. The following discussion examines an approach centered around cyber risk management practices. The areas of examination include:

- Characterizing the attack surface.
- Identifying vulnerabilities within the attack surface.
- Identify mitigations to the vulnerabilities and the costs associated with said mitigations.
- Establishing the probability such vulnerabilities will be attacked by a threat relative to a time frame.
- Calculate your potential costs relative to a successful attack versus the cost of mitigation.
- Decide whether to mitigate, accept, or transfer risks.

6.2.1   The attack surface

An organization's attack surface is essentially all possible points at which a cyber actor can access a system and engage in illicit activities. In the not-so-distant past, the attack surface for an organization was limited to a few internet-facing computing assets, while the bulk of an organization's resources were protected behind defensive assets such as firewalls (Figure 6.1).

However, today, with organizations requiring Web-based access to numerous other organizations (e.g., vendors, customers, third-party billers), and email exchanges extending beyond an organization's domain, essentially any given machine engaged in communications with an external entity can be understood as belonging to the organization's attack surface (Figure 6.2).

While this assertion may seem overly broad, the bottom line is: lest a machine is sitting in a closet, unplugged, never in use, and has no means to communicate, the machine should be considered a part of your attack surface.

6.2.2   The phishing vulnerability and mitigations

Once you've identified your attack surface, the next step involves identifying the vulnerabilities to your attack surface. While vulnerabilities in cybersecurity are mostly spoken of from a technical perspective, the vulnerabilities associated with phishing are the people. As we discussed in Chapters 2–4, the situational awareness people bring to their computing experience is multilayered and complex and rarely focused on phishing. This makes mitigating the human vulnerability a considerable challenge. Therefore, unless people are entirely restricted from receiving emails from entities outside the organization, and cannot access websites from their work machines, people should be considered the primary vulnerability relative to phishing attacks.

While some phishing threats may be addressed through traditional vulnerability management techniques, such as patching, or keeping defensive technologies up to date, for the most part, mitigating the phishing vulnerability will take a technical and managerial solutions and includes:

- Developing acceptable use policies.
- Deploying defensive technology solutions.
- Deploying a training and awareness program.
- Investigating and implementing new phishing mitigation approaches emerging from research.

Figure 6.1   Legacy architecture with net assets

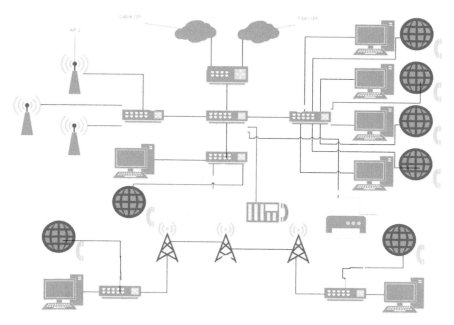

Figure 6.2   Modern architectures with multiple internet-facing access points

6.2.3   *Mitigating phishing attacks: acceptable use policies*

The first step in managing the phishing vulnerability is to define what is and is not acceptable use of organizational resources. In other words, an organization should have cybersecurity policies that address acceptable use of organizational technologies.

While it is unlikely an acceptable use policy in and of itself will protect an organization from phishing attacks, having such a policy is an important tool for employers to establish expectations, responsibilities, and enforcement procedures. The key sections of an acceptable use policy include:

- The Preamble: which describes the policy goals, and the scope of the policy.
- The Definition Section: describes terms used in the policy.
- The Policy Statement: summarizes the guidelines under which users can use the computing assets of the organization.

- The Acceptable Use Section: defines the conduct users should follow when using computing assets belonging to the organization.
- The Unacceptable Use Section: defines what actions are prohibited when using organizational computing assets.

An acceptable use policy should include:

- Who owns and is issuing the policy.
- What laws and organizational policies are being enforced through the policy (e.g. protected trademarks, patent property rights, transmission of confidential material).
- Who the policy applies to.
- What is the scope of the policy (only organizational assets, or if the organization permits personal devices to connect, what are the minimum requirements for those assets to connect).
- Expectations of the users affected by the policy (i.e. "dos" and "don'ts").
- Duration of the policy (is it reviewed annually and updated?).
- How the policy is enforced.
- How compliance with the policy is monitored.
- Consequences for breaching the policy.

While policies are designed to articulate the position of management, they are also a key element of an organization's due diligence processes, codifying the organization's commitment to good cybersecurity practices.

An acceptable use policy should address not only the organization's user population but also how vendor's and guests are managed when connecting to organizational assets.

- As it relates to phishing, an acceptable use policy should include language that addresses:
- Handling of emails from unknown senders.
- Links embedded in emails from unknown senders should not be clicked on.
- Emails from unknown senders should not be forwarded to other people in the organization.
- Emails from unknown senders should be flagged as spam.
- Handling of email suspected to be a phish.
- Phishing emails should not be forwarded to others in the organization.
- Nothing on a suspected phish should be clicked on.
- Suspected phishing emails should be immediately reported using an easy-to-access "Report" button featured on the email application (see Technology Section, Chapter 9).
- Precautions users should take when accessing personal emails on organizational machines.
- Users should use the same level of caution accessing their personal emails when using organizational emails.

After the acceptable use policy is written, reviewed by legal counsel, and the first draft approved by management, it should be socialized throughout the organization via a deliberate information campaign with feedback collected from all levels of users. Management should be interested in areas of the policy that may not be enforceable, or other blockers that could render the policy useless.

This campaign should involve groups of employees working with managers to review the policy, address questions, and possible implementation challenges. Policy owners should adjust the policy as needed.

Once the policy is released in its final official version, managers should develop procedures and guidelines designed to implement the policy. Policy owners should design assessment procedures to evaluate the implementation of the policy. It is also advisable that users receive a brief training session on the policy and how it affects their work experience going forward.

The policy implementation should be reviewed at regular intervals and adjusted as needed. Policies often require some course corrections over time. These changes may come when the regulatory environment changes, or the policy no longer fits well within an organization's workflows. If these adjustments are not made, users could circumvent policies altogether. In other words, if users are faced with a policy-driven impediment to their job performance, users will often choose their job and disregard the policy. Regular review of policy implementations helps remove such conflicts and ultimately results in well-crafted policies that protect the organization and its users.

6.2.4   *Mitigating phishing attacks: training and technology*

Mitigating phishing attacks requires a comprehensive strategy that encompasses both human elements and technological defenses.

Training employees to recognize and respond appropriately to phishing attempts is paramount. This involves educating them about the hallmarks of phishing emails such as urgent language, unsolicited attachments, and links to unrecognized sites. Regular training sessions coupled with simulated phishing exercises can enhance their vigilance and prepare them for real-world scenarios. By turning employees into informed defenders of the network, organizations can significantly reduce the risk of successful phishing attacks.

On the technology front, advanced email filtering solutions play a crucial role in intercepting phishing emails before they reach the enduser. These solutions utilize machine learning (ML) algorithms to analyze email content, metadata, and sender reputation, effectively filtering out suspicious messages. Implementing multi-factor authentication (MFA) adds another layer of security, ensuring that access to sensitive information requires more than just a password that could be compromised in a phishing attack. Furthermore, up-to-date antivirus software and regular security patches for all systems can prevent the exploitation of vulnerabilities often targeted by attackers.

However, technology alone cannot guarantee immunity against phishing. Cybercriminals continuously evolve their tactics to bypass security measures. Therefore, the integration of training and technology is critical. By fostering a security-conscious culture, organizations encourage proactive identification and reporting of phishing attempts, which, when combined with robust technological defenses, significantly strengthens their security posture. This dual approach enables rapid adaptation to the ever-changing tactics of cybercriminals, reducing the likelihood of successful attacks.

However, training and technology are comprehensive topics in themselves and are addressed further in Chapters 7 and 9.

6.2.5   Mitigating phishing attacks: new phishing mitigation approaches emerging from research

While new insights continuously emerge from research, a recent long-term phishing study conducted by Lain *et al.* [55] made an interesting discovery: when organizations provided a method for users to report suspected phishing emails, such as a report button embedded in the email program, users were more likely to remain vigilant about phishing attacks, and increased their reporting over time.

While Lain *et al.* did not elaborate on the potential reasons a crowd-sourcing approach is yielding success in catching phishing emails, a couple of assumptions can be made: submitting a suspected phishing email to the organization for evaluation is an interactive process. It becomes even more interactive if users are given a report each month on how many phishing emails they successfully "caught" as well as how many they submitted.

Interaction has shown to be a key motivator on several cybersecurity fronts. Traditionally, if someone is doing something correctly in cybersecurity, nothing happens! No one provides feedback, nothing melts down, everything is fine. This creates a bit of a challenge for humans as we are conditioned to seek out rewards. If we get no reward for good security practices the first issue is, we don't know if what we did was good or not. We're left with an ambiguity. Second, we don't have a sense of success if we receive no feedback relative to our good cybersecurity practices.

Third, recent studies from 2021 conducted by Reeves *et al.* [5] found that users overexposed to cybersecurity-related work experience cybersecurity fatigue, a form of "burn out," which results in an aversion to all things cybersecurity. While this topic is more complex than simply not receiving positive feedback, it is highly conceivable that users operating in a cybersecurity vacuum, never knowing if what they are doing matters, or makes a difference, contributes to cybersecurity fatigue.

Given these dynamics, adding a "Report" button to an email program, and encouraging users to report suspected phishing attempts, is a simple solution to several potential obstacles in keeping a user population focused on detecting phishing attacks.

Building on the insights provided by Lain *et al.*'s study, the incorporation of a "Report" button within email programs represents a strategic pivot toward engaging users in the cybersecurity process actively. This approach leverages the natural human inclination for feedback and recognition, transforming the often solitary task of identifying phishing emails into a participatory and rewarding experience. By involving users directly in the phishing detection process, organizations can cultivate a more alert and proactive cybersecurity culture. This method effectively addresses the challenge of cybersecurity complacency by providing immediate and tangible feedback for users' vigilance against phishing attempts.

Moreover, the concept of gamification can further enhance this strategy. By introducing elements such as leaderboards, badges, or rewards for reporting phishing attempts, organizations can tap into competitive and reward-based motivations. This not only makes the process of reporting phishing emails more engaging but also helps in sustaining long-term interest and participation among users. Gamification has been proven to increase motivation and engagement in various contexts, and its application in cybersecurity awareness and training programs is no exception.

Another emerging technique involves the use of artificial intelligence (AI) and ML to analyze the reported phishing attempts in real time, providing feedback not just on the accuracy of the reports but also on the sophistication of the phishing attempt. This can help in educating users about the evolving nature of phishing threats, making them more adept at identifying subtle signs of phishing in the future. Additionally, AI can aggregate and analyze data from reported attempts to identify patterns, predict future attacks, and automatically update defense mechanisms across the network, further strengthening the organization's security posture.

The effectiveness of these emerging phishing mitigation techniques underscores the importance of adopting a multifaceted approach that combines technology, user engagement, and continuous feedback. By doing so, organizations can transform their users from the weakest link in cybersecurity into their strongest asset, creating a dynamic and adaptive defense against phishing attacks. This shift not only mitigates the risk of successful phishing attacks but also addresses the broader challenges of cybersecurity fatigue and user engagement, thereby fostering a resilient and vigilant cybersecurity culture.

6.3   Managing the risk

In this section, we'll be looking at Bayesian methods for quantifying cybersecurity risks. However, to do this topic justice, a comprehensive treatment of this approach is needed. However, that is beyond the scope of this book. To explore this approach in more depth and obtain step-by-step guidance on how to conduct risk assessments using Bayesian methods, I would direct the reader to Hubbard and Seiersen's and their book of 2016 entitled *How to Measure Anything in Cybersecurity Risk*.

Earlier in Chapter 5, we discussed a variety of scoring methods used to characterize risks as well as approaches involving the aggregation of vulnerabilities into 1 score. There were also scores that involve the likelihood of occurrence and impacts that are also aggregated into a single score in some frameworks.

As popular as these frameworks are, unfortunately, there is no evidence scoring methods used in frameworks such as the NIST Risk Management Framework improve the decision-making process of risk managers. However, there is evidence these scores add "noise" to the decision-making process [54]. In other words, while these processes resulting in these scores give organizations a sense of "doing something," in the end they add nothing to the overall security posture of the organization (and the increase in cyberattacks at all levels bears this point out).

One of the main concerns around the risk management approaches discussed in Chapter 5 is that values such as "impact" or "likelihood" are estimates based on a practitioner's point of view. In other words, these are subjective values. So, while the result is articulated in a heuristic value, such as "High," "Moderate," or "Low," the individual values that make up that final score are subjectively determined by subject matter experts (SMEs).

For the most part, organizations would not be concerned about an SME's characterize of risk variables. After all, they are SMEs, trained and educated in their fields, and have experience, correct? Well, the situation is unfortunately not that simple.

There is a mature area of research in psychology that produced numerous studies demonstrating that even simpler forms of statistical models will outperform an SME in predicting potential future events.

This does not mean the SME is not trained or does not have experience. It means humans process data differently than computers. Humans rely on a form of deterministic reasoning, a reliance on anecdotes and individual events that stand out versus a compilation of data points. This holds even true for scientific researchers who understand probabilistic calculations. These are ancient survival instincts. The section of the forest where we encountered a bear while gathering food is a memory that will stick in our mind and save us in the future! Our ancestors did not envision a statistical model of potential bear encounters to make those determinations.

The validity of an SME's judgment is dependent upon said expert's cognitive representation of the subject matter and his/her communication skills [54]. That means understanding how an SME assesses risk comes into play particularly when assessing the likelihood of potential occurrences and the potential costs. Studies going back to 1970 have shown experts in all fields are generally overconfident, thus making them unable to objectively state an estimate on a given matter.

However, this does not mean SMEs cannot be taught (or better said calibrated) to process data more accurately. Aligning SME's estimates more closely with actual data and statistic models could go a long way in improving the initial starting points SME's often evaluate.

Overwhelmingly, published research posits that quantitative, probabilistic methods are effective in assessing cyber risks. Even though there are security practitioners who believe it is not possible to quantify cybersecurity risks, that position can be challenged with a simple statement: If something is detectable in some amount, it must be measurable [54].

The most resistance security practitioners have faced when proposing quantifying cyber risks revolves around the argument that organizations do not have "enough data" to formulate a quantifiable statement of risk. While your organization individually may not have enough data about the number of times you've been a target of phishing, chances are your organization belongs to a sector, a sector consisting of various organizations that share your size, mission, user population, etc. It is the inclusion of such information in your risk calculations that helps bridge some of the initial data gaps. Collecting information about your sector writ large from sources such as the annual Verizon Data Breach Investigations Report (DBIR) you will find enough data to populate a statistical model using Bayesian methods.

While this information represents your sector, it obviously doesn't address the specifics of your organization. There is a degree of uncertainty about this approach. That degree of uncertainty is quantified using the Bayesian methods and is expressed as a level of confidence.[1] For example, let's imagine we are managing the cybersecurity risks for a medium-sized hospital, and our concerns are centered around phishing attacks that result in ransomware being injected into our system. We collect the data from a national report, such as the Verizon DBIR, and insert this data into a statistical model, maybe something like a Monte Carlo simulation.[2] From this simulation, we will obtain several possible outcomes and the probability associated with each outcome. These calculations could provide us the basis to state:

We are 85 percent confident there is 90 percent probability hospitals of our size, and located in our state will be attacked within 1 year by phishing emails containing ransomware.

This is a very specific statement that allows to estimate the potential costs if such an attack were successful, and the costs of mitigations. We can do a tradeoff analysis from these data points that allow us to determine whether we wish to accept this risk, mitigate the risk, or transfer the risk (or do a combination of these).

If we were interested in understanding the relationship between the exposure of one variable relative to the occurrence of another, we could use a model based upon a log odds ratio. A log odds ratio is simply the natural logarithm of the odds ratio between an exposed variable and an occurrence variable.

There are a variety of Bayesian methods which can be used in cybersecurity risk management. Making a risk decision becomes significantly easier if you can state something like this:

We are 90 percent confident that within the next year we will experience a ransomware attack from a successful phish resulting in a loss between $50,000 and $1,000,000.

This statement not only provides your potential loss scenario, but it also provides you with a confidence level relative to the probabilistic prediction. Making an

---

[1] The confidence level is related to the number of data points we used in our calculations. As we increase our data points, our confidence levels rise. Data points are added to our models as we learn more about the problem.

[2] A Monte Carlo simulation is a probabilistic model that can include elements of uncertainty in its predictions. It provides multiple possible outcomes and the probability of each from a large pool of random data samples.

investment decision based upon such calculations should provide more investment certainty than making an investment based upon a statement that reads:

> *Your cybersecurity risk is:*
> *Confidentiality: High*
> *Integrity: High*
> *Availability: Medium*
> *Overall risk: High*

Given this information you could easily find yourself putting a steel door on a paper house. In other words, your cybersecurity investment could become misaligned as you do not have specifics relative to your potential loss, the costs needed to protect your organization, and where your protections will best protect you from an incoming threat.

Assessing risk using statistical modeling, such as Bayesian methods, is by no means a new form of assessing risks. Other disciplines that assess risks such as insurances and banks have been using statistical modeling since their inception. It was only through some specific drivers within the cybersecurity community that we arrived at these risk frameworks which did not include statistical modeling.

Since the phishing problem involves vulnerabilities directly related to the complexities of human nature, it is our recommendation that statistical modeling be used to predict as accurately as possible the likelihood your organizations will be attacked, within what time frame, and the potential results of such an attack. In other words, understanding the attack patterns of your adversary and targeting mitigations specifically toward the adversary's actions is more straightforward and workable than attempting to mitigate the contextual mind set of your users in one department, versus another etc.

To accurately characterize the phishing risk of an organization using AI, the process involves a comprehensive analysis of past data, predictive modeling, and simulations. This approach not only identifies existing vulnerabilities but also forecasts potential future threats, enabling organizations to proactively mitigate risks.

To conduct a cyber risk assessment using AI, several types of AI technologies and methodologies are employed to enhance the accuracy, efficiency, and comprehensiveness of the assessment process.

One of the primary AI technologies used in cyber risk assessment is ML. ML algorithms can analyze vast amounts of data to identify patterns and anomalies that may indicate potential threats. These algorithms are particularly effective in processing data from various sources such as system logs, network flows, endpoint data, cloud API calls, and user behaviors. By continuously learning from new data, ML models can adapt to evolving threats, making them indispensable for identifying both known and unknown vulnerabilities and attack vectors.

Another critical AI technology is natural language processing (NLP). NLP can be used to analyze unstructured data such as audit reports, incident logs, and threat intelligence feeds. This capability allows AI systems to extract relevant information and identify trends that might not be immediately apparent through manual analysis. By converting unstructured data into structured insights, NLP enhances the predictive capabilities of AI in cyber risk assessment.

User and event behavior analytics (UEBA) is another AI-driven approach that plays a significant role in cyber risk assessment. UEBA tools leverage AI to monitor and analyze user activities and system events to detect anomalies that could indicate a security breach. By establishing baselines of normal behavior, these tools can identify deviations that suggest malicious activities such as unauthorized access or data exfiltration. This real-time detection capability is crucial for mitigating risks before they escalate into significant incidents. AI-powered risk assessment tools also utilize advanced data aggregation and interpretation techniques. These tools can compile data from a risk library and other sources to generate comprehensive risk reports. For instance, solutions like Secureframe Comply AI for Risk can produce detailed insights into risks by evaluating the likelihood and impact of potential threats, suggesting treatment plans, and defining residual risks after mitigation efforts. This holistic approach helps organizations understand the full spectrum of their risk landscape and prioritize their response strategies effectively.

Predictive analytics is another essential component of AI in cyber risk assessment. By analyzing historical data and identifying patterns related to past incidents, AI can construct forward-looking scenarios to predict future risks. This capability allows organizations to proactively address potential threats and implement preventive measures. Predictive analytics also enhances the transparency of the link between business processes and risk, ensuring that risk controls are adequate and corrective actions are timely. In summary, the AI technologies used in cyber risk assessment include ML for pattern recognition and anomaly detection, NLP for analyzing unstructured data, UEBA for real-time threat detection, advanced data aggregation and interpretation tools for comprehensive risk reporting, and predictive analytics for forecasting future risks. These AI-driven approaches collectively enhance the accuracy, efficiency, and effectiveness of cyber risk assessments, enabling organizations to better protect their assets and data from evolving cyber threats.

6.3.1   Data collection and preprocessing

Historical data analysis: AI starts by analyzing an organization's past data on security incidents, email traffic, web access logs, and previous phishing attempts. This historical data is crucial for understanding the patterns and tactics previously used against the organization.

Data enrichment: The data is enriched with external threat intelligence, including details on the latest phishing schemes and indicators of compromise. This broader dataset provides a more comprehensive view of potential threats.

The data elements required can be broadly categorized into several key areas. First, you need to gather empirical data related to your network devices and IP addresses. This includes detailed information about the hardware and software components, their configurations, and any known vulnerabilities. This data helps AI models to understand the current state of your network infrastructure and identify potential weak points.

Second, historical data on past incidents and threat patterns is crucial. AI can analyze this data to identify trends and predict future risks. This includes data on previous cyberattacks, their nature, and the impact they had on your organization. By understanding past incidents, AI can develop risk predictors and construct forward-looking scenarios to anticipate and mitigate future threats.

Additionally, user behavior data is vital for building models that detect anomalies. This involves collecting data on user activities, access patterns, and any deviations from normal behavior. AI can use this data to identify potential insider threats or compromised accounts by recognizing unusual activities that may indicate a security breach.

Risk signals and threat intelligence data are also important. This includes information from external sources such as threat intelligence feeds, cybersecurity vendor reports, and government-issued alerts. These sources provide insights into emerging threats and vulnerabilities that could affect your organization. AI can aggregate and interpret this data to provide a real-time assessment of your risk exposure.

Furthermore, data on third-party vendors and their security postures is necessary. This involves evaluating the number of third parties linked to your organization, the data they have access to, and their security measures. AI can assess the risks posed by these external partners and incorporate them into the overall risk score.

Lastly, it is important to collect data on your organization's business priorities and critical assets. This includes identifying high-value assets, their importance to business operations, and the potential impact of a cyberattack on these assets. AI can use this information to prioritize risks and recommend appropriate security measures.

By collecting and analyzing these diverse data elements, AI can provide a holistic and accurate assessment of your organization's cyber risk, helping you to better protect your data and assets.

6.3.2   Predictive modeling

Risk assessment models: Using ML techniques, AI develops models that assess the organization's vulnerability to phishing attacks. These models consider factors such as the frequency of phishing attempts, the success rate of past attacks, employee susceptibility to phishing, and the effectiveness of existing defense mechanisms.

User behavior modeling: AI analyzes patterns in user behavior to identify risk factors such as susceptibility to social engineering tactics or the tendency to bypass security protocols. Understanding user behavior is key to assessing overall organizational risk.

As such, predictive modeling has emerged as a transformative tool in the realm of cyber risk assessments, offering a proactive approach to identifying and mitigating potential threats. This advanced analytical technique leverages historical and real-time data to forecast future cyber risks, enabling cybersecurity professionals, organizations, and technology researchers to enhance their defensive strategies and maintain robust security postures.

At its core, predictive modeling involves the use of statistical algorithms and ML techniques to analyze patterns from past cyber incidents and current network behaviors. By processing vast amounts of data, these models can identify trends and anomalies that may indicate potential vulnerabilities or impending attacks. This capability is particularly valuable in the dynamic and ever-evolving landscape of cybersecurity, where new threats and attack vectors continuously emerge.

For cybersecurity professionals, predictive modeling offers a significant advantage in terms of risk assessment and management. Traditional methods of cyber risk assessment often rely on historical data and reactive measures, which can leave organizations vulnerable to new and sophisticated attacks. Predictive modeling, on the other hand, enables a shift from reactive to proactive defense strategies. By anticipating potential threats before they materialize, cybersecurity teams can implement preventive measures, thereby reducing the likelihood of successful attacks and minimizing potential damage.

One of the key benefits of predictive modeling in cyber risk assessments is its ability to enhance the accuracy and efficiency of threat detection. ML algorithms can sift through large datasets, identifying subtle patterns and correlations that may be missed by human analysts. For instance, predictive AI can analyze data from various sources, such as system logs, network traffic, and user behaviors, to detect anomalies that could indicate a security breach. This real-time detection capability is crucial for mitigating risks before they escalate into significant incidents.

Moreover, predictive modeling can help organizations prioritize their cybersecurity efforts. By assessing the likelihood and potential impact of different threats, these models can generate risk scores that highlight the most critical vulnerabilities. This allows cybersecurity teams to allocate resources more effectively, focusing on the areas that pose the greatest risk to the organization. For example, predictive risk profiles can be used to evaluate third-party vendors, identifying those that present the highest risk and necessitating more stringent security measures.

In addition to enhancing threat detection and prioritization, predictive modeling supports the development of more robust incident response plans. By simulating various attack scenarios, these models can help organizations understand the potential consequences of different types of cyber incidents and prepare accordingly. This includes identifying the most effective response strategies, as well as the resources and capabilities needed to implement them. For instance, predictive analytics can guide response teams by offering insights into the attack's objectives, potential targets, and likely progression, enabling a more coordinated and effective response.

For technology researchers, predictive modeling provides a valuable tool for advancing the field of cybersecurity. By analyzing large datasets and identifying emerging trends, researchers can gain a deeper understanding of the evolving threat landscape. This knowledge can inform the development of new security technologies and methodologies, as well as the refinement of existing ones. For example, researchers can use predictive models to study the evolution of cyber risks, uncovering the processes that lead to the creation of new threats and vulnerabilities.

This can lead to the development of more sophisticated and adaptive security solutions that are better equipped to handle future challenges.

Furthermore, predictive modeling can facilitate collaboration between different stakeholders in the cybersecurity ecosystem. By providing a common framework for assessing and communicating risk, these models can help bridge the gap between technical experts, business leaders, and policymakers. This is particularly important in the context of developing and implementing cybersecurity policies and regulations. For instance, predictive analytics can inform the creation of policies that proactively address potential vulnerabilities, rather than merely reacting to incidents after they occur. This can lead to a more resilient and secure digital infrastructure, benefiting society as a whole.

In practical terms, the implementation of predictive modeling in cyber risk assessments involves several key steps. First, organizations need to collect and preprocess relevant data from various sources such as network logs, threat intelligence feeds, and user activity records. This data is then used to train ML models, which can identify patterns and correlations indicative of potential threats. Once trained, these models can be deployed to monitor network activity in real-time, providing continuous risk assessments and alerting security teams to any anomalies.

To maximize the effectiveness of predictive modeling, it is essential to ensure the quality and integrity of the data used. This involves implementing robust data governance practices, such as regular data validation and cleansing, as well as ensuring that data is collected from reliable and diverse sources. Additionally, organizations should invest in building or sourcing skilled teams with expertise in data science, ML, and cybersecurity. These teams are responsible for developing, maintaining, and refining predictive models, as well as interpreting the insights generated and translating them into actionable security measures.

Predictive modeling represents a powerful and innovative approach to cyber risk assessment, offering numerous benefits for cybersecurity professionals, organizations, and technology researchers. By leveraging historical and real-time data to forecast future threats, these models enable a proactive and strategic approach to cybersecurity, enhancing threat detection, prioritization, and incident response. As the threat landscape continues to evolve, the adoption of predictive modeling will be crucial for maintaining robust security postures and ensuring the resilience of digital infrastructures.

6.3.3   *Predictive modeling and assessing the phishing risk*

As it relates to phishing, predictive modeling has become an essential tool in identifying phishing risks during cyber risk assessments. By leveraging historical data, ML algorithms, and advanced analytics, predictive models can forecast potential phishing attacks, enabling organizations to proactively mitigate these threats. This approach not only enhances the accuracy of phishing detection but also improves the overall cybersecurity posture of an organization.

Phishing attacks have evolved significantly over the years, becoming more sophisticated and targeted. Traditional methods of phishing detection, which often rely on static rules and signature-based approaches, are increasingly inadequate in the face of these advanced threats. Predictive modeling, however, offers a dynamic and adaptive solution. By analyzing patterns in historical data, predictive models can identify indicators of phishing attempts and predict future attacks with a high degree of accuracy.

One of the primary components of predictive modeling in phishing risk assessment is the use of ML algorithms. These algorithms are trained on large datasets of known phishing and legitimate emails, learning to distinguish between the two based on various features. For instance, features such as the presence of suspicious URLs, the use of urgent language, and the structure of the email can all be used to train the model. Once trained, the model can analyze new emails in real-time, flagging those that exhibit characteristics similar to known phishing attempts.

A study by Orunsolu *et al.* [57] highlights the effectiveness of ML-based predictive models in phishing detection. The researchers developed a model that uses Support Vector Machine and Naïve Bayes algorithms, trained on a dataset consisting of 2,541 phishing instances and 2,500 benign instances. The model achieved a remarkable accuracy rate of 99.96 percent with a false positive rate of just 0.04 percent. This high level of accuracy demonstrates the potential of predictive modeling to significantly enhance phishing detection capabilities.

In addition to ML, NLP plays a critical role in predictive modeling for phishing risk assessment. NLP techniques can analyze the content of emails, identifying linguistic patterns that are indicative of phishing. For example, phishing emails often use urgent or threatening language to prompt immediate action from the recipient. By analyzing the semantics and syntax of the email content, NLP models can detect these subtle cues and flag suspicious emails for further investigation.

Anomaly detection is another important aspect of predictive modeling in phishing risk assessment. This technique involves monitoring user behavior and email traffic patterns to identify deviations from the norm. For instance, if an employee who typically receives a few emails per day suddenly receives a large number of emails from an unknown source, this could be indicative of a phishing attack. By establishing baselines of normal behavior, anomaly detection models can identify unusual activities that may signal a phishing attempt.

The integration of predictive analytics into phishing detection systems offers several benefits for cybersecurity professionals and organizations. First, it enables a proactive approach to cybersecurity. Instead of reacting to phishing attacks after they occur, organizations can anticipate and prevent them. This shift from reactive to proactive defense strategies is crucial in minimizing the impact of phishing attacks and protecting sensitive information. Second, predictive modeling enhances the efficiency of phishing detection. Traditional methods often generate a high number of false positives, which can overwhelm security teams and lead to alert fatigue. Predictive models, however, can significantly reduce false positives by accurately distinguishing between legitimate and malicious emails. This allows security teams to focus their efforts on genuine threats, improving their overall effectiveness.

Moreover, predictive modeling can help organizations prioritize their cyber-security efforts. By assessing the likelihood and potential impact of different phishing threats, predictive models can generate risk scores that highlight the most critical vulnerabilities. This enables organizations to allocate resources more effectively, focusing on the areas that pose the greatest risk. For example, predictive models can identify high-risk users who are more likely to be targeted by phishing attacks, allowing organizations to provide additional training and support to these individuals. For technology researchers, predictive modeling offers valuable insights into the evolving threat landscape. By analyzing large datasets and identifying emerging trends, researchers can gain a deeper understanding of the tactics and techniques used by phishers. This knowledge can inform the development of new security technologies and methodologies, as well as the refinement of existing ones. For instance, researchers can use predictive models to study the evolution of phishing attacks, uncovering the processes that lead to the creation of new threats and vulnerabilities. This can lead to the development of more sophisticated and adaptive security solutions that are better equipped to handle future challenges. The practical implementation of predictive modeling in phishing risk assessment involves several key steps. First, organizations need to collect and preprocess relevant data from various sources such as email logs, threat intelligence feeds, and user activity records. This data is then used to train ML models, which can identify patterns and correlations indicative of phishing attempts. Once trained, these models can be deployed to monitor email traffic in real-time, providing continuous risk assessments and alerting security teams to any anomalies.

To maximize the effectiveness of predictive modeling, it is essential to ensure the quality and integrity of the data used. This involves implementing robust data governance practices, such as regular data validation and cleansing, as well as ensuring that data is collected from reliable and diverse sources. Additionally, organizations should invest in building or sourcing skilled teams with expertise in data science, ML, and cybersecurity. These teams are responsible for developing, maintaining, and refining predictive models, as well as interpreting the insights generated and translating them into actionable security measures.

The deployment of predictive models in phishing detection systems also requires continuous monitoring and fine-tuning. As the threat landscape evolves, predictive models need to be updated and retrained to maintain their effectiveness. This involves regularly evaluating the model's performance, identifying any areas for improvement, and incorporating new data to enhance its predictive capabilities. By continuously refining predictive models, organizations can stay ahead of emerging threats and ensure that their phishing detection systems remain robust and effective.

Predictive modeling represents a powerful and innovative approach to identifying phishing risks during cyber risk assessments. By leveraging historical data, ML algorithms, and advanced analytics, predictive models can forecast potential phishing attacks with a high degree of accuracy. This proactive approach enables organizations to enhance their phishing detection capabilities, prioritize their cybersecurity efforts, and stay ahead of emerging threats. As the threat landscape continues to evolve, the adoption of predictive modeling will be crucial for maintaining robust security postures and ensuring the resilience of digital infrastructures.

6.3.4 *Predictive modeling and assessing the user population as a cyber risk*

As unsettling as it may sound, predictive modeling has revolutionized the way organizations assess their workforce, providing a data-driven approach to understanding and forecasting employee behaviors, performance, and potential risks. By leveraging historical data, statistical algorithms, and ML techniques, predictive modeling enables organizations to make informed decisions that enhance workforce management, improve employee retention, and optimize overall performance. This comprehensive approach to workforce assessment is invaluable for human resources (HR) professionals, business leaders, and technology researchers, offering insights that were previously unattainable through traditional methods.

At the heart of predictive modeling in workforce assessment is the ability to analyze vast amounts of data to identify patterns and trends. This process begins with the collection and preprocessing of relevant data from various sources such as employee records, performance reviews, engagement surveys, and exit interviews. By integrating this data into a centralized system, organizations can create a robust dataset that serves as the foundation for predictive modeling. The quality and integrity of this data are crucial, as they directly impact the accuracy and reliability of the predictive models.

One of the primary applications of predictive modeling in workforce assessment is predicting employee turnover. High turnover rates can be costly for organizations, leading to increased recruitment and training expenses, as well as disruptions in productivity. Predictive models can analyze factors such as job satisfaction, engagement levels, tenure, and performance metrics to identify employees who are at risk of leaving the company. For example, a methodology used by Visier [58] highlights how predictive HR analytics can forecast when an employee is likely to quite, allowing HR professionals to take proactive measures to retain valuable talent. By understanding the underlying causes of turnover, organizations can implement targeted interventions, such as career development programs, improved compensation packages, or enhanced work-life balance initiatives, to reduce attrition rates and retain top performers.

In addition to predicting turnover, predictive modeling can also enhance the recruitment process by identifying candidates who are most likely to succeed in a given role. Traditional recruitment methods often rely on subjective assessments and gut feelings, which can lead to biased hiring decisions and suboptimal outcomes. Predictive analytics, on the other hand, uses data-driven insights to evaluate candidates based on their skills, experience, and potential fit with the organization's culture. By analyzing historical data on successful hires, predictive models can identify the characteristics and attributes that correlate with high performance and long-term retention. This enables recruiters to make more informed decisions, reducing the likelihood of bad hires and improving overall recruitment efficiency.

Another critical application of predictive modeling in workforce assessment is performance management. Predictive models can analyze various performance-related metrics, such as productivity, quality of work, and goal attainment, to identify factors that influence employee performance. For instance, predictive analytics can reveal how different process and motivation schemes impact team productivity and job satisfaction. By understanding these correlations, HR professionals can design targeted interventions to enhance performance such as personalized training programs, performance incentives, or changes in work processes. This data-driven approach to performance management ensures that employees receive the support and resources they need to excel in their roles, ultimately driving organizational success.

Employee engagement is another area where predictive modeling can provide valuable insights. Engaged employees are more productive, have lower absenteeism rates, and are less likely to leave the organization. Predictive models can analyze engagement survey data to identify trends and patterns that influence employee engagement. For example, Best Buy found that a 0.1 percentage point increase in engagement led to a $100,000 increase in revenue per store. By understanding the drivers of engagement, organizations can implement targeted initiatives to boost employee morale and satisfaction such as recognition programs, flexible work arrangements, or opportunities for career advancement. This proactive approach to engagement not only improves employee well-being but also enhances overall organizational performance.

Predictive modeling also plays a crucial role in workforce planning and talent management. By forecasting future workforce needs, organizations can ensure they have the right talent in place to meet business objectives. Predictive models can analyze factors such as market trends, employee turnover rates, and company growth projections to anticipate talent requirements and workforce gaps. This enables HR professionals to develop strategic recruitment and training plans, ensuring that the organization is well-prepared to address future challenges. For example, predictive analytics can help HR specialists forecast an organization's talent requirements and workforce needs in advance, allowing for timely planning and resource allocation.

Moreover, predictive modeling can support succession planning by identifying high-potential employees who are ready to take on leadership roles. By analyzing data on employee experience, career trajectories, and skillsets, predictive models can identify individuals with the potential to become future leaders. This allows organizations to develop targeted development programs to groom these employees for leadership positions, ensuring a smooth transition and continuity in leadership. This proactive approach to succession planning not only mitigates the risk of leadership gaps but also fosters a culture of growth and development within the organization.

For technology researchers, predictive modeling offers a wealth of opportunities to advance the field of workforce analytics. By analyzing large datasets and identifying emerging trends, researchers can gain a deeper understanding of the factors that influence employee behavior and performance. This knowledge can inform the development of new predictive models and algorithms, as well as the refinement of existing ones. For example, researchers can use predictive models to study the impact of different HR interventions on employee outcomes, uncovering the most effective strategies for improving retention, engagement, and performance. This continuous cycle of research and development drives innovation in workforce analytics, leading to more sophisticated and accurate predictive models.

Despite the numerous benefits of predictive modeling in workforce assessment, there are also challenges that organizations must address to fully realize its potential. One of the primary challenges is ensuring data quality and integrity. Predictive models rely on accurate and comprehensive data to generate reliable insights. Therefore, organizations must implement robust data governance practices, such as regular data validation and cleansing, to maintain the quality of their datasets. Additionally, organizations must ensure that data is collected from reliable and diverse sources to avoid biases and inaccuracies.

Another challenge is the need for skilled professionals who can develop, maintain, and interpret predictive models. While predictive modeling offers powerful insights, it requires expertise in data science, ML, and HR analytics to effectively implement and manage these models. Organizations may need to invest in training and development programs to build these competencies within their HR teams or consider hiring data scientists to support their workforce analytics initiatives.

To wit, employee predictive modeling can be a powerful tool in assessing users relative to cyber risks, offering a nuanced and data-driven approach to identifying potential vulnerabilities and threats within an organization. By leveraging historical data, ML algorithms, and advanced analytics, predictive models can provide valuable insights into employee behaviors, performance, and potential risks, enabling organizations to proactively mitigate cyber threats and enhance their overall security posture.

One of the primary ways predictive modeling can be used to assess users relative to cyber risks is through the analysis of user behavior and activity patterns. By collecting and analyzing data on employee activities, such as login times, access to sensitive information, and usage of network resources, predictive models can identify deviations from normal behavior that may indicate potential security threats. For instance, if an employee who typically accesses certain files during regular business hours suddenly begins accessing sensitive data late at night, this could be a red flag for a potential insider threat or compromised account. Predictive models can flag such anomalies in real-time, allowing security teams to investigate and take appropriate action before a breach occurs.

UEBA is a key component of this approach, as it involves monitoring and analyzing user activities to detect behaviors that fall outside the range of normal activities. By combining data on user identity attributes, access rights, and historical activities, UEBA systems can create detailed profiles of typical user behavior. Predictive models can then use these profiles to identify suspicious activities that may indicate misuse of privileges or malicious intent. For example, if an employee with access to financial records suddenly starts downloading large volumes of data, predictive models can flag this behavior as potentially malicious, prompting further investigation.

Predictive modeling can also enhance the detection of phishing risks by analyzing employee interactions with emails and other communication channels.

Phishing attacks often rely on social engineering techniques to trick employees into divulging sensitive information or clicking on malicious links. By analyzing patterns in email communication such as the frequency of emails from unknown senders, the use of urgent language, and the presence of suspicious URLs, predictive models can identify emails that are likely to be phishing attempts. This proactive approach allows organizations to warn employees about potential phishing threats and reduce the likelihood of successful attacks.

Another important application of predictive modeling in assessing cyber risks is the identification of high-risk users. By analyzing factors such as job role, access to sensitive information, and historical security incidents, predictive models can generate risk scores for individual employees. These risk scores can help security teams prioritize their efforts, focusing on users who pose the greatest risk to the organization. For instance, employees with access to critical systems or sensitive data may be assigned higher risk scores, prompting more stringent monitoring and security measures. This targeted approach ensures that resources are allocated effectively, reducing the overall risk to the organization.

Predictive modeling can also support the development of personalized security training programs. By analyzing data on employee behavior and performance, predictive models can identify areas where employees may be vulnerable to cyber threats. For example, if an employee frequently falls for phishing attempts or exhibits risky online behavior, predictive models can recommend targeted training programs to address these vulnerabilities. Personalized training can help employees develop better security practices, reducing the likelihood of human error and enhancing the organization's overall security posture.

In addition to identifying potential threats, predictive modeling can help organizations understand the underlying causes of cyber risks. By analyzing data on employee behavior, performance, and engagement, predictive models can identify factors that contribute to risky behavior. For instance, employees who are disengaged or dissatisfied with their job may be more likely to engage in risky behavior, such as ignoring security protocols or attempting to steal sensitive information. By understanding these underlying factors, organizations can implement targeted interventions to address the root causes of cyber risks such as improving employee engagement or addressing workplace grievances.

For technology researchers, predictive modeling offers valuable insights into the evolving threat landscape and the factors that influence employee behavior. By analyzing large datasets and identifying emerging trends, researchers can gain a deeper understanding of the tactics and techniques used by cybercriminals. This knowledge can inform the development of new security technologies and methodologies, as well as the refinement of existing ones. For example, researchers can use predictive models to study the impact of different security interventions on employee behavior, uncovering the most effective strategies for reducing cyber risks.

Despite the numerous benefits of predictive modeling in assessing users relative to cyber risks, there are also challenges that organizations must address to fully realize its potential. One of the primary challenges is ensuring data quality and integrity. Predictive models rely on accurate and comprehensive data to generate reliable insights. Therefore, organizations must implement robust data governance practices, such as regular data validation and cleansing, to maintain the quality of their datasets. Additionally, organizations must ensure that data is collected from reliable and diverse sources to avoid biases and inaccuracies.

Another challenge is the need for skilled professionals who can develop, maintain, and interpret predictive models. While predictive modeling offers powerful insights, it requires expertise in data science, ML, and cybersecurity to effectively implement and manage these models. Organizations may need to invest in training and development programs to build these competencies within their security teams or consider hiring data scientists to support their predictive analytics initiatives.

Privacy and ethical considerations are also critical when implementing predictive modeling in workforce assessment. Organizations must ensure that employee data is collected, stored, and analyzed in compliance with relevant data protection regulations. Additionally, they must consider the ethical implications of using predictive analytics, such as the potential for bias in security decisions. By adopting transparent and ethical practices, organizations can build trust with their employees and stakeholders, ensuring that predictive modeling is used responsibly and effectively.

Predictive modeling represents a powerful and innovative approach to assessing users relative to cyber risks. By leveraging historical data, ML algorithms, and advanced analytics, predictive models can provide valuable insights into employee behaviors, performance, and potential risks. This data-driven approach enables organizations to proactively mitigate cyber threats, enhance their overall security posture, and ensure the resilience of their digital infrastructures. As the field of cybersecurity continues to evolve, the adoption of predictive modeling will be crucial for maintaining robust security measures and protecting sensitive information from emerging threats.

6.3.5   Simulation and scenario analysis

Phishing attack simulations: AI can simulate phishing attacks based on current threat intelligence and historical data. These simulations test the organization's defenses and employee reactions to various phishing scenarios, helping to identify potential weaknesses.

Scenario analysis: By creating scenarios based on different attack vectors, target departments, or timeframes, AI helps organizations understand how changes in tactics or organizational structure might impact their vulnerability to phishing.

By leveraging AI for modeling and simulating phishing attack scenarios, organizations can enhance their defensive strategies, improve employee awareness, and mitigate the risks associated with these pervasive threats. This comprehensive approach involves the use of ML algorithms, NLP, and generative AI to create realistic phishing simulations, analyze user behavior, and develop targeted training programs.

Phishing attacks are a form of social engineering where attackers deceive individuals into divulging sensitive information, such as login credentials or financial details, by impersonating trustworthy entities. These attacks have become increasingly sophisticated, making it challenging for traditional security measures to detect and prevent them effectively. AI, with its ability to analyze vast amounts of data and identify patterns, offers a powerful solution to this problem.

One of the primary ways AI is used in modeling phishing attack scenarios is through the creation of realistic phishing simulations. These simulations are designed to mimic real-world phishing attempts, allowing organizations to assess how their employees respond to such threats. AI-powered phishing simulations utilize ML algorithms to generate emails that closely resemble those used in actual phishing attacks. By analyzing historical data on phishing emails, these algorithms can identify common characteristics and patterns such as the use of urgent language, suspicious URLs, and specific formatting styles. This enables the creation of highly convincing phishing emails that can effectively test employees' ability to recognize and respond to phishing attempts.

For instance, Guardz.com offers an AI-powered phishing simulation platform that generates simulated phishing emails tailored to specific departments within an organization. This dynamic approach ensures that the content is relevant and up-to-date, reflecting the latest phishing trends. By conducting these simulations, organizations can gather detailed data on employee performance, identifying individuals who successfully identify phishing attempts and those who do not. This information is crucial for developing targeted training programs to address specific vulnerabilities and improve overall security awareness.

Generative AI, particularly large language models (LLMs) like OpenAI's GPT-3, has further enhanced the realism and effectiveness of phishing simulations. These models can generate highly sophisticated and contextually relevant phishing emails by absorbing real-time information from various sources such as news outlets and corporate websites. This capability allows for the creation of phishing emails that incorporate current events or organizational details, making them more believable and increasing the likelihood of successful deception. For example, a generative AI model could craft an email that appears to be from a company's HR department, referencing a recent policy update and requesting employees to verify their login credentials.

The use of AI in phishing simulations also extends to spear phishing attacks, which are highly targeted and personalized phishing attempts aimed at specific individuals. Spear phishing attacks often leverage information gathered from social media, data breaches, and other sources to create convincing messages that are difficult to distinguish from legitimate communications. AI-generated spear phishing emails can be particularly effective, as they can incorporate detailed personal information and adopt a professional writing style that eliminates common indicators of phishing such as spelling errors and grammatical mistakes. This level of sophistication makes it challenging for both individuals and traditional security systems to detect these attacks.

A notable example of the effectiveness of AI-generated spear phishing emails was demonstrated at Black Hat USA 2021, where Singapore's Government Technology Agency conducted an experiment comparing human-crafted and AI-generated phishing emails. The results showed that more people clicked on the AI-generated phishing emails than the human-written ones, highlighting the potential of AI to create highly convincing phishing attacks. As LLM technology continues to advance, the ability to generate realistic and targeted phishing emails will only improve, necessitating the use of AI-powered defenses to counter these threats.

In addition to generating phishing emails, AI can be used to analyze user behavior and identify patterns that may indicate susceptibility to phishing attacks. By monitoring user activities, such as email interactions, login times, and access to sensitive information, AI models can detect anomalies that suggest potential security risks. For example, if an employee who typically accesses certain files during regular business hours suddenly begins accessing sensitive data late at night, this could be a red flag for a potential insider threat or compromised account. Predictive models can flag such anomalies in real-time, allowing security teams to investigate and take appropriate action before a breach occurs.

UEBA is a key component of this approach, as it involves creating detailed profiles of typical user behavior and identifying deviations from these norms. By combining data on user identity attributes, access rights, and historical activities, UEBA systems can detect behaviors that fall outside the range of normal activities, such as unusual login locations or unexpected data transfers. This proactive approach enables organizations to identify and mitigate potential threats before they can cause significant harm.AI-powered phishing simulations also play a crucial role in employee training and awareness programs. By exposing employees to realistic phishing scenarios, organizations can assess their ability to recognize and respond to phishing attempts. This hands-on experience is invaluable for reinforcing security best practices and improving overall vigilance. For example, Microsoft Defender for Office 365 offers an attack simulation training feature that allows administrators to create and launch simulated phishing attacks. These simulations test employees' security awareness and provide insights into their susceptibility to phishing, enabling organizations to tailor their training programs accordingly.

Moreover, AI can enhance the effectiveness of security awareness training by personalizing the content based on individual performance and learning preferences. Generative AI models can adapt training curricula to address each employee's weak spots, providing targeted lessons and exercises that focus on specific vulnerabilities. Additionally, AI can identify the learning modality that best serves each employee, whether it be in-person, audio, interactive, or video-based training. By maximizing the effectiveness of security awareness training at a granular level, AI can significantly reduce overall cyber risk and improve organizational resilience. The integration of AI into phishing simulations and training programs also supports continuous improvement and adaptation to emerging threats. As cybercriminals develop new tactics and techniques, AI models can be updated and retrained to reflect the latest threat landscape. This ensures that phishing simulations remain relevant and effective, providing organizations with up-to-date insights into their security posture. For example, AI-powered platforms like Guardz.com continuously update their phishing simulation content to stay aligned with the latest phishing trends, ensuring that employees are prepared to recognize and respond to current threats.

Despite the numerous benefits of AI in modeling and simulating phishing attack scenarios, there are also challenges that organizations must address to fully realize its potential. One of the primary challenges is ensuring data quality and integrity. Predictive models rely on accurate and comprehensive data to generate reliable insights. Therefore, organizations must implement robust data governance practices, such as regular data validation and cleansing, to maintain the quality of their datasets. Additionally, organizations must ensure that data is collected from reliable and diverse sources to avoid biases and inaccuracies.

Another challenge is the need for skilled professionals who can develop, maintain, and interpret AI models. While AI offers powerful insights, it requires expertise in data science, ML, and cybersecurity to effectively implement and manage these models. Organizations may need to invest in training and development programs to build these competencies within their security teams or consider hiring data scientists to support their AI initiatives. Privacy and ethical considerations are also critical when implementing AI in phishing simulations. Organizations must ensure that employee data is collected, stored, and analyzed in compliance with relevant data protection regulations. Additionally, they must consider the ethical implications of using AI-generated content, such as the potential for bias in security decisions. By adopting transparent and ethical practices, organizations can build trust with their employees and stakeholders, ensuring that AI is used responsibly and effectively.

AI has the potential to significantly enhance the modeling and simulation of phishing attack scenarios, providing organizations with powerful tools to assess and mitigate cyber risks. By leveraging ML algorithms, generative AI, and user behavior analytics, organizations can create realistic phishing simulations, analyze employee responses, and develop targeted training programs. This proactive approach enables organizations to stay ahead of emerging threats, improve security awareness, and enhance their overall cybersecurity posture. As the field of AI continues to evolve, its integration into phishing simulations and training programs will be crucial for maintaining robust defenses and protecting sensitive information from increasingly sophisticated cyber threats.

6.3.6 Risk scoring and prioritization

Risk scoring: AI assigns risk scores to various components of the organization's security posture, including IT infrastructure, user groups, and data assets. This scoring system helps in prioritizing areas that require immediate attention or improvement.

Dynamic risk assessment: The risk models are continuously updated with new data, allowing for dynamic risk assessments that reflect the organization's current threat landscape. This ongoing process ensures that the organization can adapt its defenses in line with evolving threats.

6.3.7 Recommendations for mitigation and improvement

Mitigation strategies: Based on the risk characterization, AI provides tailored recommendations for mitigating phishing risk. This might include suggestions for enhancing email filtering, improving user training programs, or updating incident response protocols.

Improvement plans: AI can also suggest long-term strategies for reducing phishing susceptibility such as changes in IT infrastructure, adoption of new security technologies, or modifications to organizational policies.

By assessing past data, developing predictive models, and conducting simulations, AI offers powerful tools for characterizing the phishing risk of an organization. This proactive approach enables organizations to identify vulnerabilities, anticipate potential threats, and implement effective defenses. The continuous learning capability of AI ensures that the risk assessment remains accurate and relevant, helping organizations stay one step ahead of phishing attackers.

The following graph is a quantitative assessment of phishing risk across various departments within an organization. Each department is scored with a precise risk score, clearly labeled above each bar for a more detailed analysis.

Departments above the green dashed line (0.8) are considered at **high risk** of phishing attacks. For instance, the Finance department, with a score of 0.85, indicates a significant vulnerability and necessitates urgent measures to enhance security and training.

Departments between the red (0.5) and green dashed lines are at **medium risk**. These departments, such as HR and Sales, with scores of 0.72 and 0.78 respectively, should be closely monitored, and employees there should receive targeted awareness training to reduce susceptibility to phishing attacks.

Departments below the red dashed line are considered at **lower risk** (e.g., Operations with a score of 0.48). While these areas are currently less susceptible, maintaining awareness and vigilance is crucial to preventing future vulnerabilities.

This quantitative approach allows an organization to prioritize cybersecurity initiatives more effectively, focusing resources on areas of greatest need to mitigate the risk of phishing attacks.

The creation of a graph involves several AI technologies and techniques, particularly those within the realms of ML, data analysis, and NLP. Here's a breakdown of how different AI components could contribute to creating such a graph:

6.3.8 Data visualization

Graph generation: Once the data is analyzed and risk levels are assessed, AI can employ data visualization libraries (in programming languages like Python, using libraries such as Matplotlib or Seaborn) to generate informative graphs. These graphs quantitatively represent the risk assessment across different organizational departments, making it easier for stakeholders to understand and act upon the findings.

6.3.9   Automation and reporting

Automated reporting: AI systems can automate the generation of reports, including the creation of graphs like the phishing risk assessment. This not only saves time but also ensures that the information is updated regularly and remains accurate.

6.3.10   Implementing AI for phishing risk assessment recommendations for mitigation and improvement

To implement such an AI system, an organization would typically integrate various AI technologies:

- Data analysis tools: For processing and analyzing large datasets to identify trends and patterns.
- ML platforms: Such as TensorFlow or PyTorch, for building and training models to assess risk and predict future phishing attempts.
- NLP libraries: Like NLTK or spaCy, for processing and analyzing text data.
- Visualization software: Tools and libraries that support the creation of graphs and charts to represent the analyzed data visually.

The integration of these technologies enables the development of an AI system capable of assessing phishing risks, predicting future threats, and visually representing the findings to aid in decision-making and resource allocation for cybersecurity efforts.

6.4   A sample outline for a phishing risk management plan

While each organization should create a tailored risk mitigation plan relative to phishing, this outline provides a comprehensive framework that can support the development of a tailored plan. Since threats are not static in nature, this plan includes phases for adaptions to growing threats, particularly those posed by AI-enhanced phishing attacks.

To make the best use of this outline, consider the following:

1. Customization: Review each section of the outline and adapt it to your organization's unique context, including its size, industry, risk appetite, and existing cybersecurity infrastructure. Tailor the content to address your organization's specific phishing risks and align with its overall risk management strategy.
2. Collaboration: Involve key stakeholders from across your organization in the development and implementation of the risk management plan. This may include IT, security, HR, legal, and communications departments. Ensure that everyone understands their roles and responsibilities in mitigating phishing risks.

3. Prioritization: Use the risk assessment section to identify and prioritize the phishing risks that pose the greatest threat to your organization. Focus your resources and efforts on addressing the most critical risks first, while still maintaining a comprehensive approach to phishing risk management.
4. Integration: Integrate the phishing risk management plan into your organization's overall cybersecurity strategy. Ensure that the plan aligns with existing policies, procedures, and incident response processes. Regularly review and update the plan to maintain its relevance and effectiveness.
5. Training and awareness: Place a strong emphasis on employee training and awareness. Use the outlines to develop targeted training programs that educate employees about the latest phishing techniques, the risks associated with AI-enhanced phishing, and the importance of maintaining a vigilant and proactive approach to cybersecurity.
6. Metrics and reporting: Use the monitoring and reporting section to establish clear metrics for measuring the success of your phishing risk management efforts. Regularly report on these metrics to key stakeholders and use the insights gained to drive continuous improvement.
7. Continuous improvement: Treat the phishing risk management plan as a living document. Regularly review and update the plan based on new threats, changes in your organization's risk landscape, and lessons learned from actual phishing incidents and simulations.

By following these guidelines and leveraging the comprehensive outlines provided, organizations can develop robust and effective cybersecurity risk management plans that address the evolving threats posed by phishing and AI-enhanced phishing attacks.

Cybersecurity risk management plan
A focus on phishing template

(I)   Introduction
In this section, provide an overview of the purpose and scope of your organization's cybersecurity risk management plan. Clearly define phishing and its potential impact on your organization, including financial losses, reputational damage, and data breaches. Tailor the introduction to your organization's specific context, such as its size, industry, and the types of data it handles.
(A)   Purpose of the risk management plan
(B)   Scope of the plan
(C)   Definition of phishing and its impact on organizations

(II) Risk assessment
Conduct a thorough risk assessment by identifying the potential phishing attack vectors that your organization may face. Consider the unique risks associated with your organization's communication channels such as email,

SMS, voice calls, and social media. Evaluate the likelihood and potential impact of each attack vector based on your organization's risk appetite and historical data. Prioritize the identified risks using a risk assessment matrix that aligns with your organization's risk management framework.

(A)   Identify potential phishing attack vectors
1.   Email-based phishing
2.   SMS/text message phishing (smishing)
3.   Voice phishing (vishing)
4.   Social media-based phishing
5.   AI-enhanced phishing attacks

(B)   Evaluate the likelihood and impact of each attack vector
(C)   Prioritize risks based on the risk assessment matrix

(III)   AI-enhanced phishing risks
Discuss the emerging threats posed by AI-enhanced phishing attacks. Explain how deepfake technology can be used to create synthetic voice and video impersonations of trusted individuals within your organization. Highlight the risks associated with NLP-based phishing, such as highly personalized and convincing phishing emails and chatbots used for social engineering. Also, address the dangers of ML-based phishing, including automated target selection, profiling, and adaptive phishing campaigns based on user behavior. Tailor this section to the specific AI-related phishing risks your organization may face based on its technology stack and the sophistication of potential attackers.

(A)   Deepfake technology
1.   Synthetic voice impersonation
2.   Synthetic video impersonation

(B)   Natural language processing (NLP) based phishing
1.   Personalized and convincing phishing emails
2.   Chatbots used for social engineering

(C)   ML-based phishing
1.   Automated target selection and profiling
2.   Adaptive phishing campaigns based on user behavior

(IV)   Risk mitigation strategies
Outline the technical, administrative, and human controls your organization will implement to mitigate the identified phishing risks. Discuss the use of email filtering, anti-spam solutions, multi-factor authentication, DMARC,[3] and regular software updates and patching. Consider implementing AI-based phishing detection tools that can help identify and block sophisticated phishing attempts. Develop policies and procedures for

---

[3]DMARC stands for "Domain-based Message Authentication, Reporting, and Conformance." It is an email authentication protocol designed to help email domain owners protect their domain from unauthorized use, commonly known as email spoofing.

handling suspicious emails and messages, and create an incident response plan specific to phishing attacks. Emphasize the importance of employee awareness training, phishing simulation exercises, and fostering a culture of reporting suspicious activities. Tailor the mitigation strategies to your organization's existing security infrastructure, budget, and risk tolerance.

- (A) Technical controls
    1. Email filtering and anti-spam solutions
    2. Multi-factor authentication (MFA)
    3. Domain-based Message Authentication, Reporting, and Conformance (DMARC)
    4. Regular software updates and patching
    5. AI-based phishing detection tools

- (B) Administrative controls
    1. Policies and procedures for handling suspicious emails and messages
    2. Incident response plan for phishing attacks
    3. Regular risk assessments and audits

- (C) Human controls
    1. Employee awareness training on phishing and social engineering
    2. Phishing simulation exercises
    3. Encouraging a culture of reporting suspicious activities

- (V) Incident response
    Establish a dedicated incident response team responsible for handling phishing incidents. Develop a comprehensive incident response plan that outlines the steps to be taken during a phishing attack, including containment measures, eradication steps, recovery procedures, and post-incident analysis. Regularly test and update the incident response plan to ensure its effectiveness and relevance to your organization's evolving threat landscape. Tailor the incident response section to your organization's existing incident management processes and the roles and responsibilities of key stakeholders.
    - (A) Establish an incident response team
    - (B) Develop an incident response plan for phishing attacks
        1. Containment measures
        2. Eradication steps
        3. Recovery procedures
        4. Post-incident analysis and lessons learned
    - (C) Regularly test and update the incident response plan

- (VI) Monitoring and reporting
    Implement monitoring tools that can help detect phishing attempts in real-time. Define metrics to measure the effectiveness of your risk management

plan, such as the number of reported phishing incidents, the time taken to detect and respond to incidents, and the success rate of phishing simulations. Establish regular reporting mechanisms to keep stakeholders informed about the status of phishing risks and the progress of mitigation efforts. Tailor the monitoring and reporting section to your organization's existing security monitoring infrastructure and the information needs of your stakeholders.

(A)   Implement monitoring tools for early detection of phishing attempts
(B)   Establish metrics to measure the effectiveness of the risk management plan
(C)   Regular reporting to stakeholders on the status of phishing risks and mitigation efforts

(VII)   Continuous improvement
Emphasize the importance of staying informed about the latest phishing techniques and trends. Encourage your organization to participate in industry forums, subscribe to threat intelligence feeds, and collaborate with peers to share best practices. Regularly review and update your risk management plan to ensure it remains relevant and effective. Incorporate lessons learned from actual phishing incidents and simulations to refine your plan continuously. Tailor the continuous improvement section to your organization's culture of learning and its commitment to maintaining a robust cybersecurity posture.

(A)   Stay informed about the latest phishing techniques and trends
(B)   Regularly review and update the risk management plan
(C)   Incorporate feedback from incidents and simulations to refine the plan

(VIII)   Conclusion
Summarize the key points of your organization's cybersecurity risk management plan for phishing. Reiterate the importance of ongoing commitment from all stakeholders in mitigating phishing risks. Issue a call to action for everyone to actively participate in the risk management efforts by remaining vigilant, reporting suspicious activities, and adhering to the organization's security policies and procedures. Tailor the conclusion to your organization's specific goals and priorities in managing phishing risks.

(A)   Recap of the key points in the risk management plan
(B)   Importance of ongoing commitment to mitigating phishing risks
(C)   Call to action for all stakeholders to participate in the risk management efforts.

Chapter 7

Training efficacy versus direct experience and introspection

While most studies dating back to 2009 demonstrate how ineffective cybersecurity training programs are in developing a phishing resistant workforce, there is an emerging body of evidence that points toward the benefits of direct experience and introspection as possible training approach that yields success. This chapter will look at cybersecurity training with a specific focus on phishing, evaluate the research data, and present some possible approaches for application.

7.1 Cybersecurity training—the ongoing debate

When it comes to developing specific skill sets, training is the golden standard for many organizations. Whether organizations want their workforce to understand the proper handling of equipment, chemicals, or machinery, workforce training is often the first recommendation. Therefore, it stands to reason training was the logical choice for organizations to reach for when the computing world found its self-confronted with cyber criminals using deception to breach systems, steal data or disrupt operations.

However, motor vehicle safety, or operating machinery safely is not the same as resisting a phishing attack. As we discussed in Chapters 2–4, the computing experience, particularly today, introduces numerous variables that enable the personal experience to be highly individualized. Variables are not consistent between users, and often involve an individual's personal dispositions and context. These differences may be responsible for the efficacy (or lack thereof) relative to a cybersecurity training program centered around developing a phishing-resistant workforce.

As early as 2009, studies emerged pointing out how cybersecurity training was falling short in changing user behaviors. Studies continued to evolve in this area discussing a myriad of potential issues relative to these failures.

Since the vulnerability being exploited in a phishing attack is the user, most studies evaluate various aspects of users, often looking for a psychological trait that could be responsible for the failure of a cybersecurity training program to produce a phishing-resistant workforce.

A working draft paper developed by Bada and Sasse [59] addressed the problem directly and found: computer security awareness campaigns were failing to change user behaviors. These researchers also pointed directly at user behaviors and stated that changes in behavior would require changes in attitudes and intentions. However, if an individual falls prey to a well-crafted spear phishing email that appears to be coming from someone they know, and could expect an email from, is this really a statement relative to an individual's intentions and attitudes?

Since then, however, the landscape of cybersecurity training has undergone significant transformation over the years. The transition has gone away from often rudimentary programs, consisting of basic instructions on password management and general Internet safety. Luckily, as cyber threats have become more sophisticated, so too have the training programs designed to combat them. Modern cybersecurity training encompasses a wide range of topics, from advanced threat detection to the nuances of social engineering attacks.

One of the most notable advancements in cybersecurity training is the integration of artificial intelligence (AI) and machine learning (ML). These technologies are being used to create more dynamic and responsive training programs. AI-driven training platforms can analyze user behavior and adapt training content to address specific vulnerabilities. This personalized approach ensures that employees receive training that is relevant to their roles and the specific threats they are likely to encounter.

However, phishing remains a significant challenge for organizations worldwide. Despite advancements in technology, phishing attacks continue to be highly effective, primarily because they exploit human behavior. Cybercriminals use various tactics to deceive individuals into revealing sensitive information or downloading malicious software. To combat this, organizations have turned to comprehensive training programs that focus on phishing awareness and prevention.

Effective phishing training programs typically include several key components. First, they provide employees with the knowledge to recognize phishing attempts. This includes understanding common phishing tactics such as email spoofing, malicious links, and fraudulent attachments. Training programs often use real-world examples to illustrate these tactics, helping employees to identify red flags in their daily communications.

Second, phishing training emphasizes the importance of reporting suspicious emails. Employees are encouraged to report potential phishing attempts to their IT or security teams, enabling the organization to respond quickly and mitigate any potential damage. This reporting mechanism is crucial, as it allows organizations to gather intelligence on phishing trends and adapt their defenses accordingly.

One of the most effective methods for reinforcing phishing training is through simulated phishing exercises. These exercises involve sending fake phishing emails to employees to test their ability to recognize and respond to phishing attempts. Simulated phishing exercises provide several benefits. They offer a practical, hands-on experience that theoretical training cannot match. Employees learn to apply their knowledge in real-world scenarios, which helps to reinforce their training and improve their ability to detect phishing attempts.

Simulated phishing exercises also serve as a valuable tool for measuring the effectiveness of training programs. By tracking how employees respond to simulated phishing emails, organizations can identify areas where additional training is needed. This data-driven approach allows for continuous improvement of training programs, ensuring that they remain effective in the face of evolving threats.

The integration of AI into phishing training programs has brought about significant improvements in their effectiveness. AI-driven training platforms can analyze vast amounts of data to identify patterns and trends in phishing attacks. This information can be used to create more targeted and relevant training content. For example, if a particular type of phishing attack is becoming more common, the training platform can automatically update its content to address this new threat. AI also enables the creation of more personalized training experiences. By analyzing individual user behavior, AI-driven platforms can identify specific vulnerabilities and tailor training content to address them. This personalized approach ensures that employees receive training that is relevant to their roles and the specific threats they are likely to encounter.

In addition, AI can enhance the delivery of training content. Traditional training methods, such as lengthy presentations or static documents, are often ineffective at engaging employees. AI-driven platforms can deliver training content in more interactive and engaging formats, such as gamified modules or interactive simulations. This not only makes the training more enjoyable but also helps to improve retention and application of the knowledge.

Despite the advancements in cybersecurity training, there are still several challenges and limitations that need to be addressed. One of the primary challenges is ensuring that training programs are consistently updated to reflect the latest threats. The cyber threat landscape is constantly evolving, and training programs must keep pace with these changes to remain effective. This requires a significant investment of time and resources, which can be a barrier for some organizations.

Another challenge is overcoming the inherent limitations of human behavior. Even with the best training, there is always the risk that an employee will fall victim to a phishing attack. Human error is a significant factor in many cyber incidents, and it is impossible to eliminate this risk entirely. Organizations must therefore adopt a multilayered approach to cybersecurity, combining training with other technical and procedural defenses.

Additionally, there is the challenge of measuring the effectiveness of training programs. While simulated phishing exercises provide valuable data, they are not a perfect measure of an employee's ability to detect and respond to real phishing attempts. Organizations must use a combination of metrics, including incident reports and user feedback, to assess the effectiveness of their training programs and make necessary adjustments.

Addressing the problem of phishing requires a comprehensive approach that goes beyond training. While training is a critical component, it must be complemented by other measures to create a robust defense against phishing attacks. This includes implementing advanced email filtering systems, multi-factor authentication, and continuous monitoring of network activity.

Email filtering systems use AI and ML algorithms to analyze email content and identify potential phishing attempts. These systems can block many phishing emails before they reach the user's inbox, reducing the risk of an employee falling victim to an attack. Multi-factor authentication adds an extra layer of security by requiring users to provide two or more verification factors to gain access to a system. This makes it more difficult for cybercriminals to gain access to accounts even if they have obtained the user's credentials.

Continuous monitoring of network activity is also essential for detecting and responding to phishing attacks. By monitoring network traffic and analyzing patterns, organizations can identify suspicious activity and take action to mitigate potential threats. This proactive approach helps to reduce the impact of phishing attacks and improve overall security.

Leadership plays a crucial role in the success of cybersecurity training programs. Senior leaders must prioritize cybersecurity and create a culture of vigilance within the organization. This involves ensuring that cybersecurity is a key consideration in strategic decision-making and that there is alignment between the Chief Information Officer (CIO), Chief Information Security Officer (CISO), and other leaders within the organization.

Leaders must also ensure that adequate resources are allocated to cybersecurity training and other defensive measures. This includes investing in advanced training platforms, hiring skilled cybersecurity professionals, and implementing robust security policies and procedures. By demonstrating a commitment to cybersecurity, leaders can foster a culture of vigilance and encourage employees to take their training seriously.

With that said, cybersecurity training has become an essential component of organizational defense strategies in today's digital landscape. As cyber threats continue to evolve, the importance of equipping employees with the knowledge and skills to recognize and respond to these threats cannot be overstated. Phishing remains one of the most prevalent and damaging forms of cybercrime, and addressing this problem requires a comprehensive approach that includes advanced training programs, simulated phishing exercises, and other technical and procedural defenses.

The integration of AI and ML into phishing training programs has brought about significant improvements in their effectiveness. AI-driven platforms can create more targeted and relevant training content, deliver it in more engaging formats, and provide valuable data for continuous improvement. However, there are still challenges and limitations that need to be addressed, including keeping training programs up to date, overcoming human error, and measuring effectiveness.

Ultimately, the success of cybersecurity training programs depends on a combination of factors, including the commitment of leadership, the allocation of resources, and the adoption of a multi-layered approach to cybersecurity. By investing in advanced training programs and other defensive measures, organizations can better protect themselves and their stakeholders from the persistent threat of phishing.

In addition, as discussed in previous chapters, the efficacy of cybersecurity training must be evaluated through the lens of behavioral psychology. One promising approach is the application of the theory of planned behavior (TPB), which suggests that behavior is directly influenced by attitudes, subjective norms, and perceived behavioral control. In the context of cybersecurity, this means that training programs should not only inform but also aim to adjust attitudes toward phishing, establish norms of vigilance, and empower individuals with the belief that they can effectively avoid phishing attacks. By addressing these psychological components, training can foster a deeper, intrinsic motivation to act securely, rather than relying solely on the external motivation of avoiding potential threats.

Incorporating personalized feedback mechanisms into training programs can significantly enhance their effectiveness. Tailoring feedback to individual experiences and recognizing successful detection of phishing attempts can reinforce positive behaviors. Additionally, utilizing realistic simulations and role-playing scenarios can help bridge the gap between theoretical knowledge and practical application. These methods can provide individuals with hands-on experience in identifying and responding to phishing attempts in a controlled environment, thereby enhancing their confidence and skills.

Finally, ongoing evaluation and adaptation of training programs are essential to cater to the evolving nature of phishing tactics and the diverse needs of the workforce. Cybersecurity training should be a continuous process, incorporating the latest research findings and threat intelligence to remain relevant and effective. By adopting a more holistic and psychologically informed approach, organizations can significantly improve the efficacy of their training programs, ultimately developing a more resilient and phishing-resistant user population.

7.2   Methods of training delivery

Cybersecurity training (which often integrates phishing training in lieu of treating phishing as a separate topic) is delivered in a variety of formats. These can include live, online instructor lead training, videotaped OnDemand training, animated-interactive training, or training blends consisting of a variety of methods combined. Training is often completed by individuals on their own time, watching an animation, case studies, or printed material on screen, answering intermittent questions between the training segments. Interestingly, however, in many cases, even though the primary attack vector for many cyberattacks involves phishing, many cybersecurity programs either do not address phishing at all, or touch on it only lightly [60].

Cybersecurity awareness training is also very big business. Cybersecurity Ventures predicts the global security awareness training market will exceed $10 billion annually by 2027, which will be up from $5 billion today.

Cybersecurity training is delivered through a variety of means. There are organizations that conduct their own cybersecurity awareness and training campaigns, to government training programs, such as those delivered by the U.S. Department of Homeland Security (DHS). Training program providers, such as Udemy offer courses in cybersecurity focused on the workforce population, and then there are companies, often small businesses, that focus very specifically on delivering cybersecurity training to numerous organizations. In all, cybersecurity training is in high demand and is delivered in a variety of forms.

Despite the focus placed on training the user population, phishing remains a viable attack vector for a wide variety of perpetrators. In fact, launching a phishing campaign today by someone with little to no experience in developing a phishing campaign could not be easier. Today, it is no longer necessary to have considerable computing skills to execute a phishing campaign. The technical resources to execute a phishing campaign are readily available packaged with easy-to-use graphical user interfaces and sets of attacks ready to customize and use [61].

7.3   Anti-phishing training using simulated phishing attacks

The most visible trend in phishing training programs today involves simulated phishing attacks with an embedded training component. This involves organizations executing a phishing campaign against their own workforce using a phish generating tool. If a user handles the phish incorrectly, the user's workflow is either immediately halted and their system launches a remedial phishing training program, or they are scheduled for remedial phishing training.

A variety of studies have produced conflicting information as it relates to the efficacy of such programs. For example, a study conducted by Sumner *et al.* [60] demonstrated an 8 percent improvement in user's ability to detect a phish after receiving the training. However, it is unknown whether this improvement remains in effect over time, and whether context-related factors influence the user's ongoing ability to detect phishing attacks.

In another study conducted by Rizzoni *et al.* [62] it was found that phishing simulation is useful but has limitations. Rizzoni *et al.* also found that executing such training programs can induce workforce backlash as well as trigger several ethical concerns that may put the organization at risk of legal actions.

When deploying an interactive phishing training program, it's essential to consider the psychological impact on employees and the organization's culture. Simulated phishing campaigns can indeed offer immediate feedback and learning opportunities, yet how these simulations are conducted and the subsequent training provided can significantly influence the program's overall effectiveness and reception.

7.3.1   *Psychological impact on employees*

While effective for instant learning, immediately interrupting an employee's workflow to launch remedial training may not always result in long-term behavioral change. The interruption approach can sometimes lead to negative emotions, such as frustration or anxiety, which may hinder rather than enhance the learning process.

Furthermore, if employees feel they are being "tested" or "tricked" by their employers, this could potentially erode trust within the organization, affecting the workplace atmosphere and employee morale.

Incorporating psychological insights into the design and execution of simulated phishing training could mitigate the negative impacts associated with such interruptions. Recognizing the diversity in individual learning styles and stress thresholds is essential. For some employees, immediate, interactive training sessions following a simulated phishing attempt might serve as a powerful wake-up call, effectively embedding the necessary vigilance for future threats. However, for others, this method could be overwhelming, especially if the individual is already navigating a high-stress workload. This discrepancy highlights the need for a more tailored approach to cybersecurity education, one that offers flexibility in how and when remedial training is delivered.

An adaptive training framework could allow employees to engage with the learning material at a time that suits their workflow and mental bandwidth. Such flexibility respects the individual's current state and can facilitate a more receptive learning environment, ultimately enhancing the retention of critical information. Additionally, integrating elements of choice, such as selecting from various learning modules or interactive simulations, can empower employees, making them active participants in their cybersecurity education rather than passive recipients of mandated training.

Moreover, establishing a supportive feedback loop is crucial. Beyond immediate remedial actions, follow-up sessions that provide constructive feedback and personalized coaching can help reinforce learning outcomes. This ongoing support system should emphasize growth and improvement, celebrating milestones achieved in recognizing and averting phishing attempts. By shifting the focus from punitive measures to a more constructive and supportive approach, organizations can foster a culture of continuous learning and improvement.

Lastly, involving employees in the development and refinement of anti-phishing strategies can contribute significantly to reducing feelings of distrust and alienation. By soliciting feedback on the training process and actively incorporating employee suggestions, organizations can demonstrate a genuine commitment to the welfare and professional development of their workforce. This collaborative approach not only enhances the effectiveness of cybersecurity measures but also reinforces a sense of communal responsibility and trust between employees and management, contributing to a more secure and harmonious workplace environment.

7.3.2   *Organizational culture and acceptance*

The success of phishing training programs is also contingent upon the organization's culture. A culture that fosters openness encourages questions and views mistakes as learning opportunities can enhance the effectiveness of these programs. In contrast, a culture that penalizes mistakes harshly may lead employees to hide or underreport phishing incidents, counteracting the program's objectives.

The foundational values and practices within a company significantly influence the outcomes of such educational initiatives. A positive, inclusive culture— one that prioritizes psychological safety—allows employees to feel secure in expressing concerns, admitting mistakes, and seeking help without fear of retribution or judgment. This environment is crucial for encouraging proactive engagement with phishing training programs, as employees are more likely to participate actively and honestly when they trust that their vulnerability will be met with support rather than criticism.

Moreover, cultivating a culture of continuous improvement and collective responsibility toward cybersecurity can drive a more engaged and informed workforce. When the entire organization, from leadership down, champions cybersecurity as a shared goal, this collective mindset can elevate the importance of training outcomes beyond individual performance metrics. It encourages a holistic approach to security, where successes are celebrated as team achievements and failures are analyzed constructively for organizational learning.

In addition, implementing a reward system for positive behaviors observed during phishing simulations can further enhance this culture. Recognizing individuals or teams that demonstrate exceptional vigilance or improvement in phishing detection not only bolsters morale but also serves as a tangible reinforcement of the organization's commitment to valuing cybersecurity awareness and action. Such recognition can motivate others to elevate their participation and adherence to security protocols, embedding these practices into the daily routine.

Lastly, leadership's role in modeling the behaviors expected of their teams cannot be overstated. Leaders who actively participate in phishing training, openly discuss their experiences with simulated attacks, and advocate for the importance of vigilance in cybersecurity set a powerful example. This visible commitment can inspire greater engagement and trust throughout the organization, reinforcing the message that cybersecurity is a priority at all levels and that everyone has a crucial role to play in safeguarding the organization's digital assets.

7.3.3   Ethical and legal considerations

The ethical concerns highlighted by Rizzoni *et al.* raise important questions about consent, privacy, and the psychological welfare of employees. Organizations must navigate these issues carefully, ensuring that simulated phishing exercises do not cross ethical boundaries or violate privacy laws. Transparency about the nature and purpose of these simulations can mitigate some of these concerns, as can ensuring that the training is designed with respect for the employees' psychological well-being.

The ethical quandary at the core of these simulations often revolves around the balance between the necessity of preparing employees for real-world cyber threats and respecting their rights and psychological integrity. The practice of simulating phishing attacks, without prior explicit consent, treads a fine line between educational intent and the potential for perceived deception by employees. This perceived deception can lead to feelings of betrayal and distress, undermining the trust between employees and the organization.

Moreover, the question of consent is intricately linked to the legal implications of conducting such training exercises. In jurisdictions with stringent data protection and privacy laws, like those under the General Data Protection Regulation in the European Union, the unauthorized use of personal information—even for training purposes—can result in significant legal repercussions. Therefore, organizations must ensure that their phishing simulation practices comply with all relevant laws and regulations, potentially necessitating explicit consent from employees before involving them in such exercises.

Transparency becomes a critical ethical pillar in this context. By clearly communicating the objectives, methods, and benefits of phishing simulations, organizations can foster an atmosphere of mutual respect and understanding. This transparency extends to the aftermath of simulations, where providing constructive feedback and support to employees who were "caught" by the test becomes a moment for learning rather than punishment. Such an approach respects the psychological welfare of employees, recognizing that the goal is to empower, not to entrap.

Additionally, the design of phishing simulations must be carefully considered to avoid unnecessary stress or anxiety. This involves tailoring scenarios to be realistic yet not overly alarming, avoiding the use of manipulative or threatening content that could exacerbate stress or lead to fear-based responses. Ethically responsible simulations should aim to educate and build resilience, rather than to shock or scare employees into compliance.

In summary, while simulated phishing attacks offer a valuable tool in the cybersecurity training arsenal, their implementation is fraught with ethical and legal challenges that demand careful consideration. Balancing the need for effective training with respect for individual rights and well-being is paramount. By adhering to principles of consent, transparency, and respect for the psychological welfare of employees, organizations can navigate these challenges successfully, creating a secure, informed, and ethical workplace culture.

7.3.4 Long-term efficacy and continuous learning

Organizations should consider integrating phishing training into a broader, continuous learning program rather than a one-off or annual event to address long-term efficacy. This could involve regular updates on new phishing tactics, refresher courses, and incorporating gamification elements to keep the training engaging. Understanding that the landscape of cyber threats is constantly evolving, the training programs should evolve accordingly, adapting to new phishing techniques and technologies.

Integrating continuous learning about phishing into the fabric of an organization's continuous learning culture is critical. This means going beyond traditional training models to embrace a more dynamic and responsive approach to cybersecurity education. Continuous learning programs benefit from including interactive and real-time feedback mechanisms, such as phishing simulations that provide immediate insights into an employee's decision-making process and areas for improvement. By receiving instant feedback, employees can quickly understand the consequences of their actions in a controlled environment, reinforcing learning in a powerful and memorable way.

Moreover, leveraging data analytics to personalize the learning experience can significantly increase the effectiveness of training programs. By analyzing employees' responses to simulations and identifying common vulnerabilities, organizations can tailor their training to address specific weaknesses within their workforce. This personalized approach not only makes the training more relevant to each individual but also more effective in fortifying the organization against phishing attacks.

Incorporating social learning elements, such as peer discussions and shared experiences, into the training program can also contribute to its long-term success. Creating spaces for employees to discuss their experiences with phishing attempts, share tips, and offer support can foster a collaborative learning environment. This social aspect of learning can demystify cybersecurity, making it more accessible and less intimidating for employees who may not have a technical background.

Finally, committing to continuous innovation in training methods is essential. This could mean exploring new technologies such as virtual reality (VR) to simulate phishing scenarios in an immersive and interactive manner, making the learning experience even more engaging and effective. Keeping abreast of educational technology trends and applying them to cybersecurity training ensures that programs remain fresh, appealing, and aligned with the ways in which employees learn best.

The long-term efficacy of phishing training hinges on its integration into a continuous, dynamic learning program that is personalized, engaging, and leverages the latest technological innovations. By adopting this approach, organizations can create a culture of cybersecurity awareness that is both resilient and adaptable to the ever-changing landscape of cyber threats.

7.3.5   Incorporating contextual learning

Considering the influence of context-related factors on an individual's ability to detect phishing attempts, training programs could benefit from incorporating more personalized and contextualized scenarios. By simulating phishing attacks that are closely related to an employee's role or daily tasks, the training could become more relevant and practical, enhancing the ability to recognize and respond to phishing attempts in real-life situations.

Integrating adaptive learning algorithms can further refine the relevance and effectiveness of simulated phishing training programs. These algorithms can analyze an employee's interaction with training modules, adjusting the difficulty level and content based on their performance and learning pace. This ensures that each employee is challenged appropriately, preventing both under-stimulation and overwhelming experiences. Moreover, adaptive learning can identify patterns in mistakes or common vulnerabilities, allowing for targeted interventions that address specific areas of weakness.

The role of storytelling and scenario-based learning in cybersecurity training cannot be overstated. By crafting narratives that mirror potential real-life situations an employee might face, training can tap into the power of storytelling to evoke empathy and engagement. This method helps in grounding abstract cybersecurity concepts into tangible experiences, making the lessons more memorable and impactful. When employees see themselves in the stories, they are more likely to internalize the behaviors and strategies being taught, leading to stronger and more intuitive defenses against phishing attempts.

Furthermore, fostering a culture of peer learning and mentorship within the organization can significantly amplify the benefits of training programs. Encouraging more experienced employees to share their insights and experiences with newer team members can help disseminate practical knowledge and strategies for identifying and mitigating phishing threats. This culture of knowledge sharing, and collaboration not only enhances learning outcomes but also strengthens the social fabric of the organization, promoting a unified front against cyber threats.

Lastly, the evaluation and feedback loop of simulated training programs is critical for continuous improvement. Regularly assessing the effectiveness of training through metrics such as reduced phishing susceptibility rates, increased reporting of suspicious emails, and feedback surveys can provide valuable insights into the program's strengths and areas for enhancement. This ongoing evaluation ensures that the training remains aligned with the latest cyber threats and organizational needs, fostering a resilient, security-aware culture that evolves in tandem with the cybersecurity landscape.

In essence, by embracing a comprehensive approach that emphasizes personalization, storytelling, peer learning, and continuous evaluation, organizations can develop a robust phishing training program. Such a program not only equips employees with the necessary skills to combat phishing threats but also embeds a culture of cybersecurity awareness and vigilance throughout the organization.

In summary, while simulated phishing attacks with embedded training components are a valuable tool in cybersecurity education, their design and implementation must carefully consider the psychological, cultural, ethical, and legal aspects to maximize their effectiveness and minimize potential drawbacks. Continuous, contextual, and respectful training programs represent a more holistic approach to cultivating an organization's security-aware culture.

7.4   A proposed blended approach

In Chapters 2–4, we found several variables can affect an individual's perception of their computing experience. These variables can include an individual's momentary context, their perception of the technology they are using, organizational culture, the way an individual's brain processes inputs. We also found that only up to 40 percent of an individual's decision-making process can be attributed to their psychology (my psychologists would peg that number closer to 28 percent). That means at any given time, in addition to someone's psychological disposition, how they ultimately behave could be dependent upon a variety of things.

Additionally, what is influencing one person, may not be influencing another (even though there may be numerous things individuals have in common, such as location, or technology being used). What this data points to is that there is no one single silver bullet that can magically produce a phishing-resistant workforce.

However, it does suggest a blended approach to user training may increase the user population's resistance to phishing. Such a blended approach could include aspects that have, in the research, proven to increase phishing resistance:

- Conventual/theoretical training centered around explaining phishing and the threat it poses.

- Crowd sourcing phishing reporting used in combination with simulated phishing attacks.
- Mindfulness techniques to develop context awareness, and individual experiences with technology.

A blended approach to phishing training necessitates combining various instructional methods and psychological strategies to cater to the diverse needs of the workforce. This holistic strategy acknowledges that while conventional theoretical training lays the foundational knowledge about phishing and cybersecurity threats, it must be complemented with practical, interactive experiences to truly enhance phishing resistance among employees.

Crowdsourcing phishing reporting, as part of this blended approach, leverages the collective vigilance of the organization. By encouraging employees to share and report phishing attempts, an environment of collective security consciousness is fostered. This method not only aids in quickly identifying and mitigating threats but also reinforces the training by putting theory into practice. It turns passive recipients of knowledge into active participants in the organization's cybersecurity posture. Furthermore, integrating this with simulated phishing attacks provides a safe, controlled environment for employees to apply their knowledge, learn from mistakes, and improve their detection skills without the risk of real-world consequences.

Mindfulness techniques introduce a novel dimension to phishing training, focusing on enhancing context awareness and refining individual interactions with technology. Mindfulness can help employees become more aware of their moment-to-moment experiences, thoughts, and feelings while engaging with digital communication. This increased awareness can make it easier to recognize when something feels "off" about a particular email or request, even if it's not immediately obvious why. Training that incorporates mindfulness practices can improve focus and reduce the likelihood of oversight, particularly in high-pressure or fast-paced work environments. It encourages a more thoughtful and discerning engagement with emails and digital requests, crucial for spotting subtle phishing cues.

To further enrich the blended training approach, incorporating elements of gamification and social learning can enhance engagement and retention. Gamification introduces competitive and fun elements to the learning process, making it more engaging and memorable. Leaderboards, achievements, and rewards can motivate employees to participate actively and improve their skills. Meanwhile, social learning, facilitated through discussion forums, team-based learning challenges, and peer-to-peer feedback sessions, leverages the social dynamics of the workplace to foster a culture of learning and knowledge sharing.

This blended approach, combining conventional training, practical simulations, crowdsourcing, mindfulness, gamification, and social learning, addresses the multifaceted nature of phishing threats and the diverse learning preferences of employees. It acknowledges that no single method is sufficient on its own. By adopting a varied and integrated strategy, organizations can develop a more resilient, aware, and proactive workforce capable of responding effectively to the evolving landscape of phishing and cyber threats.

7.4.1 Conventional training

While in this section we've focused on the ineffectiveness of conventional cybersecurity training as a standalone solution to phishing, conventional training plays an important role in providing information and educating a user population about the dangers of phishing.

In the context of this blended framework, conventional training should be understood as providing users with background information on the cyber risks posed by phishing. Users should understand what it is, how it could appear, what techniques are used, what the risk is, and methods they can use to protect themselves. This form of training lays the groundwork for a user's understanding of the nature of phishing and what they need to do to protect themselves. In other words, conventional training provides the theoretical background which will then be reinforced using experiential training.

To that end, conventional training in this context refers to the traditional methods of educating employees about the principles, risks, and best practices related to securing information systems and data. This training typically takes the form of structured educational programs that aim to equip individuals with the knowledge and skills necessary to understand and mitigate cybersecurity threats. Conventional training methods often include in-person seminars, workshops, online courses, webinars, and educational materials such as handbooks or instructional videos.

The core objective of conventional cybersecurity training is to raise awareness among employees about the various types of cyber threats, including phishing, malware, ransomware, and social engineering attacks, among others. It also covers the policies, procedures, and best practices for protecting sensitive information and responding to incidents. The training emphasizes the critical role that every employee plays in maintaining cybersecurity, reinforcing the concept that security is not solely the responsibility of the IT department but a shared organizational responsibility.

Conventional training methods are typically characterized by their formal, one-way communication from the instructor or training material to the learner. This approach provides a foundational understanding of cybersecurity concepts and protocols in a systematic and controlled manner. Participants are often required to complete modules or sessions covering different topics, followed by assessments or quizzes to evaluate their comprehension and retention of the material.

A blended approach to cybersecurity training would leverage the positive aspects of conventional training and supplement it with more interactive and experiential learning approaches, such as simulated phishing exercises, gamification, and role-playing scenarios, to enhance engagement and improve the effectiveness of their cybersecurity training programs.

Conventional training forms the bedrock of cybersecurity education within organizations, providing a comprehensive overview of the cyber threat landscape and foundational security practices. However, to address the complexities of human behavior and the dynamic nature of cyber threats, a more holistic training approach that includes interactive and practical experiences is increasingly being recognized as essential for developing a truly resilient and aware workforce.

7.4.2   Bridging theory and practice: the need for experiential learning

Integrating experiential learning with conventional learning methodologies offers a multi-dimensional approach to cybersecurity education, particularly in the realm of phishing training. This fusion not only reinforces theoretical knowledge but also sharpens practical skills through direct experience and reflection, a process that is critical for effective learning in the context of rapidly evolving cyber threats.

While conventional training lays the theoretical groundwork, it must be seamlessly integrated with experiential learning to cement the users' understanding and foster practical skills. In other words, there should be a clear connection between theory and application.

To achieve these outcomes, trainers and educators can leverage the varied forms experiential learning can take. These include simulated phishing attacks, gamified learning experiences, and role-playing exercises. These methods help translate theoretical knowledge into practical skills, allowing users to recognize and react to phishing attempts in a safe, controlled environment.

One innovative form of experiential learning is the use of VR simulations for phishing training. VR environments can immerse users in highly realistic scenarios that mimic everyday digital interactions, where they might encounter phishing attempts. This level of immersion enables users to practice their response to phishing in a variety of contexts, from email communications to social media interactions, enhancing their ability to transfer these skills to real-world situations. The immersive nature of VR also significantly increases engagement and retention of information, making it a powerful tool for cybersecurity training.

Another aspect of integrating experiential learning into phishing training is the incorporation of real-time feedback mechanisms. Unlike conventional training methods, where feedback may be delayed or generalized, experiential learning environments can provide immediate, specific feedback on a user's actions. For instance, in a simulated phishing exercise, if a user clicks on a malicious link, the system can instantly provide feedback on the cues missed and offer tips for recognizing similar threats in the future. This immediate reinforcement helps in correcting misconceptions and reinforces learning through direct experience.

Collaborative learning experiences also play a vital role in enhancing the efficacy of phishing training. By engaging in team-based challenges or competitions, individuals can learn from the experiences and perspectives of their peers, fostering a collective learning environment. This collaborative approach not only leverages the diverse knowledge and skills within a group but also builds a sense of community and shared responsibility among employees toward maintaining cybersecurity.

Moreover, integrating experiential learning with conventional training methods allows for the personalization of learning experiences. Based on individual performance in simulations or exercises, training programs can adapt to focus more on areas where a user needs improvement, providing a tailored learning path that addresses specific vulnerabilities. This level of customization ensures that each user gains the most from the training, making it more effective and efficient.

The integration of experiential learning with conventional training approaches creates a comprehensive and dynamic educational framework for phishing training. By combining the strengths of both methodologies, organizations can equip their employees with not only the necessary theoretical knowledge but also the practical skills and reflexes needed to navigate the complex landscape of cybersecurity threats effectively.

7.4.3   Customization and continuous education

To enhance cybersecurity training's effectiveness, programs should be tailored to the organization's and users' specific needs and vulnerabilities. Customized training scenarios that reflect the most relevant and recent phishing techniques can make the learning experience more engaging and applicable. Additionally, cybersecurity education should not be a one-time event but a continuous process that adapts to the evolving landscape of cyber threats. Regular updates, refresher sessions, and feedback loops can help ensure that the user population remains vigilant and informed.

The call for customization and continuous education in cybersecurity training, particularly in phishing awareness, is a response to the dynamic and sophisticated nature of cyber threats. Customizing training to the specific context of an organization and its users involves an in-depth understanding of the unique risk profile, industry-specific threats, and common user behaviors within the organization. This personalized approach ensures that training scenarios are not only relevant but also resonate with the users' daily experiences, thereby increasing their alertness to phishing tactics directly applicable to their work environment.

For instance, financial institutions might face different phishing strategies compared to healthcare organizations, necessitating tailored content that reflects the specific phishing threats each sector is more likely to encounter. This could include simulated phishing emails that mimic those used by cybercriminals targeting these industries, thus preparing employees for the types of attacks they are most likely to face.

Furthermore, leveraging data analytics and user feedback to continually refine and update training content is crucial for maintaining its relevance. Cyber threats evolve rapidly, and what was a cutting-edge training module one year might be outdated the next. By analyzing which training exercises users find most challenging or engaging and tracking the latest phishing schemes, organizations can update their training programs to focus on current vulnerabilities and emerging threats.

The principle of continuous education extends beyond periodic updates to the curriculum. It encompasses a culture of security awareness where learning about cyber threats is woven into the fabric of daily work life. This could be facilitated through a variety of channels, such as regular security bulletins, phishing awareness weeks, and interactive security awareness portals. These platforms can provide ongoing tips, updates on the latest phishing scams, and forums for employees to share their experiences and strategies for avoiding phishing attempts.

Integrating feedback loops into the training process is also pivotal. This means not just collecting feedback from users about the training experience but also providing them with feedback on their performance in simulated phishing tests. Such feedback should be constructive, highlighting both strengths and areas for improvement, and it should be delivered in a timely manner to reinforce learning effectively.

In crafting a culture of continuous education, recognition and rewards can play a significant role in motivating and engaging employees. Acknowledging those who excel in identifying and reporting phishing attempts or show significant improvement in their awareness levels can serve as a powerful incentive for others.

Customizing cybersecurity training to align with the specific needs and threats facing an organization, coupled with a commitment to continuous education and improvement, can significantly enhance the efficacy of phishing training programs. By creating an engaging, relevant, and adaptive learning environment, organizations can foster a vigilant, informed workforce capable of responding adeptly to the evolving landscape of cyber threats.

7.4.4   Cultivating a culture of cybersecurity awareness

Beyond individual training programs, there's a need to foster a culture of cybersecurity awareness within organizations. This involves leadership endorsement, policy development, and creating an environment where cybersecurity is everyone's responsibility. Encouraging open dialogue about cyber threats and sharing experiences of phishing attempts can further reinforce the importance of vigilance among all users.

While conventional training provides the essential theoretical foundation for understanding phishing and other cyber threats, it should be considered the first step in a comprehensive, experiential, and continuous cybersecurity education strategy. By combining theoretical knowledge with practical experience, customizing training to meet specific needs, and fostering an organizational culture of cybersecurity awareness, organizations can significantly enhance their resilience against phishing attacks.

Cultivating a culture of cybersecurity awareness extends beyond structured training sessions and into the daily operational fabric of the organization. This culture is characterized by a proactive stance on security, where every employee, from the newest intern to the CEO, is informed and engaged in the collective effort to protect the organization's digital assets.

A key component of fostering this culture is the normalization of cybersecurity discussions within the workplace. Regularly scheduled meetings, updates, or newsletters that highlight recent security incidents, discuss the latest phishing tactics, and share success stories of averted threats can keep cybersecurity at the forefront of everyone's mind.

Leadership plays a critical role in setting the tone for a cybersecurity-aware culture. When leaders prioritize cybersecurity in their communications and decision-making, it sends a powerful message about its importance to the entire organization. Leaders can demonstrate their commitment by participating in the same cybersecurity training as their employees, sharing their experiences with cybersecurity, and recognizing individuals or teams who contribute significantly to the organization's cybersecurity posture.

Policy development is another cornerstone of cultivating a cybersecurity-aware culture. Clear, comprehensive policies provide a framework for expected behaviors and responses to cybersecurity threats. These policies should be accessible, easy to understand, and regularly reviewed to ensure they remain relevant to the current threat landscape. Moreover, policies should be complemented by procedures that outline specific steps to be taken in response to cybersecurity incidents, ensuring that every employee knows how to act quickly and effectively when faced with a potential threat.

Creating an environment where cybersecurity is everyone's responsibility also means empowering all employees to take action. This can be facilitated through tools and systems that make it easy to report suspicious activities and through open channels of communication that encourage questions and discussions about cybersecurity concerns. By removing barriers to reporting and fostering an atmosphere of support rather than blame, organizations can enhance their early warning systems against cyber threats.

Furthermore, integrating cybersecurity awareness into the very ethos of the organization can be achieved through innovative approaches such as appointing cybersecurity ambassadors within different departments. These ambassadors can serve as go-to resources for their peers, offer informal guidance on cybersecurity best practices, and help disseminate important security updates and information. This peer-led approach not only decentralizes cybersecurity efforts but also enhances their penetration throughout the organization.

Cultivating a culture of cybersecurity awareness is a multifaceted endeavor that requires active participation from everyone within the organization. Through leadership endorsement, comprehensive policy development, open dialogue, and the empowerment of all employees to take action, organizations can create a proactive and informed workforce capable of confronting the challenges posed by phishing and other cyber threats. This culture not only enhances the organization's resilience against attacks but also contributes to a more secure and vigilant global digital community.

7.5   Simulated phishing attacks and crowd sourcing reporting

The basis for this element of the phishing training program addresses experiential learning, positive reinforcement, and shared experiences.

Numerous studies found humans learn comprehensively if there is an actual experience associated with the learning experience. Additionally, studies have also found humans respond well to positive reinforcement, particularly if it centers around a sense of success. Lastly, in the study conducted by Lain *et al.* [55], it was found that users who had a shared experience relative to phishing developed more phishing resistance which also lasted over time.

To execute this segment, the organization would need a simulated phishing tool and an integrated "Report" button on the organization's email application.

In this segment the suggestion is to provide this experiential training following conventional training. This training involves forming teams of users who run through various exercises attempting to distinguish the phish from the non-phish email. The exercise should include various forms of phishing emails and should be executed multiple times.

The objective is for each team to work collaboratively, identify the phish and report the phish. At the end of each session, teams are given their scores. While it is understood users work individually at their desks, the learning experience is enhanced when given the opportunity to work collaboratively in a team. In addition, the team experience gives users the opportunity to process their experience amongst each other and learn from each other.

Receiving the score at the end of each round allows users to assess their phishing resistance skills immediately and improve upon them during each round.

7.6   Implementation strategies

To effectively implement a comprehensive cybersecurity training program that addresses the multifaceted nature of phishing threats, organizations must adopt a holistic approach that integrates conventional training, simulated phishing attacks, AI, and mindfulness training. This multi-pronged strategy ensures that employees are not only aware of the theoretical aspects of cybersecurity but are also equipped with practical skills and mental resilience to recognize and respond to phishing attempts.

Conventional cybersecurity training forms the foundation of any robust security awareness program. This type of training typically involves educating employees about the basics of cybersecurity, including the different types of cyber threats, the importance of strong passwords, and the principles of safe internet usage. However, to address the specific challenge of phishing, conventional training must go beyond these basics.

Effective conventional training programs should include detailed modules on phishing, explaining what phishing is, how it works, and the various forms it can take, such as email phishing, spear phishing, and whaling. Employees should be taught to recognize common indicators of phishing, such as suspicious email addresses, unexpected attachments, and urgent requests for personal information. Real-world examples of phishing attacks can be particularly effective in illustrating these concepts and making the training more relatable.

Conventional training should emphasize the importance of reporting suspicious emails. Employees need to understand that reporting potential phishing attempts is a critical part of the organization's defense strategy. To facilitate this, organizations should ensure that the "Report" button is seamlessly integrated into their email systems, making it easy for employees to report suspicious emails with a single click. This not only encourages active participation but also helps the IT and security teams to quickly identify and respond to potential threats.

While conventional training provides the necessary theoretical knowledge, simulated phishing attacks offer a practical, hands-on experience that is crucial for reinforcing this knowledge. Simulated phishing exercises involve sending fake phishing emails to employees to test their ability to recognize and respond to phishing attempts. These simulations should be designed to closely mimic actual phishing attempts, incorporating the latest tactics used by cybercriminals.

To create realistic phishing scenarios, organizations must stay up-to-date with the latest phishing trends and techniques. This involves continuously monitoring the threat landscape and incorporating new phishing tactics into the simulations. For example, if a new type of phishing attack is gaining popularity, the organization should quickly adapt its simulations to include this new threat. This ensures that employees are always prepared to deal with the most current and relevant phishing tactics.

Simulated phishing exercises also provide valuable data on employee performance. By tracking how employees respond to simulated phishing emails, organizations can identify areas where additional training is needed. For instance, if a significant number of employees fall for a particular type of phishing email, this indicates a gap in their knowledge that needs to be addressed. This data-driven approach allows for continuous improvement of the training program, ensuring that it remains effective in the face of evolving threats.

The integration of AI into cybersecurity training programs represents a significant advancement in the fight against phishing. AI-driven training platforms can analyze vast amounts of data to identify patterns and trends in phishing attacks. This information can be used to create more targeted and relevant training content, ensuring that employees receive training that is tailored to their specific needs and vulnerabilities.AI can also enhance the delivery of training content. Traditional training methods, such as lengthy presentations or static documents, are often ineffective at engaging employees. AI-driven platforms can deliver training content in more interactive and engaging formats, such as gamified modules or interactive simulations. This not only makes the training more enjoyable but also helps to improve retention and application of the knowledge.

AI can be used to personalize the training experience. By analyzing individual user behavior, AI-driven platforms can identify specific vulnerabilities and tailor training content to address them. For example, if an employee frequently falls for phishing simulations, the AI system can provide additional training modules focused on phishing awareness. This personalized approach ensures that employees receive the training they need to improve their cybersecurity behavior. AI can also be used to automate certain aspects of cybersecurity training. For instance, AI algorithms can automatically generate and send simulated phishing emails to employees, track their responses, and provide immediate feedback. This automation not only saves time and resources but also ensures that training is delivered consistently and at scale.

Mindfulness training is an emerging approach in cybersecurity that focuses on improving employees' mental resilience and decision-making abilities. The concept of mindfulness involves being fully present and attentive to one's surroundings, which can help individuals to better recognize and respond to phishing attempts.

Research has shown that mindfulness can mitigate automatic responses to phishing attempts by improving rational decision-making. For example, a study by CybSafe found that a brief mindfulness practice can help employees to detect more phishing cues, particularly in emails that are difficult to discern and originate from familiar sources. This suggests that mindfulness training can be a valuable tool in enhancing email security awareness.

Incorporating mindfulness techniques into cybersecurity training programs can help employees to be more present and less likely to click on malicious content. Techniques such as four-square or "box" breathing, which involves inhaling for four counts, holding the breath for four counts, exhaling for four counts, and holding the breath again for four counts, can help employees to concentrate and focus. Encouraging employees to engage in physical activity, such as taking a short walk, can also boost their concentration and focus.

Mindfulness training can also help employees to recognize and manage their emotional responses to phishing attempts. Cybercriminals often exploit human emotions, such as fear and urgency, to trick individuals into clicking on malicious links. By teaching employees to be attuned to their body's signals and to slow down and think before acting, mindfulness training can help to reduce the likelihood of falling for these tactics.

To effectively implement a comprehensive cybersecurity training program that includes conventional training, simulated phishing attacks, AI, and mindfulness training, organizations must adopt a strategic and integrated approach. This involves several key steps:

- First, organizations must conduct a thorough needs assessment to identify the specific cybersecurity risks and vulnerabilities they face. This assessment should consider factors such as the organization's industry, workforce demographics, and the types of data and systems that need to be protected. Based on this assessment, organizations can develop a tailored training strategy that addresses their unique needs and objectives.
- Second, organizations must secure executive buy-in for the training program. Strong executive support is crucial for ensuring that the training program receives the necessary resources and is prioritized across the organization. Executives should be involved in communicating the importance of cybersecurity training to employees and in setting the tone for a culture of vigilance.
- Third, organizations must develop and deliver engaging and relevant training content. This involves using a variety of formats, such as live training sessions,

on-demand videos, interactive modules, and gamified training. The content should be tailored to the specific needs and learning styles of employees, and should be regularly updated to reflect the latest threats and best practices.

- Fourth, organizations must implement continuous monitoring and evaluation of the training program. This involves tracking employee performance in simulated phishing exercises, gathering feedback from employees, and analyzing data to identify areas for improvement. Regularly updating the training content and delivery methods based on this feedback ensures that the program remains effective and relevant.

- Finally, organizations must foster a culture of cybersecurity mindfulness. This involves integrating mindfulness techniques into the training program and encouraging employees to practice mindfulness in their daily work. By promoting a culture of mindfulness, organizations can help employees to be more present, focused, and resilient in the face of phishing threats.

Implementing a comprehensive cybersecurity training program that includes conventional training, simulated phishing attacks, AI, and mindfulness training is essential for effectively addressing the problem of phishing. Conventional training provides the necessary theoretical knowledge, while simulated phishing exercises offer practical, hands-on experience. AI enhances the delivery and personalization of training content, and mindfulness training improves employees' mental resilience and decision-making abilities. By adopting a strategic and integrated approach, organizations can create a robust training program that equips employees with the knowledge, skills, and mindset needed to recognize and respond to phishing attempts. This multi-pronged strategy not only improves the organization's overall cybersecurity posture but also fosters a culture of vigilance and resilience that is crucial for defending against the ever-evolving threat of phishing.

7.6.1   Enhancing team collaboration

While we've discussed the importance of teamwork in identifying phishing emails, further strategies can be employed to enhance collaboration and engagement among team members. For instance, incorporating role-playing elements where team members assume specific roles within the scenario can add depth to the exercise. Roles could include a security analyst, an end user, and a supervisor, each providing different perspectives on the phishing attempt.

Role-playing in team collaboration training can significantly enhance the effectiveness of a phishing training program by providing a dynamic, interactive, and immersive learning experience. This method leverages the power of experiential learning, where participants actively engage in simulated scenarios that mimic real-world phishing attacks. By doing so, role-playing not only reinforces theoretical knowledge but also builds practical skills and fosters a deeper understanding of the tactics used by cybercriminals.

One of the primary benefits of role-playing in phishing training is its ability to enhance engagement and retention. Traditional training methods, such as lectures or static presentations, often fail to capture the attention of participants and may not effectively convey the complexities of phishing attacks. In contrast, role-playing involves active participation, which has been shown to improve learning outcomes. According to research, people learn and retain information more effectively when they have hands-on experience rather than just receiving information passively. By immersing participants in realistic phishing scenarios, role-playing helps them internalize new concepts and strategies more effectively.

For instance, a role-playing exercise might involve participants taking on the roles of both the attacker and the victim. In the role of the attacker, participants design and execute phishing emails, gaining insights into the techniques used by cybercriminals. As victims, they must identify and respond to these phishing attempts, applying the knowledge they have gained from conventional training. This dual perspective not only deepens their understanding of phishing tactics but also enhances their ability to recognize and respond to real phishing attacks in the future.

Role-playing provides an excellent opportunity for participants to practice their communication, problem-solving, and decision-making skills in a safe, controlled environment. These skills are crucial for effectively responding to phishing attacks. For example, participants can practice how to communicate with their IT department when they suspect a phishing attempt, or how to verify the authenticity of an email before clicking on a link or downloading an attachment. By simulating these scenarios, role-playing helps participants develop the confidence and competence needed to handle phishing threats in real-world situations.

Role-playing exercises can be tailored to reflect the specific challenges and contexts of the participants' work environment. This relevance is critical for ensuring that the training is practical and applicable. For example, a role-playing scenario for a financial institution might involve a phishing email that appears to come from a trusted client, while a scenario for a healthcare organization might involve a phishing attempt disguised as an urgent request for patient information. By customizing the scenarios to the participants' roles and responsibilities, role-playing ensures that the training is directly applicable to their daily work.

Phishing attacks often target multiple individuals within an organization, making it essential for employees to work together to identify and respond to these threats. Role-playing in team collaboration training fosters a sense of teamwork and collective responsibility, which is crucial for effective phishing defense. By working together in role-playing scenarios, participants learn how to collaborate effectively, share information, and support each other in identifying and mitigating phishing threats.

For example, a role-playing exercise might involve a simulated phishing attack that targets several team members simultaneously. Participants must work together to identify the phishing emails, report them to the appropriate authorities, and take steps to mitigate the threat. This collaborative approach not only reinforces the importance of teamwork in cybersecurity but also helps to build a culture of vigilance and mutual support within the organization.

Phishing attacks are constantly evolving, with cybercriminals developing new tactics to bypass security measures and deceive their targets. To effectively defend against these threats, employees must be able to think critically and adapt to new situations. Role-playing exercises encourage participants to think on their feet, make quick decisions, and adapt their strategies based on the evolving nature of the threat.

A role-playing scenario might involve a phishing email that uses a new technique not covered in the conventional training. Participants must analyze the email, identify the signs of phishing, and decide on the best course of action. This process helps to develop their critical thinking and problem-solving skills, making them more resilient to new and emerging phishing tactics.

The integration of AI and mindfulness training into role-playing exercises can further enhance their effectiveness. AI-driven training platforms can analyze participants' performance in role-playing scenarios, providing personalized feedback and identifying areas for improvement. This data-driven approach ensures that the training is continuously adapted to meet the needs of the participants and address the latest phishing threats. Mindfulness training, on the other hand, helps participants to stay focused and present, reducing the likelihood of falling for phishing attempts. By incorporating mindfulness techniques into role-playing exercises, participants can learn to recognize and manage their emotional responses to phishing emails, such as fear or urgency. This mental resilience is crucial for making rational decisions and avoiding impulsive actions that could lead to a security breach.

For example, a role-playing exercise might begin with a brief mindfulness practice, such as deep breathing or visualization, to help participants center themselves and focus on the task at hand. During the exercise, participants can use mindfulness techniques to stay calm and composed, even when faced with a convincing phishing email. This combination of practical skills and mental resilience makes them better equipped to handle real phishing threats.

To ensure the effectiveness of role-playing in phishing training, organizations must implement continuous monitoring and evaluation. This involves tracking participants' performance in role-playing exercises, gathering feedback, and analyzing data to identify areas for improvement. Regularly updating the training content and delivery methods based on this feedback ensures that the program remains effective and relevant.

For instance, organizations can use AI-driven platforms to analyze participants' responses to simulated phishing emails, identifying common mistakes and areas where additional training is needed. This data can be used to refine the role-playing scenarios, making them more challenging and realistic. Additionally, organizations can gather feedback from participants on the effectiveness of the role-playing exercises, using this information to make continuous improvements to the training program.

Ultimately, the goal of role-playing in phishing training is to create a culture of vigilance within the organization. By engaging employees in realistic and interactive training exercises, role-playing helps to build a sense of collective responsibility and awareness of cybersecurity threats. This culture of vigilance is crucial for defending against phishing attacks, as it ensures that all employees are actively engaged in identifying and reporting potential threats. Organizations can encourage employees to share their experiences and best practices from role-playing exercises with their colleagues. This peer-to-peer learning helps to reinforce the training and build a sense of community and mutual support. Additionally, organizations can recognize and reward employees who demonstrate exceptional vigilance and responsiveness in role-playing exercises, further reinforcing the importance of cybersecurity awareness.

Role-playing in team collaboration training can significantly enhance the effectiveness of a phishing training program by providing a dynamic, interactive, and immersive learning experience. By engaging participants in realistic phishing scenarios, role-playing helps to enhance engagement and retention, build practical skills, foster collaboration and teamwork, encourage critical thinking and adaptability, and integrate AI and mindfulness training. By implementing continuous monitoring and evaluation, organizations can ensure that the training remains effective and relevant, ultimately creating a culture of vigilance that is crucial for defending against phishing attacks.

7.6.2   Gamification elements

Introducing gamification elements can significantly increase participants' engagement and motivation. Leaderboards, badges, and rewards for the teams or individuals with the most improvement or highest accuracy in identifying phishing attempts can foster a competitive yet collaborative environment. This not only makes the training more enjoyable but also enhances learning retention.

One of the primary benefits of gamification in phishing training is its ability to enhance engagement and motivation. Traditional training methods often fail to capture the attention of participants, leading to low participation rates and poor retention of information. In contrast, gamification leverages the inherent appeal of games to make training more enjoyable and engaging. According to a study by Inspired eLearning, 88 percent of participants reported that gamifying the training experience improved their mood and made them more likely to complete the training. This increased engagement is crucial for ensuring that employees actively participate in the training and absorb the necessary information to recognize and respond to phishing attempts.

Gamification elements such as points, badges, and leaderboards introduce a competitive aspect to the training, motivating employees to perform better. For example, participants can earn points for correctly identifying phishing emails or for completing training modules. These points can be displayed on a leaderboard, fostering a sense of competition among employees. This competitive element encourages employees to engage with the training content more thoroughly, as they strive to improve their scores and outperform their peers. The desire to achieve higher rankings on the leaderboard can drive continuous engagement with the training program, leading to better retention and application of cybersecurity knowledge.

Gamification also enhances knowledge retention by making the learning process more interactive and immersive. Research has shown that people retain information more effectively when they actively participate in the learning process rather than passively receiving information. A study by Delinea found that the knowledge retention rate for employees with traditional learning methods is only about 5 percent, whereas experiential learning methods, such as gamification, can achieve retention rates as high as 90 percent. By incorporating interactive scenarios and simulations into phishing training, gamification helps employees to better understand and remember the concepts being taught.

Interactive scenarios and simulations are particularly effective in phishing training because they allow employees to practice their skills in a safe, controlled environment. A gamified training program might include a simulation where participants must identify and respond to a series of phishing emails. These simulations can mimic real-world phishing attempts, providing employees with hands-on experience in recognizing and mitigating phishing threats. By actively engaging with these scenarios, employees can develop a deeper understanding of phishing tactics and improve their ability to detect and respond to phishing attempts in real-world situations.

One of the key goals of phishing training is to encourage positive behavioral change among employees. Gamification can play a crucial role in achieving this by providing immediate feedback and rewards for desired behaviors. Participants can receive badges or certificates for successfully completing training modules or for consistently identifying phishing emails. These rewards serve as positive reinforcement, encouraging employees to continue practicing good cybersecurity habits.

Gamification can also help create a sense of ownership and accountability among employees. By tracking their progress and achievements in the training program, employees can see the tangible results of their efforts. This sense of ownership can motivate employees to take their cybersecurity responsibilities more seriously and to apply what they have learned in their daily work. For instance, employees who have earned high scores in phishing simulations may feel more confident in their ability to recognize and report phishing attempts, leading to a more vigilant and proactive approach to cybersecurity.

Phishing attacks often target multiple individuals within an organization, making it essential for employees to work together to identify and respond to these threats. Gamification can foster a sense of collaboration and teamwork by incorporating multiplayer elements and team-based challenges into the training program. Participants can work together in teams to complete phishing simulations or to solve cybersecurity challenges. This collaborative approach not only reinforces the importance of teamwork in cybersecurity but also helps to build a culture of vigilance and mutual support within the organization.

By working together in gamified training exercises, employees can share their knowledge and learn from each other's experiences. This peer-to-peer learning can enhance the overall effectiveness of the training program, as employees gain insights and perspectives that they might not have encountered on their own. Additionally, team-based challenges can create a sense of camaraderie and collective responsibility, motivating employees to support each other in maintaining a secure work environment.

The cybersecurity landscape is constantly evolving, with cybercriminals developing new tactics to bypass security measures and deceive their targets. To effectively defend against these threats, phishing training programs must be adaptable and continuously updated. Gamification can facilitate this adaptability by incorporating AI and ML technologies into the training program. AI-driven platforms can analyze participants' performance and provide personalized feedback, ensuring that the training content is continuously adapted to meet the needs of the participants and address the latest phishing threats. An AI-driven gamified training platform can automatically generate new phishing scenarios based on emerging threats and trends. Participants can then engage with these scenarios, gaining experience in recognizing and responding to the latest phishing tactics. This continuous adaptation ensures that employees are always prepared to deal with the most current and relevant phishing threats, enhancing their overall resilience to phishing attacks.

One of the challenges of cybersecurity training is overcoming the negative perception that many employees have toward mandatory training programs. Traditional training methods are often seen as boring and monotonous, leading to low participation rates and poor engagement. Gamification can transform this perception by making the training experience more enjoyable and rewarding. By incorporating elements of fun and competition, gamification can create a positive training experience that employees look forward to participating in.

To wit, a gamified phishing training program might include a storyline or narrative that immerses participants in a cybersecurity adventure. Participants can take on the role of a cybersecurity hero, completing missions and challenges to protect their organization from cyber threats. This narrative approach can make the training more engaging and memorable, helping employees to better retain and apply the information they have learned. Additionally, the use of rewards and recognition can create a sense of accomplishment and satisfaction, further enhancing the overall training experience.

Lastly, gamification elements in a phishing training program can significantly improve phishing resistance by enhancing engagement and motivation, improving knowledge retention, encouraging positive behavioral change, fostering collaboration and teamwork, adapting to evolving threats, and creating a positive training experience. By transforming the learning process into an interactive and immersive experience, gamification helps employees to better understand and respond to phishing threats, ultimately enhancing the organization's overall cybersecurity posture. As cyber threats continue to evolve, the integration of gamification into phishing training programs will be crucial for building a resilient and vigilant workforce capable of defending against the ever-changing landscape of cyber threats.

7.6.3   *Continuous learning and adaptation*

To ensure that the learning from these exercises is not transient, the program should include follow-up sessions where teams revisit past phishing scenarios and discuss what was learned. This could be supplemented with updates on new phishing techniques and how to counter them. Continuous adaptation of the training material to reflect the evolving nature of phishing attacks is essential for maintaining relevance and effectiveness.

One potential challenge is ensuring that all participants are actively engaged and not relying on more knowledgeable team members to carry the workload. This can be mitigated by randomly assigning team members to different groups for each session to encourage broader participation and knowledge sharing. Additionally, setting up scenarios that require input from all team members can ensure active participation.

While scores provide immediate feedback, qualitative measures such as confidence levels in identifying phishing attempts, the perceived value of the team experience, and willingness to report potential phishing attempts in real life should also be considered. Surveys and interviews can be employed to gather this data, providing insights into the program's impact on the organization's cybersecurity culture.

Simulated phishing attacks combined with crowd-sourced reporting represent a powerful method for enhancing phishing resistance within organizations. Organizations can create a dynamic and effective training program by focusing on experiential learning, positive reinforcement through teamwork, and the incorporation of gamification and continuous adaptation. Addressing potential challenges and measuring success through both quantitative and qualitative metrics will further ensure the program's long-term effectiveness and sustainability.

7.7   Working with the sub-conscious experiences—a last word on mindfulness training

As discussed in Chapters 2–4, there are many variables involved with the human experience. These variables involve how we perceive our technology, motivations we experience relative to our individual contexts, how we interact and relate to each other, and perceptions related to the way our minds process inputs. While this may seem like an endless array of variables that could detract us from processing events clearly, a study from 2017 along with additional studies coming from 2020 offer some insights into techniques that may offer some mitigating tools: using mindfulness training to affect behavior changes.

In 2017, Jensen *et al.* investigated mindfulness theory as a potential mitigating solution for phishing attacks [61]. The mindfulness approach used by the Jensen *et al.* team involved mindfulness theory which taught participants to dynamically allocate attention when evaluating incoming emails.

To evaluate the efficacy of the mindfulness training, the researchers contrasted the phishing detection abilities of a study population after receiving conventional rules-based training to the detection abilities of a population after receiving mindfulness training. The study validated that students who received mindfulness training were better able to detect phishing attacks than those who only received conventional, rule-based training. These results are consistent with the anecdotal evidence I found in my own classroom after introducing mindfulness techniques to detect phishing attacks. In fact, up to six months after taking my class, students continued to report phishing attacks, suggesting there may be long-term benefits to this approach.

In a *Harvard Review of Psychiatry* paper involving a review of mindfulness and behavior changes, Schuman-Olivier *et al.* [63] attention and cognitive control as:

a fundamental capacity of human cognition that regulates access to specific goal-relevant information to facilitate the performance of specific behaviors.

The authors go on to explain that executive functions are cognitive processes which are required to initiate and maintain changes in behavior. Given these data points, it would appear our goal of making users aware of their individual context, and the relationship they experience with their technologies, and how these could be impairing their ability to detect a phish, a strong case could be made for adding mindfulness training to the conventional training-experiential training repertoire.

In addition to controlling executive functions, mindfulness training includes an emotional regulation component that would help users resist an immediate emotional reaction to a phish (which is often designed to specifically illicit an emotional response). This means that mindfulness training could mitigate the contextual and emotional experiences that often lead to a user falling victim to a phish.

While individual studies point to a combination of theoretical training, experiential training, and mindfulness training as a potent approach to developing a more phishing-resistant user population, research is required to determine how well such an approach works and how long its effects last.

Mindfulness training transcends the mere act of detecting phishing emails; it fundamentally alters how individuals engage with their digital environments. By fostering heightened awareness, mindfulness encourages users to critically evaluate digital communications, making them less susceptible to the urgency and emotional manipulation often employed in phishing attacks. This conscious engagement can lead to a more deliberate and skeptical interaction with technology, where users take a moment to reflect before acting.

To effectively incorporate mindfulness training into phishing awareness programs, organizations can adopt a multifaceted approach that includes:

Mindfulness workshops: Conduct workshops that introduce employees to mindfulness principles and practices, emphasizing the application of these techniques to enhance cybersecurity practices.

Mindful email practices: Encourage the development of mindful email practices, such as taking a moment to breathe and assess before opening emails, especially those from unknown sources or those that invoke a sense of urgency.

Regular mindfulness exercises: Incorporate regular mindfulness exercises into the workday, such as guided meditations or breathing exercises, to cultivate a general state of mindfulness that can extend to digital interactions.

7.7.1   Evaluating the impact of mindfulness training

Organizations should employ quantitative and qualitative measures to assess the efficacy of integrating mindfulness into cybersecurity training. This could involve pre-and post-training assessments of employee phishing detection rates, as well as surveys or interviews to gauge changes in attitudes toward email security and emotional responses to potential phishing attempts. Monitoring the long-term sustainability of these behavioral changes is crucial to understanding the lasting impact of mindfulness training.

Measuring the efficacy of mindfulness training in creating a phishing-resistant workforce involves a multifaceted approach that combines quantitative and qualitative assessments. The goal is to determine whether mindfulness training can effectively enhance employees' ability to recognize and respond to phishing attempts, thereby reducing the organization's vulnerability to such attacks. This discussion explores various methods and metrics for evaluating the impact of mindfulness training on phishing resistance, drawing on insights from existing research and practical applications. While most of the following discussion centers around the studies researchers would conduct, organizations should also consider some form of measuring the efficacy of a training program involving mindfulness training as well.

Quantitative assessments provide objective data on the effectiveness of mindfulness training. One of the primary methods for measuring the efficacy of mindfulness training is through controlled experiments and randomized controlled trials. In these studies, participants are typically divided into two groups: one group receives mindfulness training, while the other group receives either no training or an alternative form of training. The performance of both groups is then compared to assess the impact of mindfulness training on phishing resistance.

A study might involve a pre-test where both groups are exposed to a series of simulated phishing emails. The number of phishing emails successfully identified and reported by each participant is recorded. After the pre-test, the experimental group undergoes mindfulness training, which may include techniques such as breath meditation, body scan, and mindful awareness exercises. Following the training, both groups are subjected to a post-test involving another set of simulated phishing emails. The improvement in phishing detection and reporting rates from pre-test to post-test is then analyzed to determine the efficacy of the mindfulness training.

Another quantitative method involves the use of self-report questionnaires and scales to measure changes in mindfulness and related psychological constructs. Instruments such as the Mindful Attention Awareness Scale (MAAS) and the Five Facet Mindfulness Questionnaire (FFMQ) can be administered before and after the training to assess changes in participants' mindfulness levels. Higher scores on these scales post-training would indicate an increase in mindfulness, which can be correlated with improved phishing resistance. Additionally, measures of self-efficacy, such as the Self-Efficacy for Mindfulness Meditation Practice (SEMMP) scale, can be used to assess participants' confidence in their ability to apply mindfulness techniques in real-world situations, including recognizing and responding to phishing attempts.

Qualitative assessments provide deeper insights into the subjective experiences and perceptions of participants, which can complement quantitative data. One common qualitative method is conducting interviews or focus groups with participants to gather feedback on their experiences with the mindfulness training. Participants can be asked about their perceptions of the training's relevance and effectiveness, as well as any changes they have noticed in their ability to recognize and respond to phishing emails. This feedback can provide valuable information on the strengths and weaknesses of the training program and suggest areas for improvement.

Another qualitative approach involves analyzing participants' written reflections or journals. Participants can be encouraged to keep a journal during the training period, documenting their experiences and any changes they observe in their behavior and mindset. These reflections can provide rich, detailed accounts of how mindfulness training influences their awareness and decision-making processes, particularly in the context of phishing resistance.

Behavioral metrics are crucial for assessing the real-world impact of mindfulness training on phishing resistance. One effective method is to track the number of phishing emails reported by employees over a specified period before and after the training. An increase in the number of reported phishing emails post-training would indicate that employees are more vigilant and proactive in identifying potential threats. Additionally, the accuracy of these reports can be evaluated to determine whether employees are correctly identifying phishing emails or generating false positives.

Simulated phishing exercises can also be used to measure behavioral changes. These exercises involve sending fake phishing emails to employees and tracking their responses. Metrics such as the click-through rate (the percentage of employees who click on a phishing link), the report rate (the percentage of employees who report the phishing email), and the response time (the time it takes for employees to report the phishing email) can provide objective data on the effectiveness of the training. A decrease in the click-through rate and an increase in the report rate and response time post-training would indicate improved phishing resistance.

Psychophysiological measures offer an innovative approach to assessing the impact of mindfulness training on phishing resistance. These measures involve monitoring physiological responses, such as brain activity, heart rate variability (HRV), and skin conductance, to assess changes in participants' stress levels and cognitive processes. For example, event-related brain potentials (ERPs) can be used to measure changes in attention and emotional regulation, which are key components of mindfulness. Studies have shown that mindfulness training can lead to changes in ERPs that are associated with improved attention and reduced emotional reactivity, which can enhance phishing resistance.

HRV is another psychophysiological measure that can be used to assess the impact of mindfulness training. HRV reflects the balance between the sympathetic and parasympathetic nervous systems and is an indicator of stress and emotional regulation. Higher HRV is associated with better stress management and emotional regulation, which can help employees remain calm and focused when evaluating potential phishing emails. By measuring HRV before and after mindfulness training, researchers can assess whether the training has led to improvements in stress management and emotional regulation, which can enhance phishing resistance.

Longitudinal studies are essential for assessing the long-term impact of mindfulness training on phishing resistance. These studies involve tracking participants' performance and behavior over an extended period, allowing researchers to assess the sustainability of the training's effects. For example, a longitudinal study might involve periodic assessments of phishing resistance, mindfulness levels, and self-efficacy over several months or even years. This approach can provide valuable insights into the durability of the training's impact and identify any factors that influence the long-term effectiveness of mindfulness training.

Measuring the efficacy of mindfulness training in creating a phishing-resistant workforce requires a comprehensive approach that combines quantitative and qualitative assessments, behavioral metrics, psychophysiological measures, and longitudinal studies. By employing a variety of methods and metrics, organizations can gain a holistic understanding of the impact of mindfulness training on phishing resistance. This multifaceted approach not only provides objective data on the training's effectiveness but also offers deeper insights into the subjective experiences and behavioral changes of participants. Ultimately, these assessments can inform the development and refinement of mindfulness training programs, ensuring that they effectively enhance employees' ability to recognize and respond to phishing attempts, thereby reducing the organization's vulnerability to cyber threats.

7.7.2 Addressing challenges and limitations

While the benefits of mindfulness training are promising, organizations must also consider potential challenges and limitations. These might include employee skepticism toward mindfulness practices, the time investment required to see tangible benefits, and the difficulty of measuring mindfulness's direct impact on phishing detection rates. To address these challenges, it's important to communicate the relevance of mindfulness to cybersecurity, provide flexible training options, and establish clear metrics for success.

7.7.3 Future research directions

Further research is needed to explore the optimal integration of mindfulness training with conventional and experiential phishing training. This could involve comparative studies to identify the most effective training combinations, longitudinal studies to assess the durability of training effects and research into the applicability of mindfulness training across different demographics and organizational cultures.

Including mindfulness training in phishing awareness programs represents an innovative approach to bolstering cybersecurity defenses. Mindfulness training enhances individuals' attentional control and emotional regulation, offering a complementary tool to traditional cybersecurity education efforts. As the digital landscape evolves, adopting comprehensive, human-centered training strategies will be key to developing resilient, phishing-resistant user populations.

Key takeaways

- Cybersecurity training is an important tool in mitigating phishing attacks.
- However, theoretical training in and of itself is not sufficient to develop a phishing-resistant workforce.
- A blended training program combining theoretical training, experiential training, and mindfulness training could yield a more phishing-resistant user population.

Chapter 8
Addressing social media presence

In today's social media environment, perpetrators find a rich body of information about individuals, enabling them to construct a profile of a potential victim with a high degree of information fidelity. This profile ultimately provides a perpetrator with sufficient information about a target, enabling them to craft highly personalized phishing emails. However, in addition to capturing information about a potential target, phishing involving social media accounts is expanding as rapidly as organizational use of social media is expanding. In today's social media world, it is not uncommon for companies and other organizations to have a broad social media presence, and with that comes the phish. This chapter explores the unique phishing risks that come with an organization's social media presence and how these attacks provide an opportunity for the escalation of an adverse cyber event.

8.1   How do we define social media?

Social media, a term that has become ubiquitous in modern discourse, refers to a collection of online platforms and technologies that facilitate the creation, sharing, and exchange of information, ideas, and content through virtual communities and networks. These platforms have revolutionized the way we communicate, interact, and consume information, profoundly impacting various aspects of society and individual behavior.

At its core, social media encompasses a wide range of websites and applications designed to enable users to create and share content or to participate in social networking. The essence of social media lies in its interactive nature, allowing users to engage in dialogues, share personal experiences, and build communities around shared interests. Unlike traditional media, which operates on a one-to-many model of communication, social media thrives on a many-to-many model, where information flows freely among users, creating a dynamic and participatory environment.

The origins of social media can be traced back to the early days of the internet, with the development of platforms that allowed for electronic information exchange. The evolution of social media has been rapid, moving from simple communication tools to complex ecosystems that integrate various forms of media, including text, images, videos, and live streams. Early examples of social media include bulletin board systems and internet relay chat, which laid the groundwork for more sophisticated platforms like MySpace, Facebook, and Twitter.

The history of social media is marked by significant milestones that reflect its growing influence and complexity. In the late 1990s, platforms like Six Degrees and Friendster emerged, allowing users to create profiles and connect with friends. These early social networks were rudimentary compared to today's standards but were instrumental in shaping the concept of online social networking.

The launch of Facebook in 2004 marked a turning point in the evolution of social media. Founded by Mark Zuckerberg, Facebook introduced features that allowed users to share status updates, photos, and links, creating a more immersive and interactive experience. The platform's success spurred the development of other social media sites, such as Twitter, which focused on microblogging, and LinkedIn, which catered to professional networking.

As social media platforms proliferated, they began to diversify in terms of functionality and user base. Instagram, launched in 2010, capitalized on the growing popularity of photo sharing, while platforms like Snapchat introduced ephemeral content that disappeared after a short period. The rise of video-sharing platforms like YouTube and TikTok further expanded the scope of social media, enabling users to create and consume video content on an unprecedented scale.

Social media has fundamentally altered the way we communicate, both on a personal and societal level. One of the most significant changes is the shift from one-to-one or one-to-many communication to a more interactive and immediate form of engagement. Social media platforms allow users to share their thoughts and experiences with a global audience in real time, fostering a sense of connectedness and immediacy that was previously unattainable.

This transformation has had profound implications for how we consume and disseminate information. In the past, traditional media channels such as newspapers, television, and radio were the primary sources of news and information. These channels operated under a gatekeeping model, where information was curated and controlled by a select few. Social media, on the other hand, democratizes information dissemination, allowing anyone with an internet connection to become a content creator and share news with the world.

While this democratization has many positive aspects, such as increased access to diverse perspectives and the ability to mobilize social movements, it also presents significant challenges. The lack of gatekeepers means that misinformation and fake news can spread rapidly, often with real-world consequences. For example, during the COVID-19 pandemic, social media platforms were inundated with false information about the virus and vaccines, leading to confusion and mistrust among the public.

The impact of social media extends beyond communication, influencing various aspects of society, including politics, business, and personal relationships. In the political realm, social media has become a powerful tool for activism and advocacy. Movements such as Black Lives Matter and #MeToo have leveraged social media to raise awareness, organize protests, and effect change. These platforms provide a space for marginalized voices to be heard and for communities to rally around shared causes.

However, the political impact of social media is not without controversy. The same features that enable activism can also be exploited for malicious purposes.

Social media has been criticized for its role in spreading hate speech, radicalizing individuals, and interfering in democratic processes. The 2016 U.S. presidential election, for instance, saw widespread use of social media by foreign actors to influence public opinion and sow discord.

In the business world, social media has transformed marketing and customer engagement. Companies use social media platforms to reach their target audiences, build brand loyalty, and gather consumer insights. The ability to create highly personalized marketing campaigns has made social media an essential component of modern business strategies. However, this also raises concerns about privacy and data security, as companies collect vast amounts of personal information to tailor their advertising efforts.

On a personal level, social media has reshaped how we form and maintain relationships. Platforms like Facebook and Instagram allow users to stay connected with friends and family, share life events, and engage in social interactions. While this can enhance a sense of community and belonging, it can also lead to negative outcomes such as social isolation, cyberbullying, and mental health issues. Studies have shown that excessive social media use can contribute to feelings of anxiety, depression, and low self-esteem, particularly among young people.

In all, social media is likely to become even more integrated into our daily lives. Emerging technologies such as artificial intelligence (AI) and virtual reality (VR) are poised to further transform the social media landscape. AI can enhance user experiences by providing personalized content recommendations and improving moderation of harmful content. VR, on the other hand, has the potential to create immersive social experiences that blur the lines between the virtual and physical worlds.

However, the future of social media also raises important ethical and regulatory questions. Issues such as data privacy, misinformation, and the impact of social media on mental health will require ongoing attention and intervention from policymakers, technology companies, and society at large. Balancing the benefits of social media with its potential harms will be a critical challenge in the years to come.

Social media is a multifaceted phenomenon that has profoundly impacted the way we communicate, interact, and live our lives. Its evolution from simple communication tools to complex ecosystems reflects the dynamic nature of technology and its ability to shape human behavior. As we navigate the opportunities and challenges presented by social media, it is essential to remain mindful of its potential to both connect and divide, inform and mislead, empower and exploit. By understanding the complexities of social media, we can better harness its power for positive change while mitigating its negative effects.

8.2   What does social media have to do with phishing?

Since social media is clearly an integral part of our daily lives, connecting us with friends, family, and communities across the globe as the popularity of these platforms continues to rise, so does the risk of falling victim to social media phishing attacks. These attacks exploit the trust and familiarity we have with social media, making them particularly insidious and challenging to detect.

Social media phishing is a form of cyberattack where criminals impersonate legitimate individuals or organizations on social media platforms to trick users into revealing sensitive information or clicking on malicious links. These attacks can take various forms, such as fake friend requests, compromised accounts, or fraudulent advertisements.

One of the most common tactics used by cybercriminals is to create fake social media profiles that mimic those of well-known companies or public figures. These profiles are designed to appear legitimate, often using official logos, branding, and language. Once a user interacts with the fake profile, the attacker can then attempt to extract personal information, login credentials, or financial data through carefully crafted messages or links.

One prevalent technique is the hijacking of legitimate accounts. Cybercriminals may gain access to an individual's or organization's social media account through various means, such as phishing attacks or malware infections. Once they have control of the account, they can use it to spread malicious links or messages to the account's followers, exploiting the trust and familiarity that already exists between the account holder and their connections.

Social media phishing attacks can be launched through various channels and employ a wide range of techniques to deceive users. One common vector is the use of malicious links or attachments shared through direct messages, posts, or comments. These links may lead to fake login pages designed to steal credentials or websites that attempt to install malware on the user's device.

Another technique involves the use of social engineering tactics to manipulate users into revealing sensitive information. Cybercriminals may create fake profiles or accounts that appear to be from trusted sources, such as banks, government agencies, or well-known companies. They may then send messages or posts claiming that the user's account has been compromised or that they need to verify their personal information to resolve an issue.

Phishing attacks on social media can also take advantage of current events, trends, or popular topics to increase their chances of success. For example, during the COVID-19 pandemic, cybercriminals created fake accounts and posts offering information about vaccines or relief funds, luring unsuspecting users into providing their personal data or clicking on malicious links.

The consequences of falling victim to a social media phishing attack can be severe and far-reaching. On a personal level, individuals may suffer financial losses, identity theft, or damage to their reputation if their sensitive information is compromised. Additionally, the emotional and psychological impact of being targeted by cybercriminals can be significant, leading to feelings of violation, anxiety, and distrust.

For businesses and organizations, the impacts of social media phishing can be even more devastating. A successful attack can result in the theft of valuable data, such as trade secrets, customer information, or intellectual property. This can lead to significant financial losses, damage to the company's reputation, and potential legal liabilities.

Compromised social media accounts can be used to spread misinformation, defame the organization, or launch further attacks on customers, partners, or employees. This can erode trust in the brand and undermine the company's credibility, making it difficult to recover from the incident.

Social media phishing attacks can affect individuals, businesses, and organizations of all sizes and across various industries. While personal users may be targeted for financial gain or identity theft, businesses and organizations are often targeted for their valuable data, customer information, or financial resources.

Individuals have a responsibility to remain vigilant and educated about the risks of social media phishing. This includes being cautious about sharing personal information online, verifying the authenticity of messages and links, and keeping their devices and accounts secure with strong passwords and two-factor authentication.

Businesses and organizations, on the other hand, have a heightened responsibility to protect their employees, customers, and stakeholders from social media phishing attacks. This involves implementing robust cybersecurity measures, such as employee training programs, email and web filtering solutions, and incident response plans.

There is much discussion around the need for social media platforms themselves to play a crucial role in combating phishing attacks. They must continuously improve their security measures, implement advanced detection and prevention techniques, and provide users with clear guidelines and resources to identify and report suspicious activities.

Social media phishing is a growing and evolving threat that requires constant vigilance and proactive measures from individuals, businesses, and social media platforms alike. By understanding the nature of these attacks, the various attack vectors and techniques employed, and the potential impacts and consequences, we can better prepare ourselves to mitigate the risks and protect our personal and professional interests. Ultimately, combating social media phishing requires a collaborative effort involving education, awareness, and the implementation of robust security measures. Only by working together can we create a safer and more secure online environment, where the benefits of social media can be enjoyed without the constant fear of falling victim to cybercriminals.

8.2.1   How do hackers use social media to phish?

Hackers employ a variety of sophisticated techniques to create fake social media login pages, leveraging both technical skills and psychological manipulation to deceive users into divulging their credentials. The process typically begins with the selection of a target platform, such as Facebook, Instagram, or LinkedIn, which are popular due to their vast user bases and the valuable personal information they contain.

To create a convincing fake login page, hackers first need to replicate the appearance and functionality of the legitimate site. This involves downloading the HTML source code of the target's login page. By right-clicking on the login page and selecting "View Source," hackers can access and copy the HTML, CSS, and JavaScript that define the page's structure and design. This code is then saved and modified to suit the hacker's needs.

One critical modification involves changing the form's action attribute, which specifies where the form data should be sent upon submission. In a legitimate login page, this attribute points to the server that processes the login request. In a phishing page, hackers alter this attribute to direct the data to a server they control. This server is set up to capture and store the entered credentials, often in a simple text file or database.

To further enhance the deception, hackers ensure that the fake page closely mimics the real one in every detail. This includes using the same logos, fonts, colors, and layout as the legitimate site. They may also include elements like security badges or trust seals to give the page an air of authenticity. Advanced attackers might even incorporate dynamic elements, such as error messages or loading animations, to make the interaction feel more genuine.

Once the fake login page is ready, hackers need to drive traffic to it. This is typically done through phishing emails or messages that appear to come from the legitimate social media platform. These messages often contain urgent or enticing content, such as a security alert or a special offer, prompting the recipient to click on a link. This link directs the user to the fake login page, where they are asked to enter their credentials.

In some cases, hackers use URL manipulation techniques to make the link appear legitimate. This can involve using homograph spoofing, where characters in the URL are replaced with visually similar ones from different character sets, or URL shortening services to obscure the true destination of the link. Additionally, attackers might employ covert redirect methods, where the user is first taken to a legitimate-looking intermediate page before being redirected to the fake login page.

Once the user enters their credentials on the fake page, the data is captured and stored by the hacker's server. The user is often then redirected to the real login page, making them unaware that their information has been compromised. With the stolen credentials, hackers can gain unauthorized access to the victim's social media account, which can be used for various malicious activities, such as spreading further phishing attacks, stealing personal information, or conducting financial fraud.

The creation and deployment of fake social media login pages highlight the importance of vigilance and cybersecurity awareness. Users should be cautious of unsolicited messages and links, verify the authenticity of URLs, and use security features like two-factor authentication to protect their accounts. By understanding the methods hackers use to create and exploit fake login pages, individuals and organizations can better defend against these pervasive threats.

8.2.2 Other forms of cyberattacks using social engineering techniques on social media

Beyond the well-known phishing attacks, there are several other forms of cyber threats that exploit the unique features and user behaviors associated with social media. One prevalent form of cyberattack on social media is the distribution of malware. Cybercriminals often use social media platforms to spread malicious software by embedding it in links, attachments, or even advertisements. These links can be shared through direct messages, posts, or comments, enticing users to click on them. Once clicked, the malware can be downloaded onto the user's device, leading to a range of harmful outcomes, such as data theft, system damage, or unauthorized access to the user's accounts. For instance, malware like the HAMMERTOSS, which uses social media for command and control, can automatically search for commands posted by attacker profiles, allowing cybercriminals to control the malware via social media posts.

Identity theft and impersonation are significant threats on social media. Attackers can create fake profiles that mimic real individuals or brands, using publicly available information and photos to make these profiles appear legitimate. These fake accounts can then be used to scam others, steal personal information, or tarnish reputations. For example, scammers might use a fake profile to trick users into providing sensitive information or to send money under false pretenses. This type of attack is particularly harmful if the impersonator gains access to the victim's bank accounts or other critical personal data.

In all, as with other forms of social engineering, social engineering attacks on social media involve manipulating users into divulging confidential information or performing actions that compromise their security. These attacks exploit human psychology, often invoking emotions such as fear, urgency, or curiosity. For example, attackers might pose as a trusted contact or authority figure and send messages that prompt the victim to reveal passwords, click on malicious links, or download harmful attachments. Social engineering can also involve more elaborate schemes, such as creating fake personas to build trust over time before exploiting the relationship to extract sensitive information.

Account takeovers are another common cyber threat on social media. In these attacks, cybercriminals gain unauthorized access to a user's social media account, often through phishing or by exploiting weak passwords. Once they have control of the account, they can use it to spread malware, send scam messages, or post harmful content. This not only affects the account owner but also their network of friends or followers, who may trust the compromised account and fall victim to further attacks. High-profile account takeovers can have significant repercussions, as seen in the case of the Associated Press's Twitter account being hacked, which led to a temporary but substantial drop in the stock market.

Data breaches on social media can occur when cybercriminals gain access to sensitive information stored on these platforms. This can happen through direct hacking of the social media site itself or by compromising individual user accounts. The stolen data can include personal details, login credentials, and even financial information, which can then be sold on the dark web or used for further attacks. For example, the LinkedIn data breach exposed 117 million email and password combinations, highlighting the potential scale and impact of such incidents.

Disinformation campaigns are a form of cyberattack that involves spreading false or misleading information on social media to influence public opinion or disrupt societal processes. These campaigns can be orchestrated by state actors, political groups, or other malicious entities. They often use fake accounts, bots, and trolls to amplify their messages and create a false sense of consensus. Disinformation campaigns can have serious consequences, such as undermining trust in institutions, inciting violence, or interfering in elections.

Social media scams are diverse and can take many forms, including fake giveaways, fraudulent advertisements, and investment schemes. These scams often promise significant rewards or benefits to lure victims into providing personal information, making payments, or clicking on malicious links. For instance, scammers might post content that appears to offer free gift cards or other prizes, requiring users to enter their personal email addresses or other details. Instead of receiving the promised reward, victims are often subjected to spam, phishing attempts, or malware infections.

Cybercriminals use social media for reconnaissance and data mining, gathering information about potential targets to facilitate more targeted attacks. By analyzing users' profiles, posts, and interactions, attackers can collect valuable data such as contact information, personal interests, and social connections. This information can be used to craft highly personalized phishing attacks, guess security questions, or even perform identity theft. The extensive amount of personal information shared on social media makes it a rich resource for cybercriminals conducting reconnaissance.

Cross-site scripting (XSS) and cross-site request forgery (CSRF) are technical attacks that exploit vulnerabilities in web applications, including social media platforms. XSS attacks involve injecting malicious scripts into web pages viewed by other users, which can steal cookies, session tokens, or other sensitive information. CSRF attacks trick users into performing actions they did not intend, such as changing account settings or making unauthorized transactions, by exploiting the trust that a web application has in the user's browser. These attacks can lead to significant data breaches and unauthorized access to user accounts.

Phishing attacks using text messages are called Smishing. This is another attack used to gather personal information. Smishing is implemented through text messages or SMS (Short Message Service), giving the attack the name "SMiShing." These attacks may contain something like "Congrats! You've won!", or "we have been trying to contact you regarding . . . ." A common tactic is to confirm an order or charge to a credit card. The user is then instructed to click on a link to review the charge or the order. Sometimes a phone number is provided. When the user calls the provided number they will be asked to provide identifying information.

Attackers also send messages impersonating a bank or service provider. The user will be requested to click on a link or call the customer service number, included within the text message, about a recent suspicious charge or a compromised account. The user will be required to provide their account number and other personal information.

Smishing attacks exploit the widespread use and trust in mobile communication, leveraging the immediacy and personal nature of text messaging to elicit quick responses from targets. These attacks are becoming increasingly sophisticated, utilizing psychological tactics and social engineering to create a sense of urgency or fear, prompting victims to act hastily without scrutinizing the message's authenticity.

One evolving tactic in smishing involves the use of deepfake technology or voice synthesis to impersonate known contacts or authority figures in voice messages accompanying the text. Victims receiving a voice message that seemingly comes from a trusted source are more likely to believe the legitimacy of the accompanying text, increasing the likelihood of falling for the scam.

Furthermore, smishing campaigns are increasingly leveraging current events or popular trends to create compelling narratives for their attacks. For example, during tax season, messages purporting to be from tax authorities, claiming issues with tax returns or promising tax refunds, can be particularly effective. Similarly, during global events like pandemics or natural disasters, smishing messages may offer emergency funds or require personal information to receive aid, preying on the heightened emotional state and vulnerabilities of recipients.

Attackers are also employing techniques to mask the originating number of the smishing message, making it appear as though it comes from a legitimate source, such as a well-known bank or government agency. This technique, known as spoofing, makes it even more challenging for recipients to identify the message as a phishing attempt. Additionally, attackers sometimes use short links in their messages, which disguise the actual URL, making it difficult for users to ascertain the legitimacy of the link before clicking.

To counter smishing attacks, it is crucial for individuals to adopt a skeptical approach to unsolicited messages, especially those that request personal information or prompt immediate action. Verifying the identity of the sender through independent channels, such as the official website or customer service line of the purported organization, is a key step in protecting oneself against these attacks.

Organizations, on their part, can play a significant role in combating smishing by educating their employees and customers about the nature of these attacks and how to respond to them. Implementing and promoting secure communication channels for customer interactions can also help reduce the effectiveness of smishing campaigns by making it easier for customers to identify legitimate messages.

As smishing attacks continue to evolve in complexity and deceit, awareness and education remain the most powerful defenses. By staying informed about the latest smishing tactics and exercising caution with mobile communications, individuals and organizations can significantly reduce the risk of falling victim to these increasingly common cyber threats.

8.3 Examples of social media phishing tools

Attackers have developed tools to streamline their attacks and to provide the tools to other attackers that are less technically capable. This is also true with malware development and tools for a variety of system attacks. These tools enable a larger number of attackers to develop their attacks quickly and effectively.

The evolving landscape of cyber threats, the democratization of hacking tools, particularly for phishing attacks, has lowered the barrier to entry for aspiring cybercriminals. These tools not only facilitate the creation of phishing campaigns but also enhance their sophistication, making them more difficult to detect and counteract. The following are examples of social media phishing tools.

The first tool is **HiddenEye**. HiddenEye is an automated tool that was developed in the Python Language and is available on the GitHub platform. It is free to download and use. The program is commonly installed on a Linux operating system. The program is menu-driven and provides several options for the user to create a phishing site that is a clone of social platforms LinkedIn, Facebook, Twitter, Instagram, and many other legitimate sites. This provides phishing capabilities to those who do not have the requisite skills to perform an attack on their own.

The current version of HiddenEye contains forty phishing modules. The user selects a site they would like to clone from one of these modules. This simplifies the creation of a social platform site. The program also provides the ability to add a keylogger to the site. This option captures the keystrokes of the user.

Another option is the ability to add a fake Cloudfare Protection page. Cloudflare is a service that provides security for the Internet. By adding this option to the cloned site, the site looks more secure and encourages the user to trust the site.

The focus of this attack is to get the user to enter information like username, password, account number, address, and other information related to the particular attack. Once the user has entered the requested data, the attacker can then redirect the user to another website. The redirection is an attempt to not make the user aware that they have been phished.

Another tool used in phishing attacks is **Kismet**. Kismet is a sniffer tool used for capturing data packets on networks or through Wi-Fi and Bluetooth communications. The captured data packets may contain sensitive information such as account information and passwords. By placing Kissmet on a network in promiscuous mode, an attacker can capture and analyze the target's network traffic. This tool is used with phishing tools like SocialFish and ShellPhish.

The **SocialFish** tool is used to avoid detection by the target while phishing his/her credentials. Newer versions of SocialFish have an intuitive user interface used for managing and creating phishing attacks. Socialphish provides the capability of cloning a target website to be used for phishing for credentials. This tool can be integrated with open-source intelligence tools like Shodan. Shodan is a search engine that maps and gather information about internet-connected devices and systems and is often referred to as a search engine for the internet of things.

Another tool is **ShellPhish**. ShellPhish is a tool is an easy-to-use phishing tool. The tool leverages some of information generated by SocialFish. The tool offers phishing templates for 18 popular sites, which include social media and email providers. The attacker may also create a custom template.

Gophish is an open-source phishing toolkit designed for businesses and penetration testers to test the susceptibility of their organizations or clients to phishing attacks. Unlike other tools that focus solely on creating phishing sites,

Gophish offers a comprehensive platform for designing, executing, and tracking entire phishing campaigns. This includes crafting emails, setting up fake web pages, and detailed reporting on how targets interact with the emails and web pages. Gophish's user-friendly interface and detailed analytics make it a valuable tool for understanding and improving phishing awareness within an organization.

PhishX is another tool that streamlines the phishing process, focusing on ease of use and effectiveness in launching spear-phishing campaigns. PhishX uses a technique known as 'masquerading' where the phishing email appears to come from a trusted sender within the target's own contacts, increasing the likelihood of the target's engagement with the email. It supports a wide array of social media and email platforms, making it versatile for various phishing scenarios.

The development of these tools indicates a shift toward more accessible cybercrime, where the technical knowledge required to launch attacks is being abstracted away by user-friendly interfaces and automated processes. This trend not only increases the volume of phishing attacks but also allows attackers to rapidly adapt to new security measures and exploit emerging vulnerabilities.

To counteract the threats posed by these tools, organizations must adopt a proactive and multilayered approach to cybersecurity. This includes regular security training that keeps pace with the evolving threat landscape, advanced phishing detection technologies, and a culture of security that encourages vigilance and reporting of suspicious activities. Furthermore, leveraging threat intelligence platforms can provide insights into new phishing tools and tactics, enabling organizations to update their defenses accordingly.

As phishing tools become more sophisticated and accessible, the arms race between cybercriminals and cybersecurity professionals continues to intensify. Staying informed about the latest developments in phishing tactics and tools is crucial for organizations aiming to protect their assets and stakeholders in this dynamic cyber threat environment.

The landscape of cyberattacks on social media is vast and continually evolving. From malware distribution and identity theft to social engineering and disinformation campaigns, the threats are diverse and sophisticated. As social media continues to play a central role in our personal and professional lives, it is crucial to remain vigilant and adopt robust security practices to protect against these pervasive threats. Understanding the various forms of cyberattacks that can occur on social media is the first step in safeguarding our digital presence and maintaining the integrity of these platforms.

8.4   Social media in the organization

Many organizations use LinkedIn and various job-hunting services to post career opportunities and find new job candidates. Attackers can use social media to impersonate an organization to obtain detailed information on applicants. Businesses also use social media to notify customers about the latest product offerings and specials, for marketing, and to promote the business.

Attackers can also scrape social media sites for information. Scrapping is the process of copying the contents of a website and storing the retrieved data in a structured, readable format. Useful information can then be used to identify targets for phishing attacks. Many social media sites contain names, email addresses, phone numbers, names of associates, addresses, and other useful information that can produce a specific, targeted phishing attack that provides enough detail and context to reduce the suspicion of the victim of the attack.

LinkedIn is an example of a site that provides employees and employers a means of connecting. However, LinkedIn also lets attackers create fraudulent company pages and operate fake job scams. Attackers use the scams to collect job resumes and applications or lure applicants through private messages. The process gives the attackers sensitive information to further their later phishing attacks.

If the attacker can obtain enough information from social media sites to identify members for the organization and the positions they hold, they can create a phishing attack that is difficult to detect. The message can appear to come from a supervisor requesting a subordinate to review and complete an attached document. Many users would automatically open the document and follow any instructions, thinking the email message was from their supervisor.

In addition to the sophisticated tactics already mentioned, attackers are exploiting social media's advertising and promotional features to conduct phishing attacks. Social media platforms allow organizations to run targeted ad campaigns aimed at specific demographics, interests, or behaviors. Cybercriminals can misuse these features to create malicious ads that mimic legitimate business promotions. These ads might direct users to counterfeit websites that resemble the official sites of the organization, where users are prompted to enter sensitive information, such as login credentials or payment details. The deceptive nature of these ads, combined with the trust users place in the apparent endorsement of the social media platform, can make them particularly effective vectors for phishing attacks.

Another dimension to the threat landscape is the use of social media bots. Automated accounts can be programmed to carry out a variety of tasks, from posting content to interacting with users. In the context of phishing, bots can be used to amplify malicious content, automate the scraping of information, or even engage directly with users to deliver phishing links through comments or messages. The scalability of bot-driven activities means that a single attacker can target thousands of users across multiple platforms simultaneously, significantly increasing the potential reach and impact of their phishing campaigns.

The convergence of personal and professional use of social media also presents unique challenges. Employees may inadvertently expose their organization to risk through their personal social media activities. For example, sharing details about work projects, tagging their location at corporate events, or publicly discussing internal processes can provide attackers with valuable insights. This information can be used to craft highly credible phishing messages or to map out the organizational structure for more targeted attacks.

To combat these evolving threats, organizations must extend their cybersecurity measures beyond traditional defenses. This includes deploying advanced threat detection tools that can identify and block malicious social media activities and implementing comprehensive digital risk protection strategies that monitor external threats across the web, social media, and the dark web.

Educating employees about the risks associated with their personal social media use and its potential impact on organizational security is also crucial. This might involve providing guidelines on what information can be safely shared online and training employees to recognize and report suspicious social media activities.

Moreover, establishing strong collaboration between the cybersecurity team and the marketing or social media departments can ensure that promotional activities are conducted securely. This collaboration can help in vetting the security of marketing campaigns, monitoring the organization's social media presence for signs of impersonation, and quickly responding to incidents of misuse.

Ultimately, protecting against social media-based phishing attacks requires a holistic approach that combines advanced technology, employee education, and cross-departmental collaboration. By acknowledging the unique vulnerabilities introduced by social media and adapting their security practices accordingly, organizations can safeguard their data, protect their reputation, and maintain the trust of their customers and employees.

8.4.1   The tradeoffs for an organizational presence on social media

Social media has become an indispensable tool for businesses, offering a myriad of benefits that can significantly enhance brand visibility, customer engagement, and overall market reach. However, the advantages of maintaining a social media presence come with inherent risks and potential damages that organizations must carefully navigate. This high-level tradeoff analysis explores the benefits and drawbacks of social media for businesses, providing a comprehensive understanding of the landscape.

Social media platforms offer businesses unparalleled opportunities to connect with their target audiences. One of the most significant advantages is the ability to reach a larger audience. Platforms like Facebook, Instagram, Twitter, and LinkedIn boast billions of active users, providing businesses with a vast pool of potential customers. This extensive reach allows companies to increase brand awareness and attract new customers, which is particularly beneficial for small businesses looking to expand their market presence.

Another key benefit is the direct connection with the audience. Social media enables businesses to engage with their customers in real time, fostering a sense of community and loyalty. Through comments, direct messages, and interactive posts, companies can address customer inquiries, resolve issues, and gather valuable feedback. This direct line of communication helps build stronger relationships with customers, enhancing their overall experience with the brand.

Social media also offers cost-effective marketing opportunities. Compared to traditional advertising methods, social media marketing is relatively affordable and accessible. Businesses can create and share content, run targeted ad campaigns, and track performance metrics without breaking the bank. This cost-efficiency is particularly advantageous for small businesses with limited marketing budgets, allowing them to compete with larger companies on a more level playing field.

Additionally, social media provides businesses with valuable insights and analytics. Platforms like Facebook and Instagram offer in-depth performance analytics that help companies understand their audience's behavior, preferences, and engagement patterns. These insights enable businesses to make data-driven decisions, optimize their marketing strategies, and improve their overall performance. Social media analytics also allow companies to track the success of their campaigns, measure return on investment (ROI), and identify areas for improvement.

Despite the numerous benefits, maintaining a social media presence also exposes businesses to several risks and potential damages. One of the most significant drawbacks is the potential for negative feedback and public embarrassment. Social media platforms provide a public forum for customers to voice their opinions, and negative reviews or comments can quickly tarnish a company's reputation. A single negative post can go viral, causing significant damage to the brand's image and customer trust.

Another major concern is the spread of misinformation. Social media platforms are notorious for the rapid dissemination of false or misleading information. Businesses can fall victim to rumors, fake news, or malicious attacks that can harm their credibility and reputation. For example, a false rumor about a company's financial stability or product safety can lead to a loss of customer confidence and a decline in sales. Managing and mitigating the impact of misinformation requires constant vigilance and proactive communication strategies.

The time-consuming nature of social media management is another drawback. Maintaining an active and engaging social media presence requires regular content creation, monitoring, and interaction with followers. This can be particularly challenging for small businesses with limited resources and staff. The constant need for fresh and relevant content can also lead to burnout and decreased productivity among employees tasked with managing social media accounts.

Privacy and security risks are also significant concerns for businesses on social media. Sharing sensitive business information or personal data on these platforms can expose companies to cyberattacks, data breaches, and identity theft. Hackers can exploit vulnerabilities in social media accounts to gain unauthorized access to confidential information, leading to financial losses and legal liabilities. Ensuring robust security measures and educating employees about safe social media practices are essential to mitigate these risks.

Social media can also amplify customer service challenges. While it provides a platform for addressing customer concerns, it also exposes businesses to public scrutiny. Negative customer experiences shared on social media can quickly escalate, attracting widespread attention and damaging the brand's reputation. Companies must be prepared to respond promptly and professionally to customer complaints, demonstrating their commitment to excellent customer service.

Balancing the benefits and drawbacks of social media for businesses requires a strategic approach that maximizes the advantages while mitigating the risks.

The following high-level tradeoff analysis provides a comprehensive overview of the key considerations for businesses navigating the social media landscape.

Brand visibility versus reputation management. One of the primary benefits of social media is increased brand visibility. By leveraging the vast user base of platforms like Facebook, Instagram, and Twitter, businesses can reach a broader audience and enhance their brand recognition. However, this increased visibility also means that any negative feedback or public relations issues are more likely to be amplified. Effective reputation management strategies, such as monitoring social media channels, responding to negative comments, and proactively addressing customer concerns, are essential to balance this tradeoff.

Customer engagement versus resource allocation. Social media offers businesses the opportunity to engage directly with their customers, fostering loyalty and building strong relationships. However, maintaining an active social media presence requires significant time and resources. Businesses must allocate sufficient staff and budget to manage social media accounts, create content, and interact with followers. To optimize resource allocation, companies can use social media management tools, automate certain tasks, and prioritize platforms that offer the highest ROI.

Cost-effective marketing versus security risks. Social media marketing is a cost-effective way to promote products and services, especially for small businesses with limited budgets. However, the use of social media also exposes businesses to security risks, such as data breaches and cyberattacks. Implementing robust security measures, such as two-factor authentication, regular account monitoring, and employee training on safe social media practices, can help mitigate these risks while still reaping the benefits of cost-effective marketing.

Valuable insights versus privacy concerns. Social media platforms provide businesses with valuable insights and analytics that can inform marketing strategies and improve performance. However, the collection and use of customer data raise privacy concerns. Businesses must ensure compliance with data protection regulations, such as the General Data Protection Regulation (GDPR), and be transparent about how they collect, use, and protect customer data. Balancing the need for valuable insights with respect for customer privacy is crucial for maintaining trust and avoiding legal issues.

Real-time communication versus potential for misinformation. The ability to communicate with customers in real time is a significant advantage of social media. It allows businesses to address customer inquiries, resolve issues, and share updates quickly. However, the same immediacy that facilitates real-time communication also enables the rapid spread of misinformation. Businesses must be proactive in monitoring social media channels, correcting false information, and maintaining open lines of communication with their audience to mitigate the impact of misinformation.

Social media presents a complex landscape for businesses, offering both significant benefits and potential risks. The ability to reach a larger audience, engage directly with customers, and leverage cost-effective marketing opportunities makes social media an invaluable tool for business growth. However, the potential for negative feedback, misinformation, privacy and security risks, and the time-consuming nature of social media management are critical challenges that businesses must address. A strategic approach that balances these tradeoffs is essential for maximizing the benefits of social media while minimizing potential damages. By implementing effective reputation management strategies, optimizing resource allocation, ensuring robust security measures, respecting customer privacy, and proactively managing communication, businesses can navigate the social media landscape successfully.

Ultimately, the key to leveraging social media for business success lies in understanding the unique dynamics of each platform, staying informed about emerging trends and risks, and continuously adapting strategies to meet the evolving needs of the digital age. With careful planning and execution, businesses can harness the power of social media to build strong brands, foster customer loyalty, and drive sustainable growth.

8.4.2   Managing social media use

While it is recognized that in the digital age, social media has become an essential tool for businesses, enabling them to engage with customers, promote their brands, and drive growth, the widespread use of social media also introduces significant risks, particularly in terms of cybersecurity, information sharing, and company privacy. To mitigate these risks, organizations must implement comprehensive social media policies that provide clear guidelines for employees and protect the company's interests.

Cybersecurity is a critical concern for organizations that maintain a social media presence. Social media platforms are frequent targets for cyberattacks, including phishing, malware distribution, and account hijacking. To safeguard against these threats, organizations must establish robust cybersecurity policies that address the unique challenges posed by social media.

One fundamental aspect of cybersecurity policies is the implementation of strong password protocols. Employees should be required to use complex passwords that combine letters, numbers, and special characters. Additionally, organizations should mandate the use of multi-factor authentication (MFA) for accessing social media accounts. MFA adds an extra layer of security by requiring users to provide two or more verification factors, such as a password and a one-time code sent to their mobile device.

Regular training and awareness programs are also essential components of cybersecurity policies. Employees should be educated about common social media threats, such as phishing scams and social engineering attacks. Training sessions should include practical exercises that help employees recognize and respond to suspicious activities. For example, employees can be taught to verify the authenticity of links before clicking on them and to report any unusual account activity immediately.

Access control is another critical element of cybersecurity policies. Organizations should restrict access to social media accounts to authorized personnel only. This can be achieved by implementing role-based access controls (RBAC), which assign permissions based on an employee's role within the company. Regular audits should be conducted to ensure that access permissions are up-to-date and that former employees no longer have access to company accounts.

Organizations should establish protocols for monitoring and responding to security incidents. This includes setting up intrusion detection and prevention systems (IDPS) to identify and mitigate potential threats in real time. In the event of a security breach, organizations should have a clear incident response plan that outlines the steps to be taken, including notifying affected parties, containing the breach, and conducting a thorough investigation to prevent future incidents.

Information sharing on social media can be a double-edged sword. While it allows organizations to communicate with their audience and share valuable content, it also poses risks related to the disclosure of sensitive information. To manage these risks, organizations must develop information sharing protocols that govern what can and cannot be shared on social media.

One key aspect of information sharing protocols is the classification of information. Organizations should categorize information into different levels of sensitivity, such as public, internal, confidential, and restricted. Clear guidelines should be provided on what type of information falls into each category and the appropriate channels for sharing it. For example, public information, such as marketing materials and press releases, can be shared freely on social media, while confidential information, such as financial data and trade secrets, should be restricted to internal communication channels.

Employees should be trained to recognize and handle sensitive information appropriately. This includes understanding the potential consequences of sharing confidential information on social media and the importance of maintaining discretion. For instance, employees should avoid discussing proprietary information in public forums or sharing details about ongoing projects that have not been officially announced.

Organizations should also implement approval processes for social media posts. This involves designating specific individuals or teams responsible for reviewing and approving content before it is published. The approval process ensures that all posts align with the company's information sharing protocols and do not inadvertently disclose sensitive information. Additionally, organizations can use social media management tools that provide workflow features to streamline the approval process and maintain a record of all published content.

Another important consideration is the use of disclaimers and disclaimers. Employees should be instructed to include disclaimers in their social media profiles, stating that their views are their own and do not necessarily reflect the views of the company. This helps to mitigate the risk of employees' personal opinions being misconstrued as official company statements. Furthermore, organizations should provide guidelines on how to handle inquiries and comments on social media, ensuring that employees respond professionally and do not disclose sensitive information.

Protecting company privacy is paramount in the age of social media, where information can be easily disseminated and accessed by a global audience.

Organizations must implement privacy policies that safeguard their data and ensure compliance with relevant privacy laws and regulations.

One of the first steps in protecting company privacy is to conduct a thorough assessment of the types of data collected and shared on social media. This includes identifying personally identifiable information, sensitive personal information, and any other data that could pose a privacy risk if exposed. Organizations should then implement data minimization practices, collecting only the data necessary for their operations and avoiding the storage of unnecessary information.

Privacy policies should also address the use of third-party applications and integrations. Many social media platforms offer third-party tools that enhance functionality, such as analytics and scheduling tools. However, these tools often require access to company data, which can pose privacy risks. Organizations should carefully vet third-party applications, reviewing their privacy policies and security measures to ensure they comply with the company's privacy standards. Additionally, organizations should limit the access of third-party applications to only the data necessary for their functionality. Data encryption is another critical component of privacy policies. Organizations should use encryption to protect data both in transit and at rest. This ensures that even if data is intercepted or accessed without authorization, it remains unreadable and secure. Encryption should be applied to all sensitive data, including login credentials, personal information, and proprietary company data.

Transparency is also essential in maintaining company privacy. Organizations should be transparent with their employees and customers about how their data is collected, used, and protected. This includes providing clear and accessible privacy policies that outline the company's data practices and the rights of individuals. Regular updates to privacy policies should be communicated to ensure that all stakeholders are aware of any changes.

Organizations should also implement data access controls to limit who can view and handle sensitive information. This includes using RBAC to assign permissions based on an employee's role and responsibilities. Regular audits should be conducted to ensure that access permissions are appropriate and that any unnecessary access is revoked.

In addition to these measures, organizations should establish protocols for responding to data breaches. This includes having a clear incident response plan that outlines the steps to be taken in the event of a breach, such as notifying affected individuals, containing the breach, and conducting a thorough investigation. Organizations should also comply with any legal requirements for reporting data breaches, such as notifying regulatory authorities and providing timely updates to affected parties.

In all, the integration of social media into business operations offers numerous benefits but also introduces significant risks related to cybersecurity, information sharing, and company privacy. To navigate these challenges, organizations must implement comprehensive social media policies that provide clear guidelines for employees and protect the company's interests.

Effective cybersecurity policies should include strong password protocols, MFA, regular training and awareness programs, access control measures, and incident response protocols. These measures help safeguard against cyberattacks and ensure the security of social media accounts.

Information sharing protocols should categorize information based on sensitivity, provide guidelines for handling sensitive information, implement approval processes for social media posts, and include disclaimers to mitigate risks. These protocols help manage the disclosure of information and protect the company's reputation.

Company privacy protections should involve data minimization practices, careful vetting of third-party applications, data encryption, transparency in data practices, data access controls, and protocols for responding to data breaches. These measures ensure the protection of sensitive data and compliance with privacy laws and regulations.

By implementing these comprehensive social media policies, organizations can leverage the benefits of social media while mitigating the associated risks, ultimately protecting their data, reputation, and overall business interests.

8.4.3   When an organization becomes a target-disinformation campaign

When a company becomes the target of a social media disinformation campaign, the repercussions can be severe, affecting its reputation, customer trust, and financial stability. Disinformation campaigns are deliberate attempts to spread false or misleading information to manipulate public perception and cause harm. These campaigns can be orchestrated by competitors, disgruntled employees, or even state actors. To effectively counter such attacks, companies must adopt a multifaceted approach that includes preparation, rapid response, legal action, and long-term strategies to rebuild trust and resilience.

Disinformation campaigns are designed to exploit the rapid dissemination capabilities of social media platforms. They often involve the creation and spread of false narratives that can quickly go viral, reaching millions of users in a short period. These campaigns can take various forms, including fake news articles, doctored images or videos, and misleading social media posts. The goal is to create confusion, sow distrust, and damage the target's reputation.

One of the first steps a company should take when targeted by a disinformation campaign is to understand the nature and scope of the attack. This involves identifying the sources of disinformation, the platforms being used, and the specific narratives being promoted. Companies should employ social listening tools to monitor mentions of their brand across social media and identify any emerging false narratives. By understanding the tactics and channels used by disinformation actors, companies can develop more effective countermeasures.

Preparation is crucial in mitigating the impact of disinformation campaigns. Companies should have an "Emergency Response Strategy" in place, which outlines the steps to be taken in the event of a disinformation attack. This strategy should involve teams from different departments, including communications, legal, IT, and customer service, to ensure a coordinated and comprehensive response.

Each team should have clearly defined roles and responsibilities to prevent confusion and ensure a swift and effective response.

Social listening is an essential component of this preparation. By continuously monitoring social media for mentions of their brand, companies can detect disinformation early and take proactive measures to counter it. This involves not only tracking direct mentions but also monitoring related keywords and hashtags that could be associated with the disinformation campaign. Early detection allows companies to respond quickly before the false narratives gain significant traction.

When a disinformation campaign is detected, rapid response is critical. Companies should address their audience directly and transparently, providing accurate information to counter the false narratives. This involves using verified social media accounts to communicate with the public and ensure that the response is seen as coming directly from the company. Transparency and honesty are key to maintaining credibility and trust with the audience.

Engagement with the audience is also important. Companies should actively respond to comments and questions on social media, providing clarifications and correcting any misinformation. This helps to reduce the spread of false information and reassures customers that the company is taking the issue seriously. For example, when Starbucks was targeted by a false "Dreamer Day" campaign, the company quickly responded on Twitter, warning users that the campaign was a hoax and providing accurate information about their policies. In addition to direct communication, companies should leverage their network of influencers and advocates to help spread accurate information. Influencers with large followings can amplify the company's message and reach a wider audience, helping to counter the disinformation more effectively. Building a community of advocates on social media before an attack occurs can provide a valuable resource in times of crisis.

In some cases, legal action may be necessary to address disinformation campaigns. Companies should work with their legal teams to identify the sources of disinformation and take appropriate legal measures. This can include filing complaints with local courts and regulatory bodies, such as the Federal Trade Commission (FTC), to seek redress and hold the perpetrators accountable. Legal action can serve as a deterrent to future attacks and demonstrate the company's commitment to protecting its reputation.

Engaging with social media platforms is also crucial. Companies should report false information to the platforms and request its removal. Many platforms, such as Facebook and Twitter, have policies in place to counter fake news and disinformation. By providing evidence of the falsehoods, companies can work with the platforms to have the misleading content taken down. This helps to limit the spread of disinformation and reduce its impact on the company's reputation.

While immediate response measures are essential, companies must also focus on long-term strategies to build resilience against disinformation. This involves fostering a culture of transparency and trust with customers, employees, and other stakeholders. By consistently communicating openly and honestly, companies can build a strong reputation that is more resistant to disinformation attacks.

One effective long-term strategy is to invest in media literacy and education for both employees and customers. Educating employees about the risks of disinformation and how to recognize and respond to it can help prevent internal vulnerabilities. Similarly, educating customers about how to identify false information and encouraging them to verify sources before sharing can reduce the spread of disinformation.

Companies should also develop and maintain a robust crisis communication plan. This plan should include detailed procedures for responding to disinformation attacks, including pre-drafted statements and messaging templates that can be quickly adapted to specific situations. Regularly testing and updating the crisis communication plan through simulations and exercises can ensure that the company is prepared to respond effectively when an attack occurs.

Building and maintaining trust with the audience is crucial in countering disinformation. Companies should focus on creating a positive narrative around their brand and consistently delivering on their promises. This involves providing high-quality products and services, engaging with customers in meaningful ways, and addressing any issues or concerns promptly and transparently.

Trust can also be reinforced through third-party endorsements and partnerships. Collaborating with reputable organizations, industry experts, and influencers can enhance the company's credibility and provide additional validation of its messages. For example, partnering with fact-checking organizations to verify and promote accurate information can help counter disinformation and build trust with the audience.

The fight against disinformation is ongoing, and companies must continuously monitor their social media presence and adapt their strategies as needed. This involves regularly reviewing and updating social media policies, monitoring the effectiveness of response measures, and learning from past incidents. By staying vigilant and proactive, companies can better protect themselves against future disinformation attacks. Monitoring should also extend to the broader information ecosystem. Companies should stay informed about emerging trends and tactics in disinformation, as well as changes in social media platform policies and regulations. Engaging with industry groups, attending conferences, and participating in discussions about disinformation can provide valuable insights and help companies stay ahead of potential threats.

Disinformation campaigns on social media pose significant challenges for companies, but with the right strategies and preparation, they can effectively counter these attacks and protect their reputation. By understanding the nature of disinformation, preparing proactive measures, responding rapidly and transparently, taking legal action when necessary, and building long-term resilience, companies can mitigate the impact of disinformation and maintain trust with their audience.

The key to success lies in a multifaceted approach that combines immediate response measures with long-term strategies to build credibility and trust. By fostering a culture of transparency, educating employees and customers, and continuously monitoring and adapting their strategies, companies can navigate the complex landscape of social media disinformation and emerge stronger and more resilient.

8.4.4   When key personnel in an organization become a target on social media—deepfakes

When members of a company's leadership team become victims of deepfake attacks on social media, the organization faces a multifaceted crisis that can severely impact its reputation, stakeholder trust, and operational stability. Deepfakes, which are AI-generated videos or audio clips that convincingly mimic real people, can be used to spread false information, create damaging narratives, and manipulate public perception. To effectively respond to such incidents, organizations must adopt a comprehensive approach that includes immediate crisis management, legal action, long-term reputation rebuilding, and proactive measures to prevent future attacks.

The first step in responding to a deepfake attack is to activate the company's crisis management plan. This plan should be pre-established and include specific protocols for handling deepfake incidents. The crisis management team, which typically includes members from communications, legal, IT, and executive leadership, must be mobilized immediately to coordinate the response.

A swift and transparent communication strategy is crucial. The company should issue a public statement acknowledging the deepfake attack and clarifying that the content in question is fabricated. This statement should be disseminated through all available channels, including the company's official social media accounts, press releases, and the corporate website. Transparency is key to maintaining credibility and trust with stakeholders. The statement should provide clear information about the nature of the deepfake, the steps the company is taking to address the situation, and assurances that the leadership team is actively working to mitigate any potential damage.

Engaging with the audience is also essential. The company should monitor social media platforms for discussions about the deepfake and respond to comments and questions in real time. This helps to control the narrative and prevent the spread of misinformation. It is important to remain calm and professional in all communications, avoiding any defensive or confrontational language that could exacerbate the situation.

Legal action is a critical component of the response to a deepfake attack. The company's legal team should work to identify the source of the deepfake and pursue appropriate legal remedies. This may involve filing complaints with local courts and regulatory bodies, such as the FTC in the United States, to seek redress and hold the perpetrators accountable. Legal action can serve as a deterrent to future attacks and demonstrate the company's commitment to protecting its reputation and the integrity of its leadership team.

In addition to legal action, the company should engage with social media platforms to have the deepfake content removed. Most social media platforms have policies in place to counter fake news and disinformation. By providing evidence that the content is a deepfake, the company can work with the platforms to ensure that the misleading material is taken down promptly. This helps to limit the spread of the deepfake and reduce its impact on the company's reputation.

While immediate response measures are essential, the company must also focus on long-term strategies to rebuild its reputation and restore stakeholder trust. This involves a sustained effort to communicate transparently and consistently with all stakeholders, including employees, customers, investors, and the general public.

One effective strategy is to leverage third-party endorsements and partnerships to reinforce the company's credibility. Collaborating with reputable organizations, industry experts, and influencers can provide additional validation of the company's messages and help to counter the negative impact of the deepfake. For example, partnering with fact-checking organizations to verify and promote accurate information can help to rebuild trust with the audience.

The company should also invest in media literacy and education for both employees and stakeholders. Educating employees about the risks of deepfakes and how to recognize and respond to them can help prevent internal vulnerabilities. Similarly, educating stakeholders about how to identify false information and encouraging them to verify sources before sharing can reduce the spread of disinformation.

To prevent future deepfake attacks, the company must implement proactive measures that enhance its overall security posture and resilience. This includes updating cybersecurity programs and incident response plans to specifically address deepfake threats. Companies should consider running tabletop exercises to test these new protocols and ensure that the crisis management team is prepared to respond effectively to deepfake incidents.

Monitoring social media for potential deepfake threats is also crucial. The company should use social listening tools to continuously track mentions of its brand and leadership team across social media platforms. Early detection of deepfake content allows the company to respond quickly and mitigate the impact before the false narratives gain significant traction.

Implementing robust data protection measures is another key component of preventing deepfake attacks. This includes using encryption to protect sensitive data, implementing RBAC to restrict access to critical information, and requiring MFA for accessing social media accounts. Regular training and awareness programs for employees can also help to reinforce best practices for data protection and cybersecurity.

Building a culture of transparency and trust within the organization is essential for maintaining resilience against deepfake attacks. This begins with leadership. Company leaders should set the tone by prioritizing transparency and demonstrating their commitment to data protection and privacy. This includes allocating sufficient resources for privacy initiatives, supporting ongoing training and awareness programs, and holding employees accountable for adhering to privacy policies.

Employees at all levels should be encouraged to take ownership of privacy and to integrate data protection practices into their daily activities. This can be achieved through regular communication, recognition of privacy champions, and the establishment of clear channels for reporting privacy concerns. By creating an environment where privacy is valued and respected, the company can ensure that its social media policies are not only compliant but also effective in protecting user data.

As deepfake technology continues to evolve, companies must leverage advanced tools and technologies to detect and respond to deepfake threats. AI and machine learning-based detection systems can help to identify manipulated media and distinguish between genuine and synthetic content. These systems can analyze various indicators, such as unnatural blinking, head positioning, and speech cadence, to detect deepfakes with a high degree of accuracy.

In addition to detection systems, companies should use social media management tools to monitor their online presence and identify potential deepfake threats in real time. These tools can provide valuable insights into social media activities, helping companies to detect and respond to deepfake content before it causes significant damage.

Responding to a deepfake attack on a company's leadership team requires a comprehensive and multifaceted approach. Immediate crisis management, legal action, long-term reputation rebuilding, and proactive measures to prevent future attacks are all essential components of an effective response strategy. By understanding the nature of deepfake threats, preparing proactive measures, responding rapidly and transparently, taking legal action when necessary, and building long-term resilience, companies can mitigate the impact of deepfake attacks and maintain trust with their stakeholders.

The key to success lies in a holistic approach that combines immediate response measures with long-term strategies to build credibility and trust. By fostering a culture of transparency, educating employees and stakeholders, leveraging advanced detection technologies, and continuously monitoring and adapting their strategies, companies can navigate the complex landscape of social media disinformation and emerge stronger and more resilient.

8.4.5   Ensuring social media policies align with emerging privacy laws

In the rapidly evolving landscape of digital communication, social media has become a cornerstone for businesses to engage with their audiences, promote their brands, and drive growth. However, the dynamic nature of social media also brings significant challenges, particularly in terms of aligning company policies with evolving privacy laws. As privacy regulations become more stringent and comprehensive, companies must ensure that their social media policies are not only compliant but also adaptable to future changes. This analysis explores how companies can develop and maintain social media policies that align with evolving privacy laws, focusing on key strategies and best practices.

The first step in aligning social media policies with evolving privacy laws is to thoroughly understand the regulatory landscape. Privacy laws vary significantly across jurisdictions, and staying informed about these differences is crucial for compliance. For instance, the GDPR in Europe sets stringent requirements for data protection and privacy, while the California Consumer Privacy Act and its successor, the California Privacy Rights Act, impose specific obligations on businesses operating in California. Additionally, new state-level privacy laws are continually being enacted in the United States, such as the Texas Data Privacy and Security Act and the Florida Digital Bill of Rights, each with unique provisions and compliance requirements.

To navigate this complex regulatory environment, companies should establish a dedicated compliance team or designate a privacy officer responsible for monitoring legal developments and ensuring that social media policies are up-to-date. This team should regularly review and interpret new laws, assess their impact on the company's social media practices, and provide guidance on necessary policy adjustments. By staying informed and proactive, companies can avoid the pitfalls of non-compliance and mitigate the risk of legal penalties.

A well-crafted social media policy serves as the foundation for aligning with privacy laws. Such a policy should clearly outline the company's expectations for social media use, including guidelines for content creation, information sharing, and user interactions. It should also address specific privacy concerns, such as the handling of personal data, consent requirements, and data security measures.

One critical aspect of a comprehensive social media policy is the inclusion of clear guidelines on data collection and processing. Companies must ensure that any personal data collected through social media is handled in accordance with applicable privacy laws. This includes obtaining explicit consent from users before collecting their data, providing clear and accessible privacy notices, and allowing users to exercise their rights, such as accessing, correcting, or deleting their personal information. For example, under the GDPR, companies must obtain informed consent from users before processing their personal data, and this consent must be freely given, specific, informed, and unambiguous.

Additionally, social media policies should address the use of third-party applications and integrations. Many social media platforms offer third-party tools that enhance functionality, such as analytics and scheduling tools. However, these tools often require access to user data, which can pose privacy risks. Companies should carefully vet third-party applications, ensuring that they comply with relevant privacy laws and that data sharing agreements are in place to protect user information. The policy should also specify the procedures for regularly auditing third-party applications to ensure ongoing compliance.

Data protection is a cornerstone of privacy compliance, and companies must implement robust measures to safeguard personal information collected through social media. This includes both technical and organizational measures designed to prevent unauthorized access, data breaches, and other security incidents. One essential technical measure is data encryption. Companies should use encryption to protect data both in transit and at rest, ensuring that even if data is intercepted or accessed without authorization, it remains unreadable and secure. Encryption should be applied to all sensitive data, including login credentials, personal information, and proprietary company data.

Access control is another critical component of data protection. Companies should implement RBAC to restrict access to social media accounts and sensitive data to authorized personnel only. This involves assigning permissions based on an employee's role and responsibilities and conducting regular audits to ensure that access permissions are appropriate and up-to-date. Additionally, MFA should be required for accessing social media accounts, adding an extra layer of security by requiring users to provide two or more verification factors.

Organizational measures include regular training and awareness programs for employees. These programs should educate employees about the importance of data protection, common social media threats, and best practices for maintaining security. Training sessions should include practical exercises that help employees recognize and respond to suspicious activities, such as phishing attempts or unauthorized access to social media accounts.

Transparency is a fundamental principle of privacy laws, and companies must ensure that their social media policies reflect this principle. This involves being open and honest with users about how their data is collected, used, and protected. Companies should provide clear and accessible privacy notices that explain their data practices in plain language, avoiding legal jargon that may confuse users.

Companies should establish mechanisms for users to exercise their privacy rights. This includes providing easy-to-use tools for users to access, correct, delete, or port their personal information. For example, under the GDPR, users have the right to access their personal data and request its deletion or correction. Companies should ensure that these rights are clearly communicated to users and that procedures are in place to handle such requests promptly and efficiently.

Accountability is also crucial for ensuring compliance with privacy laws. Companies should implement data protection impact assessments (DPIAs) to evaluate the risks associated with data processing activities and to identify measures to mitigate those risks. DPIAs are particularly important for high-risk processing activities, such as profiling or processing sensitive data. By conducting DPIAs, companies can demonstrate their commitment to privacy and their proactive approach to managing data protection risks.

Privacy laws are continually evolving, and companies must be prepared to adapt their social media policies to keep pace with these changes. This requires a proactive approach to policy management, including regular reviews and updates to ensure ongoing compliance.

One effective strategy is to establish a policy review committee comprising members from various departments, such as legal, HR, communications, and IT. This committee should meet regularly to assess the effectiveness of the social media policy, review any changes in privacy laws, and recommend updates as needed. By involving a cross-functional team, companies can ensure that their social media policies are comprehensive and aligned with the latest legal and regulatory requirements.

Additionally, companies should leverage technology to streamline policy management. Social media management tools can help automate the monitoring and enforcement of social media policies, ensuring that employees adhere to the guidelines and that any policy violations are promptly addressed. These tools can also provide valuable insights into social media activities, helping companies identify potential risks and take corrective actions before issues escalate.

Ultimately, aligning social media policies with evolving privacy laws requires more than just technical and procedural measures; it requires a cultural shift within the organization. Companies must foster a culture of privacy where data protection is ingrained in every aspect of their operations.

This begins with leadership. Company leaders should set the tone by prioritizing privacy and demonstrating their commitment to data protection. This includes allocating sufficient resources for privacy initiatives, supporting ongoing training and awareness programs, and holding employees accountable for adhering to privacy policies.

Employees at all levels should be encouraged to take ownership of privacy and to integrate data protection practices into their daily activities. This can be achieved through regular communication, recognition of privacy champions, and the establishment of clear channels for reporting privacy concerns. By creating an environment where privacy is valued and respected, companies can ensure that their social media policies are not only compliant but also effective in protecting user data.

The dynamic nature of social media and the evolving landscape of privacy laws present significant challenges for companies. To navigate these challenges, companies must develop and maintain comprehensive social media policies that align with current and future privacy regulations. This involves understanding the regulatory landscape, implementing robust data protection measures, ensuring transparency and accountability, adapting to legal changes, and fostering a culture of privacy within the organization.

By taking a proactive and holistic approach to social media policy management, companies can protect user data, maintain compliance with privacy laws, and build trust with their audiences. As privacy regulations continue to evolve, companies that prioritize data protection and privacy will be better positioned to navigate the complexities of the digital age and to thrive in an increasingly interconnected world.

Key takeaways

- Phishing is no longer limited to email attacks. Social media and text messaging can now be included in the capabilities at risk of exploitation.
- The benefits of organizations maintaining a social media presence must be done with an understanding that such a presence does create additional risks to the organization.
- Countless users continuously fall prey to social media phishing attacks. These exploitations of individuals could also put entire organizations at risk. Individuals and their respective employers should have a strategy to

ensure legitimate communications on social media can be discerned from nefarious communications.

- If organizations choose to have a social media presence, they should consider having a multi-tiered social media strategy. This should include adherence to emerging privacy laws, crisis management plan in the event of becoming the victim of a targeted disinformation campaign (this should be in place whether the company has a social media presence or not), and include a plan to address deepfakes in the event personnel are targeted with deepfakes.

Chapter 9

Technology solutions

Some users have the perception that technology is protecting them from phishing attacks. Intrusion protection devices, intrusion prevention devices, firewalls, anti-virus tools, anti-malware tools, and email monitoring tools are thought to protect the user and the enterprise from various attacks, including phishing attacks. There is also a perception that cybersecurity professionals can quickly restore data and resolve any attack or cybersecurity incident even if something were to affect the systems. This results in a diminished level of awareness and concern related to phishing attacks. Technology plays a significant role in cybersecurity; however, the human component is the most successful and frequent attack.

9.1   Current technical solutions

Cybersecurity best practices use a layered technical solutions approach, not relying on one technical solution for protection. As mentioned above, intrusion prevention devices, firewalls, antivirus tools, anti-malware tools, and email monitoring tools are commonly used together to assist in reducing phishing attacks and their affects. Most current technical protections against phishing attacks are implemented in network devices and user applications which provide content to the users. These network devices and user applications use content filtering, source identification, and provide some post-delivery protection. Many of the technical solutions combine phishing protection along with malware protection.

In Chapter 1, we described phishing as attacks masquerading as legitimate forms of electronic communications, designed to compel a user into taking a specific action that will benefit the attacker. User actions include opening an email message attachment, clicking on a link embedded in the message, or responding to the message with an electronic response providing requested information or verbal communication via a provided phone number. These common phishing attack tactics are also used through other communication-based applications like websites, and instant messages. Any medium that provides the capability of inserting a hyperlink, downloads content to the user's computer, or requests the user to respond to a request with information.

Building on the multifaceted approach to combating phishing attacks, current technological solutions have evolved to include more sophisticated mechanisms that address the ever-changing tactics of cybercriminals. Machine learning (ML) and artificial intelligence (AI) technologies have become instrumental in identifying and mitigating phishing threats. These technologies can analyze patterns and anomalies in emails and web content at a scale and speed unattainable by human oversight alone. By learning from vast datasets of known phishing attempts and malicious behaviors, ML algorithms can predict and flag potential threats before they reach the end user. This proactive stance allows for the dynamic updating of defense mechanisms in response to new and evolving phishing strategies.

Another critical advancement is the implementation of Domain-based Message Authentication, Reporting, and Conformance (DMARC), along with associated standards like Sender Policy Framework (SPF) and DomainKeys Identified Mail (DKIM). These email authentication methods help in verifying the sender's identity, significantly reducing the success rate of phishing emails that rely on spoofing legitimate email addresses. By ensuring that incoming messages are authenticated, organizations can prevent many phishing attempts from reaching their intended targets.

Blockchain technology also presents a novel approach to phishing defense, particularly in verifying the authenticity of websites and digital identities. By creating immutable records of legitimate sites and emails, blockchain can provide a secure and unforgeable reference point for verifying digital content. This approach could significantly reduce the effectiveness of phishing sites that mimic legitimate websites to capture sensitive information.

Browser extensions and web filters have become more advanced, offering users real-time alerts if they attempt to navigate to known phishing sites or download suspicious attachments. These tools rely on constantly updated databases of threat intelligence and can often integrate with other security solutions to provide a comprehensive defense against phishing.

Moreover, the integration of security information and event management (SIEM) systems offers a holistic view of an organization's security posture, correlating data from various sources to identify potential phishing campaigns and coordinate responses across different security technologies.

These technological solutions, when combined with ongoing user education and a robust organizational cybersecurity policy, form a formidable defense against phishing attacks. However, as cybercriminals continue to innovate, the development of new defensive technologies and strategies remains a critical priority for the cybersecurity community. The future of phishing defense will likely involve even more integration of AI and ML, along with the development of more intuitive and user-friendly security tools that can further minimize the risk of successful phishing attacks.

9.1.1   Using AI and ML to defend against phishing

The integration of AI and ML technologies into cybersecurity defenses represents a significant shift in the battle against phishing attacks. These sophisticated technologies offer dynamic, intelligent solutions that adapt and evolve in response to the ever-changing landscape of cyber threats, particularly phishing.

AI and ML algorithms are designed to learn from data, identifying patterns and anomalies that may indicate a phishing attempt. This learning process involves several stages, including data pre-processing, feature extraction, model training, and continuous learning. In the context of phishing defense, these technologies analyze various data points such as email headers, text content, sender reputation, website characteristics, and user behavior patterns.

The integration of AI and ML in phishing detection offers several advantages over traditional approaches. These technologies can analyze large volumes of data, identify patterns and anomalies that may be imperceptible to human analysts, and adapt to evolving tactics employed by cybercriminals. In addition, AI and ML systems can process information in real time, enabling rapid response and mitigation of potential threats.

One of the primary applications of AI and ML in phishing defense is email analysis. Sophisticated algorithms can scrutinize the content, structure, and metadata of incoming emails, detecting subtle indicators of phishing attempts. These indicators may include suspicious URLs, malicious attachments, or language patterns that deviate from the norm. By continuously learning from new data and feedback, these systems can adapt to the ever-changing tactics of cybercriminals, ensuring that they remain effective in identifying even the most sophisticated phishing attempts.

Another area where AI and ML can play a crucial role is in the analysis of user behavior. By monitoring user interactions with emails, websites, and other digital assets, these technologies can detect anomalous behavior that may indicate a potential phishing attack. For instance, if a user clicks on a suspicious link or enters sensitive information on a dubious website, AI and ML systems can flag this activity and initiate appropriate countermeasures, such as blocking access or alerting security personnel.

AI and ML can be employed in the development of advanced phishing simulations and training programs. By analyzing real-world phishing attempts and user responses, these systems can generate realistic scenarios tailored to an organization's specific needs. These simulations can help educate employees on how to identify and respond to phishing attempts, thereby strengthening the overall security posture of the organization.

AI and ML can also be leveraged in proactive threat hunting and intelligence gathering. By analyzing vast amounts of data from various sources, such as dark web forums, social media platforms, and cybercrime reports, these technologies can identify emerging threats and patterns, enabling security teams to stay ahead of potential attacks.

However, it is important to note that AI and ML are not silver bullets in the fight against phishing. These technologies must be integrated into a comprehensive security strategy that includes robust policies, user education, and continuous monitoring and adaptation. Additionally, the effectiveness of AI and ML systems is heavily dependent on the quality and quantity of the data used for training and the expertise of the security professionals responsible for their implementation and maintenance. As cybercriminals continue to evolve their tactics, the role of AI and

ML in phishing defense will become increasingly crucial. By embracing these technologies and fostering collaboration between security experts, researchers, and technology providers, we can develop more effective and resilient defense mechanisms against the ever-present threat of phishing attacks.

The integration of AI and ML in phishing defense represents a significant step forward in the ongoing battle against cybercrime. By leveraging the power of these technologies, organizations can proactively identify and mitigate phishing threats, safeguarding their data, operations, and reputation. However, it is essential to approach AI and ML as part of a comprehensive security strategy, recognizing their strengths and limitations, and continuously adapting to the evolving landscape of cyber threats.

9.1.2   Email and web content analysis

ML models are trained on vast datasets comprising both phishing and legitimate emails or websites. By analyzing the content and structure of these communications, ML algorithms learn to differentiate between benign and malicious intent. For emails, this might involve scrutinizing the presence of suspicious links, unusual sender addresses, or deceptive language patterns. For websites, ML models assess factors like URL structures, page content, and meta-data for signs of phishing.

The integration of AI and ML into email security systems has revolutionized the way organizations defend against phishing and other email-based cyber threats. These advanced technologies have proven to be highly effective in real-world scenarios, providing robust protection that adapts to the ever-evolving landscape of cyber threats. This comprehensive analysis explores the effectiveness of AI-powered email protection systems, drawing on various case studies, expert opinions, and research findings.

Email remains a primary vector for cyberattacks, with phishing being one of the most prevalent and damaging forms of attack. Traditional email security measures, such as spam filters and signature-based detection, have struggled to keep pace with the sophistication of modern phishing tactics. Cybercriminals have become adept at crafting emails that mimic legitimate communications, making it increasingly difficult for conventional security systems to detect and block these threats.

The advent of AI and ML has introduced a new paradigm in email security. These technologies leverage vast amounts of data and advanced algorithms to identify patterns and anomalies that indicate malicious intent. By continuously learning from new data and adapting to emerging threats, AI-powered email protection systems offer a dynamic and proactive defense against phishing and other email-based attacks.

One of the key advantages of AI-powered email protection systems is their ability to automate threat detection and response. These systems can analyze the content, structure, and metadata of incoming emails in real time, identifying potential threats with high accuracy. For instance, AI algorithms can detect suspicious URLs, malicious attachments, and language patterns that deviate from the norm, flagging these emails for further investigation or automatically blocking them from reaching the user's inbox.

In real-world scenarios, this capability has proven invaluable. For example, financial institutions, which are frequent targets of phishing attacks, have successfully implemented AI-powered email security solutions to protect their clients' sensitive information. By leveraging AI, these institutions can detect and mitigate phishing attempts in real-time, preventing unauthorized access to accounts and reducing the risk of financial fraud.

AI and ML algorithms excel at accurately classifying emails as legitimate or malicious, significantly reducing the incidence of false positives and false negatives. Traditional email security systems often struggle with this balance, either allowing malicious emails to slip through or mistakenly flagging legitimate emails as threats. AI-powered systems, however, continuously learn from user interactions and feedback, refining their classification models to improve accuracy over time.

This enhanced classification accuracy is particularly beneficial in high-stakes environments where the cost of a missed threat or a false positive can be substantial. For instance, healthcare organizations, which handle sensitive patient data, have adopted AI-powered email security solutions to ensure that critical communications are not disrupted by false positives while maintaining robust protection against phishing and other email-based threats.

AI-powered email protection systems also leverage behavioral analysis to detect anomalies in email communication patterns and user behavior. By establishing a baseline of normal activity, these systems can identify deviations that may indicate a potential security breach. For example, if a user suddenly starts receiving emails from unfamiliar sources or exhibits unusual email usage patterns, the AI system can flag this activity for further investigation.

In practice, this capability has helped organizations detect and respond to compromised accounts and other security incidents more effectively. For example, a multinational corporation implemented an AI-powered email security solution that monitored user behavior and detected an unusual spike in email forwarding activity. Upon investigation, the security team discovered that a phishing attack had compromised several employee accounts. The AI system's ability to detect this anomaly allowed the organization to respond swiftly, mitigating the impact of the attack and preventing further damage.

AI-powered email protection systems are not limited to reactive measures; they also offer predictive threat response capabilities. By analyzing vast amounts of data and identifying patterns indicative of emerging threats, these systems can proactively defend against new attack vectors. This predictive capability is particularly important in the context of rapidly evolving cyber threats, where traditional security measures may lag behind.

For instance, AI-powered email security solutions can integrate real-time threat intelligence feeds, staying abreast of the latest email-based threats and attack tactics. By correlating external threat data with internal telemetry, these systems can identify and mitigate emerging threats before they can compromise an organization's security. This proactive defense approach has been instrumental in helping organizations stay ahead of cybercriminals and protect their digital assets.

Numerous case studies highlight the effectiveness of AI-powered email protection systems in real-world scenarios. For example, a global financial institution implemented an AI-driven email security solution to combat phishing attacks targeting its customers. The AI system analyzed email content, sender behavior, and contextual information to identify sophisticated phishing attempts. As a result, the institution significantly reduced the incidence of successful phishing attacks, safeguarding its clients' financial information and enhancing overall trust in its services. Another notable example is a healthcare organization that adopted an AI-powered email security platform to protect patient data. The system utilized natural language processing (NLP) techniques to understand the context of emails and detect malicious content. By continuously learning from new data and adapting to evolving threats, the AI system provided robust protection against phishing, ransomware, and other email-based attacks. This implementation not only improved the organization's security posture but also ensured the uninterrupted delivery of critical healthcare services.

While AI-powered email protection systems offer significant advantages, they are not without challenges. One of the primary concerns is the reliance on high-quality data for training AI models. The effectiveness of these systems depends on the richness and volume of the data they are trained on. Organizations must ensure that their AI systems are exposed to a diverse array of real-world attack scenarios to enhance their predictive accuracy and response capabilities.

Another consideration is the need for continuous monitoring and adaptation. Cyber threats are constantly evolving, and AI systems must be regularly updated to stay effective. This requires ongoing collaboration between security experts, data scientists, and technology providers to refine AI models and incorporate the latest threat intelligence.

While AI can significantly reduce the burden on human defenders, it is not a replacement for human expertise. Security teams must remain vigilant and actively involved in the oversight and management of AI-powered systems. Human analysts play a crucial role in interpreting AI-generated alerts, investigating potential threats, and making informed decisions about incident response.

As cyber threats continue to evolve, the role of AI and ML in email security will become increasingly critical. The integration of these technologies into email protection systems represents a significant step forward in the ongoing battle against cybercrime. By leveraging the power of AI, organizations can proactively identify and mitigate phishing threats, safeguarding their data, operations, and reputation.

Looking ahead, we can expect further advancements in AI-powered email security solutions. For instance, the development of more sophisticated NLP techniques will enhance the ability of AI systems to understand and interpret email content, improving their accuracy in detecting malicious intent. Additionally, the integration of AI with other cybersecurity technologies, such as endpoint detection and response (EDR) and SIEM systems, will provide a more comprehensive and cohesive defense against cyber threats.

AI-powered email protection systems have proven to be highly effective in real-world scenarios, offering robust defense against phishing and other email-based attacks. By automating threat detection, enhancing classification accuracy, leveraging behavioral analysis, and providing predictive threat response, these systems empower organizations to stay ahead of cybercriminals and protect their digital assets. However, it is essential to approach AI as part of a comprehensive security strategy, recognizing its strengths and limitations, and continuously adapting to the evolving landscape of cyber threats. With the right combination of AI technology and human expertise, organizations can achieve a more resilient and secure email environment.

9.1.3   AI email security and zero-day attacks

AI-powered email protection systems have become a cornerstone in the defense against zero-day attacks, which are among the most challenging threats in cybersecurity. Zero-day attacks exploit previously unknown vulnerabilities in software, leaving no time for developers to patch the flaws before they are exploited. This makes traditional signature-based detection methods ineffective, as they rely on known threat signatures to identify malicious activity. AI, with its ability to learn and adapt, offers a more dynamic and proactive approach to identifying and mitigating these threats.

AI-powered email protection systems leverage several advanced techniques to detect and respond to zero-day attacks. These systems utilize ML, NLP, and deep learning to analyze vast amounts of data, identify patterns, and detect anomalies that may indicate a zero-day exploit.

One of the primary ways AI handles zero-day attacks is through anomaly detection. AI algorithms establish a baseline of normal email behavior by analyzing historical data and user interactions. When an email deviates from this established norm, the system flags it for further investigation. For instance, if an email contains unusual language patterns, unexpected attachments, or links to unfamiliar domains, the AI system can identify these anomalies as potential indicators of a zero-day attack. This proactive approach allows the system to detect threats that have not yet been cataloged in threat databases.

Deep learning models, such as neural networks, play a crucial role in processing large volumes of data and identifying complex patterns that may be indicative of zero-day attacks. These models can analyze the content and metadata of emails, including the sender's information, the structure of the email, and the context in which it was sent. By doing so, they can detect subtle signs of malicious intent that might be missed by traditional security measures. For example, a deep learning model can identify a phishing email that uses sophisticated social engineering tactics to trick the recipient into clicking on a malicious link.

NLP enhances the ability of AI systems to understand and interpret the content of emails. NLP techniques enable the system to analyze the language used in emails, detect phishing attempts, and identify other content-based threats. For instance, NLP can help detect emails that attempt to impersonate legitimate entities, such as a company's CEO or a trusted vendor, by analyzing the language and tone used in the email. This capability is particularly important in defending against spear-phishing attacks, which are highly targeted and personalized.

Behavioral analysis is another critical component of AI-powered email protection systems. By monitoring user behavior and email interactions, AI can detect unusual activities that may indicate a compromised account or an ongoing attack. For example, if an employee suddenly starts receiving emails from unfamiliar sources or exhibits unusual email usage patterns, the AI system can flag this activity for further investigation. This real-time monitoring and analysis enable organizations to respond swiftly to potential threats, mitigating the impact of zero day attacks.

While AI-powered email protection systems offer significant advantages in defending against zero-day attacks, they are not without limitations. Understanding these limitations is crucial for developing a comprehensive and effective email security strategy.

One of the primary limitations of AI in email security is the potential for false positives and false negatives. Despite their advanced capabilities, AI algorithms may sometimes misinterpret normal behavior as malicious or overlook subtle threats, resulting in unwarranted alerts or missed vulnerabilities. False positives can lead to alert fatigue among security teams, causing them to overlook genuine threats. On the other hand, false negatives can allow malicious emails to slip through the defenses, potentially leading to security breaches. AI systems are also vulnerable to adversarial attacks, where malicious actors manipulate the inputs to the AI system to cause erroneous decisions. For example, an attacker might craft an email that appears benign to the AI system but contains hidden malicious content. This type of attack exploits the AI system's reliance on specific patterns and features to identify threats, potentially bypassing the security measures in place. Adversarial attacks highlight the need for continuous monitoring and updating of AI models to ensure they remain effective against evolving threats.

The effectiveness of AI-powered email protection systems heavily depends on the quality and completeness of the data used for training. Biased or incomplete data can skew AI outcomes, potentially reinforcing existing biases in cybersecurity practices. For instance, if the training data predominantly includes certain types of phishing emails, the AI system may become less effective at detecting other types of threats. Ensuring diverse and representative training data is essential for developing robust AI models that can accurately identify a wide range of email-based threats.

Another limitation is the complexity and interpretability of AI systems. The decision-making processes of AI algorithms can be intricate and difficult for cybersecurity professionals to understand. This lack of transparency can hinder incident response, as security teams may struggle to comprehend why the AI system flagged a particular email as malicious. Enhancing the interpretability of AI models and providing clear explanations for their decisions can help address this challenge and improve the overall effectiveness of AI-powered email protection systems. AI systems also require significant computational resources and expertise to develop, deploy, and maintain. Small and medium-sized businesses, which may lack the necessary resources and expertise, can find it challenging to implement and manage AI-powered email security solutions. This limitation underscores the importance of accessible and affordable AI solutions that can cater to organizations of all sizes.

While AI can significantly reduce the burden on human defenders, it is not a replacement for human expertise. Security teams must remain actively involved in the oversight and management of AI-powered systems. Human analysts play a crucial role in interpreting AI-generated alerts, investigating potential threats, and making informed decisions about incident response. Combining AI technology with human expertise can enhance the overall effectiveness of email security measures.

AI-powered email protection systems have demonstrated remarkable effectiveness in defending against zero-day attacks, leveraging advanced techniques such as anomaly detection, deep learning, NLP, and behavioral analysis. These systems offer a dynamic and proactive approach to identifying and mitigating email-based threats, providing organizations with robust defense mechanisms that adapt to the evolving threat landscape.

However, it is essential to recognize the limitations of AI in email security. False positives and false negatives, vulnerability to adversarial attacks, reliance on high-quality training data, complexity and interpretability, and the need for significant computational resources and expertise are some of the challenges that must be addressed. By understanding these limitations and integrating AI with human expertise, organizations can develop comprehensive and effective email security strategies that leverage the strengths of AI while mitigating its weaknesses.

As cyber threats continue to evolve, the role of AI in email security will become increasingly critical. By embracing AI technology and fostering collaboration between security experts, researchers, and technology providers, organizations can enhance their email security posture and protect their digital assets from the ever-present threat of zero-day attacks.

9.1.4 *Integrating AI-powered email protection systems*

AI-powered email protection systems have become an integral part of modern cybersecurity strategies, offering advanced capabilities to detect, analyze, and mitigate email-based threats. These systems do not operate in isolation; instead, they are designed to seamlessly integrate with existing cybersecurity measures, creating a cohesive and comprehensive defense mechanism. This integration enhances the overall security posture of organizations by enabling coordinated responses to threats and leveraging the strengths of various security tools.

One of the primary ways AI-powered email protection systems integrate with existing cybersecurity measures is through their connection with endpoint security solutions. Endpoint security tools, such as antivirus software and EDR systems, are critical for protecting individual devices within an organization. By integrating with these tools, AI-powered email security systems can provide a more holistic view of potential threats.

For instance, when an AI system detects a suspicious email, it can communicate with endpoint security solutions to ensure that any malicious attachments or links are blocked at the device level. This integration allows for real-time threat mitigation, preventing malware from spreading across the network. Additionally, endpoint security tools can provide valuable telemetry data to the AI system, enhancing its ability to detect and analyze threats. This bidirectional flow of information ensures that both the email security system and the endpoint security tools are working in concert to protect the organization.

Network firewalls are another critical component of an organization's cybersecurity infrastructure. These devices monitor and control incoming and outgoing network traffic based on predetermined security rules. AI-powered email protection systems can integrate with network firewalls to enhance their ability to detect and block malicious traffic. For example, if an AI system identifies an email containing a link to a known malicious domain, it can communicate this information to the network firewall. The firewall can then block any attempts to access that domain, preventing users from inadvertently visiting harmful websites. This integration not only stops threats at the email level but also ensures that network-level defenses are aware of and can respond to emerging threats. By sharing threat intelligence between the email security system and the network firewall, organizations can create a more robust and responsive security environment.

SIEM systems play a crucial role in aggregating and analyzing security data from various sources within an organization. These systems provide a centralized platform for monitoring, detecting, and responding to security incidents. AI-powered email protection systems can integrate with SIEM platforms to enhance their threat detection and response capabilities.

When an AI system detects a potential threat in an email, it can send detailed logs and alerts to the SIEM system. This integration allows security analysts to correlate email-based threats with other security events occurring within the organization. For instance, if a phishing email is detected, the SIEM system can cross-reference this event with login attempts, network traffic, and other security logs to identify any related malicious activity. This comprehensive view enables security teams to understand the full scope of an attack and respond more effectively.

SIEM systems can provide historical data that AI-powered email protection systems can use to refine their detection algorithms. By analyzing past incidents and patterns, AI systems can improve their ability to identify similar threats in the future. This continuous learning process ensures that the email security system remains effective against evolving threats.

AI-powered email protection systems also integrate with threat intelligence feeds to stay updated on the latest cyber threats. Threat intelligence feeds provide real-time information about known malicious domains, IP addresses, and other indicators of compromise. By incorporating this data, AI systems can enhance their ability to detect and block emerging threats.

For example, if a threat intelligence feed reports a new phishing campaign targeting specific industries, the AI-powered email protection system can use this information to update its detection rules. This proactive approach allows the system to identify and block phishing emails associated with the campaign before they reach users' inboxes. Additionally, the integration with threat intelligence feeds enables the AI system to share its findings with other security tools, such as firewalls and SIEM systems, ensuring a coordinated defense across the organization.

One of the significant advantages of AI-powered email protection systems is their ability to automate incident response processes. By integrating with other cybersecurity tools, these systems can streamline the detection, analysis, and remediation of email-based threats. For instance, when an AI system identifies a malicious email, it can automatically quarantine the email, block any associated attachments or links, and trigger alerts to the security team. This automated response reduces the burden on security teams and accelerates the mitigation of threats. Additionally, the AI system can communicate with other security tools to ensure a coordinated response. For example, if a phishing email is detected, the AI system can instruct the endpoint security solution to scan affected devices for any signs of compromise. Similarly, it can update the network firewall to block any related malicious domains. This integrated approach ensures that all security measures are working together to protect the organization.

AI-powered email protection systems can also integrate with authentication and access control mechanisms to enhance security. For instance, these systems can work with multi-factor authentication (MFA) solutions to verify the identity of users before granting access to sensitive information. By analyzing email communication patterns and user behavior, AI systems can detect anomalies that may indicate a compromised account.

If an AI system identifies suspicious activity, such as an unusual login attempt or an email containing a request for sensitive information, it can trigger additional authentication steps. This integration ensures that only authorized users can access critical resources, reducing the risk of unauthorized access and data breaches. Additionally, AI systems can provide insights into user behavior that can help organizations refine their access control policies and improve overall security.

While the integration of AI-powered email protection systems with existing cybersecurity measures offers significant benefits, it is not without challenges. One of the primary challenges is the complexity of integrating AI systems with diverse security tools and platforms. Organizations must ensure that their AI systems can communicate effectively with other security measures, which may require custom configurations and ongoing maintenance.

Data privacy and compliance are also critical considerations. AI-powered email protection systems must handle sensitive information responsibly and comply with regulatory standards. Organizations must ensure that their AI systems are designed to protect user data and maintain compliance with legal requirements.

Human oversight and interpretability are essential for the effective use of AI in email security. Security teams must be able to understand and validate the decisions made by AI systems. Ensuring that AI algorithms are interpretable and providing clear explanations for their actions can help security teams make informed decisions and respond effectively to threats.

AI-powered email protection systems have transformed the landscape of email security by offering advanced capabilities to detect, analyze, and mitigate email-based threats. By integrating with existing cybersecurity measures, such as endpoint security solutions, network firewalls, SIEM systems, and threat intelligence feeds, these systems create a cohesive and comprehensive defense mechanism. This integration enhances the overall security posture of organizations, enabling coordinated responses to threats and leveraging the strengths of various security tools.

However, successful integration requires careful planning, expertise, and ongoing maintenance. Organizations must address challenges related to integration complexity, data privacy, and the need for human oversight. By understanding these challenges and leveraging the strengths of AI-powered email protection systems, organizations can enhance their email security measures and protect against evolving cyber threats. As the threat landscape continues to evolve, the integration of AI with existing cybersecurity measures will remain a critical component of effective email security strategy.

9.1.5 Challenges and considerations

While AI and ML technologies offer promising solutions to phishing defense, they are not without challenges. The quality of the underlying data is crucial; models trained on outdated or biased datasets may not perform effectively against new or sophisticated attacks. Additionally, the complexity of these technologies requires specialized expertise, posing a barrier to implementation for some organizations.

Moreover, cybercriminals are also leveraging AI and ML to enhance their phishing campaigns, creating a perpetual arms race between attackers and defenders. This dynamic underscores the need for continuous investment in research and development to maintain the effectiveness of AI and ML in phishing defense.

Looking ahead, the integration of AI and ML in phishing defense is set to become more sophisticated. Future developments may include greater personalization of security measures, where AI systems tailor their defense mechanisms to the specific risk profile of individual users or organizations. Additionally, advances in NLP could improve the detection of phishing attempts in text-based communications, further enhancing the accuracy of phishing defenses.

AI and ML technologies are transforming the landscape of phishing defense, offering adaptive, intelligent solutions to counteract the evolving threat of phishing attacks. Despite the challenges, the potential of these technologies to significantly bolster cybersecurity defenses is undeniable. As cyber threats continue to advance, the strategic deployment of AI and ML will be crucial in safeguarding digital assets and information in an increasingly interconnected world.

9.2   More technologies to protect email systems

While phishing attacks have evolved into various areas of electronic communications, email remains a critical communication tool for individuals and organizations alike. Its widespread use continues to make it a prime target for phishing attacks. To combat these threats, several email authentication technologies have been developed, including DMARC, SPF, and DKIM. These technologies work individually and collectively to validate the authenticity of email messages, thereby protecting against phishing and other email-based attacks.

9.2.1 Sender Policy Framework

SPF is an email validation system designed to prevent email spoofing by verifying the sender's IP address. It allows the domain owner to specify which mail servers are authorized to send emails on behalf of their domain. When an email is received, the receiving server checks the SPF record in the DNS to verify that the email comes from an authorized server. If the check fails, the email can be rejected, quarantined, or flagged as suspicious, depending on the receiving server's policy. SPF helps to ensure that receivers can identify and block emails that impersonate domains but does not validate the header information that users typically see, such as the "From" address.

9.2.2 DomainKeys Identified Mail

DKIM provides a way to validate a domain name identity that is associated with a message through cryptographic authentication. It enables an organization to take responsibility for a message in a way that can be validated by the recipient. DKIM uses a pair of cryptographic keys: a private key to digitally sign outgoing messages and a public key published in the DNS that recipients use to verify the signature. This verification process helps to ensure that the content has not been altered in transit, thereby protecting against tampering and spoofing. While DKIM secures the content and provides a method to verify the sender's identity, it does not specify any actions for the receiving server to take if a message fails the validation.

9.2.3 Domain-based Message Authentication, Reporting, and Conformance

DMARC builds upon SPF and DKIM, providing additional features that help domain owners control how their email is processed if it fails authentication tests. It allows domain owners to publish policies in their DNS records that dictate how receiving servers should handle emails that don't pass SPF or DKIM checks. Options include reporting the failure, quarantining the message, or outright rejection. DMARC also includes reporting capabilities, enabling domain owners to receive feedback on the number of messages passing or failing DMARC evaluation, as well as information on potential authentication issues. This feedback loop is crucial for organizations to monitor and protect their domain from unauthorized use, providing insights into attack patterns and helping to refine security measures.

9.2.4 Implementation challenges and considerations

Implementing SPF, DKIM, and DMARC effectively requires careful planning and ongoing management. For SPF, organizations must ensure that all legitimate email sources are included in their SPF records to avoid legitimate emails being flagged as spam. However, the SPF protocol has a limit on the number of DNS lookups it can perform, which can complicate configurations for organizations using multiple email services.

DKIM implementation involves managing cryptographic keys, which can be challenging, especially for organizations with many domains. Key management practices, including regular rotation and safeguarding of private keys, are critical to maintaining the integrity of the DKIM authentication process.

DMARC adoption brings its own challenges, particularly in fine-tuning policies to balance security and deliverability. Organizations often start with a "none" or "monitor" policy to assess the impact of DMARC on their email flow before moving to more restrictive policies like "quarantine" or "reject." Careful monitoring and adjustment of policies based on DMARC reports are essential to avoid disrupting legitimate email communications.

9.2.5   The way forward

The combined use of SPF, DKIM, and DMARC represents a comprehensive approach to securing email communications against phishing and spoofing attacks. While no single technology can eliminate the threat of phishing entirely, together, they significantly enhance an organization's ability to defend against email-based security threats. Continuous monitoring, management, and adjustment of these technologies, along with user education on recognizing and responding to phishing attempts, are crucial components of a robust cybersecurity strategy. As cyber threats evolve, so too must the defenses against them, with SPF, DKIM, and DMARC serving as foundational elements in the ongoing effort to secure the email ecosystem.

9.3   Blockchain technology

Blockchain technology, renowned for its robust security features inherent in its decentralized nature and cryptographic algorithms, is increasingly being explored as a novel approach to fortifying email communications against phishing attacks. By leveraging blockchain, it's possible to introduce an unprecedented level of verification and trust into email systems, thereby mitigating the risk of phishing attempts that prey on the vulnerabilities of traditional email infrastructure.

9.3.1   Blockchain fundamentals in email security

At its core, blockchain is a distributed ledger technology that records transactions across multiple computers in such a way that the registered transactions cannot be altered retroactively. This characteristic, combined with cryptographic security, makes blockchain an attractive solution for enhancing email security.

1. **Immutable record-keeping**: Blockchain's ability to maintain a tamper-proof ledger of transactions can be applied to email communications to ensure the integrity of sent and received messages. By storing hashes of emails on a blockchain, it becomes possible to verify if an email has been altered or tampered with from the time it was sent. This immutability can be crucial in detecting and preventing phishing attempts where attackers might try to modify the contents of an email to deceive the recipient.
2. **Decentralized architecture:** Traditional email systems rely on centralized servers that can be a single point of failure and a target for phishing attacks.

Blockchain's decentralized nature distributes the data across numerous nodes, making it exceedingly difficult for attackers to compromise the integrity of the email system. This decentralization not only enhances security but also improves system resilience against attacks.

3. **Enhanced authentication:** Blockchain can be utilized to create decentralized identities, enabling a more secure and verifiable way of authenticating the sender and recipient of an email. Through the use of public-private key cryptography, each participant in the blockchain network can securely prove their identity, reducing the risk of impersonation and spoofing attacks common in phishing.

9.3.2   Implementing blockchain for email security

The implementation of blockchain technology in email systems can take several forms, from enhancing existing email protocols to developing entirely new blockchain-based email platforms.

1. **Blockchain-based email platforms:** New email platforms built on blockchain technology inherently incorporate the security features of blockchain, offering end-to-end encryption, immutable storage of email data, and secure identity verification mechanisms. These platforms can significantly reduce the susceptibility of email communications to phishing, spoofing, and other cyber threats.
2. **Integrating blockchain with existing email systems:** For broader adoption, blockchain can be integrated with existing email protocols and infrastructure. This could involve using blockchain to store cryptographic hashes of emails to verify their integrity or employing blockchain-based identity management systems to authenticate email senders and recipients.

9.3.3   Challenges and considerations

While blockchain presents a promising solution to enhancing email security, several challenges need to be addressed for widespread adoption. These include scalability issues, as the size of the blockchain could grow significantly with the addition of email data. Additionally, the integration of blockchain technology with existing email systems and protocols requires careful consideration to ensure compatibility and user accessibility.

Moreover, the success of blockchain in preventing phishing attacks also depends on user awareness and behavior. Educating users on the new mechanisms and practices associated with blockchain-based email systems is essential to maximizing their potential in combating phishing.

As blockchain technology continues to mature and evolve, its application in securing email communications against phishing attacks offers a glimpse into the future of cybersecurity. Innovations in blockchain scalability, interoperability, and user interface design are likely to further enhance its suitability for email security applications.

Leveraging blockchain technology presents a novel and potentially transformative approach to safeguarding email communications against the ever-present threat of phishing attacks. By ensuring the integrity, authenticity, and confidentiality of email messages, blockchain can significantly reduce the efficacy of phishing campaigns, marking a significant step forward in the ongoing battle against cybercrime.

9.4   Browser extensions and web filters

Browser extensions add functionality to web browsers and can play a crucial role in identifying and blocking phishing attempts. These extensions work by analyzing the websites that users attempt to visit, comparing them against databases of known phishing sites, and alerting users or blocking access to these sites to prevent potential data breaches. They can also scrutinize website certificates and security features to assess their legitimacy, offering an additional layer of protection against counterfeit sites that may be used in phishing campaigns.

9.4.1   Real-time URL scanning

Browser extensions often include features that scan URLs in real time, evaluating them for patterns or characteristics common to phishing sites. This might include analysis of domain names that are slight misspellings of legitimate sites or URLs that incorporate misleading subdomains.

9.4.2   Link preview and verification

Some extensions provide previews of links found in emails, showing the actual URL and a rating of its safety. This feature helps users to visually verify links before clicking on them, reducing the risk of inadvertently accessing malicious sites.

9.4.3   Phishing site reporting

Many browser extensions allow users to report suspected phishing sites. This crowdsourced approach not only aids the individual user but also contributes to a broader community effort in identifying and neutralizing phishing threats.

9.5   Web filters for enhanced email security

Web filters, often part of broader cybersecurity solutions, regulate the content that users can access by blocking websites known to be malicious or unsuitable. When integrated with email systems, web filters scrutinize links contained within emails, preventing users from accessing phishing sites and thereby significantly reducing the risk of credential theft or malware infection.

9.5.1   Dynamic blacklisting

Utilizing continuously updated lists of known phishing and malicious websites, web filters can dynamically block access to these sites. This blacklist approach is crucial for adapting to the rapidly evolving nature of phishing threats.

9.5.2   Content analysis

Beyond simple URL filtering, advanced web filters analyze the content of web pages for suspicious elements, such as fake login forms or requests for personal information, which are indicative of phishing sites.

9.5.3   Secure email gateways

Web filters can be part of secure email gateway solutions that inspect incoming emails for malicious links and attachments, offering a comprehensive defense mechanism that filters out phishing emails before they reach the user's inbox.

9.6   Challenges and considerations

While browser extensions and web filters are indispensable tools in the fight against phishing, they are not without limitations. Cybercriminals continually devise new methods to circumvent these defenses, necessitating constant updates and refinements to these tools. Moreover, reliance on these technologies should not lead to complacency among users; education on the importance of vigilance and skepticism in digital communications remains critical.

Additionally, the effectiveness of browser extensions and web filters depends on their configurability and the accuracy of their threat detection algorithms. False positives, where legitimate sites are incorrectly blocked, can hinder productivity and user experience. Balancing security with usability is a key consideration in the deployment of these tools.

As phishing techniques become increasingly sophisticated, the role of browser extensions and web filters in safeguarding email systems will continue to grow in importance. Future advancements in AI and ML promise to enhance the capabilities of these tools, offering more personalized and proactive protection based on user behavior and risk profiles.

Browser extensions and web filters represent critical components of a multi-layered cybersecurity strategy, providing essential defenses against phishing attacks on email systems. By combining these technologies with ongoing user education and other cybersecurity measures, organizations can significantly mitigate the risks associated with phishing, protecting their data and digital assets in an increasingly perilous cyber landscape.

9.7   Security information and event management systems

SIEM systems represent a comprehensive cybersecurity solution to include phishing attacks on email systems. SIEM systems offer a sophisticated approach to detecting, analyzing, and responding to cybersecurity incidents. By aggregating and analyzing log data across an organization's technology infrastructure, SIEMs provide the visibility and intelligence needed to identify and mitigate threats before they can cause significant harm.

9.7.1   Centralized visibility and real-time monitoring

At the heart of a SIEM system's defense against phishing attacks is its ability to provide centralized visibility across the organization's entire digital environment. This includes network devices, servers, endpoints, and applications. By collecting and correlating logs from these disparate sources in real-time, SIEM systems can detect anomalous activities that may indicate a phishing attempt. For instance, if a user clicks on a phishing link and unknowingly downloads malware, the SIEM can detect the resultant unusual network traffic or unauthorized access attempts, triggering alerts for immediate investigation.

9.7.2   Advanced analytics and threat detection

SIEM systems utilize advanced analytics, incorporating both rule-based and behavioral algorithms, to identify patterns of activity that deviate from established baselines. This capability is crucial for detecting sophisticated phishing attacks that may not be caught by traditional security measures. Through the application of ML and statistical analysis, SIEMs can learn over time, continually improving their accuracy in identifying phishing-related anomalies, such as suspicious email flows or login attempts from compromised accounts.

9.7.3   Incident response and automation

When a potential phishing attack is detected, the SIEM system plays a critical role in orchestrating a swift and effective response. This often involves automating certain response actions, such as isolating affected systems, blocking malicious IP addresses, or disabling compromised user accounts. Additionally, SIEMs can provide security teams with detailed incident context and analysis, enabling them to make informed decisions and take manual intervention steps if necessary. This combination of automated and manual response capabilities ensures that phishing attacks can be contained and remediated with minimal delay.

9.7.4   Compliance and reporting

Phishing attacks often target sensitive data, putting organizations at risk of compliance violations. SIEM systems assist in maintaining regulatory compliance by providing comprehensive logging, monitoring, and reporting capabilities. They can generate detailed reports on security incidents, including phishing attacks, demonstrating to regulators and stakeholders that appropriate measures are in place to protect sensitive information. Furthermore, the insights gained from SIEM reporting can guide ongoing improvements to an organization's phishing defense strategies.

9.7.5   Challenges and future directions

While SIEM systems offer robust defenses against phishing attacks, they are not without challenges. The effectiveness of an SIEM system depends heavily on the quality of its configuration and the skills of the security analysts operating it. There is also the issue of alert fatigue, where the high volume of alerts generated by the system can overwhelm security teams, potentially causing critical warnings to be overlooked.

Looking ahead, the integration of AI and ML technologies into SIEM systems holds promise for enhancing their capabilities. These advancements could lead to more accurate threat detection, reduced false positives, and even more sophisticated automated response actions. As these technologies evolve, SIEM systems will likely become an even more integral component of organizational defenses against phishing and other cybersecurity threats.

SIEM systems represent a powerful tool in the fight against phishing attacks, offering comprehensive capabilities for monitoring, detection, response, and compliance. By leveraging the full potential of SIEM technology, organizations can significantly enhance their cybersecurity posture, protecting themselves against the damaging consequences of phishing attacks.

9.8   Remote Browser Isolation

Remote Browser Isolation (RBI) technology represents a significant advancement in cybersecurity strategies, especially in safeguarding against phishing attacks. By directing web activities through an isolated external browser, RBI ensures that any potential malware from compromised websites does not reach the organization's internal networks. This isolation mechanism significantly mitigates the risk associated with clicking on malicious links, one of the most common vectors for phishing and malware distribution.

In addition to isolating the browsing session, RBI systems often employ advanced content filtering techniques to scrutinize and sanitize the web content before it is rendered to the user. This process involves stripping out potentially harmful elements such as JavaScript, which is commonly used in web-based attacks, thereby delivering only the safe content to the user's screen. By doing so, RBI not only protects against direct malware infiltration but also shields the user from web-based exploits and phishing attempts designed to execute malicious code in the browser.

To further enhance the protective capabilities against phishing attacks, RBI solutions can be integrated with threat intelligence platforms. This integration allows for real-time analysis of URLs against known phishing sites and malicious domains. If a user attempts to access a website that is flagged by the threat intelligence feed, the RBI system can block access to the site or alert the user to the potential danger. This proactive approach helps in preventing phishing attempts from reaching the user in the first place, significantly reducing the risk of information theft or compromise.

While RBI provides a robust technical solution to reduce the risk of phishing attacks, it is also essential to complement this technology with ongoing user education and awareness programs. Educating users about the dangers of phishing and the importance of vigilance when interacting with emails and websites is crucial.

This education can be enhanced by incorporating RBI technology into cybersecurity training, demonstrating to users how RBI works and why it is an effective tool in the cybersecurity arsenal. By fostering a culture of security awareness, organizations can empower their users to act as an additional layer of defense against phishing attacks.

Implementing RBI technology, while significantly enhancing security, can sometimes impact the user experience. The process of rendering web content through an isolated browser may introduce latency, affecting the responsiveness of web applications. Moreover, certain web functionalities might be limited or behave differently in an isolated environment, potentially causing confusion or frustration for users. Addressing these challenges requires careful configuration of RBI settings to balance security and usability, ensuring that protective measures do not unduly hinder productivity.

RBI technology offers a powerful means of protecting against phishing attacks by isolating web content and reducing the attack surface. When combined with MFA, content filtering, and threat intelligence, RBI forms a comprehensive defense mechanism. However, the effectiveness of RBI and similar technologies depends on their integration into a broader cybersecurity strategy that includes user education, awareness, and attention to user experience. As cyber threats continue to evolve, leveraging advanced technologies like RBI, in concert with fostering a culture of cybersecurity awareness, will be key to maintaining robust defenses against phishing and other cyberattacks.

9.9   End-point delivery protection

End-point delivery protections play a pivotal role in fortifying an organization's defenses against phishing attacks, a prevalent threat vector in the cybersecurity landscape. As end-point devices—such as laptops, desktops, smartphones, and tablets—represent primary interfaces for users to interact with digital environments, they also emerge as prime targets for phishing schemes. Protecting these end points, therefore, is crucial in mitigating the risks associated with phishing and ensuring the integrity of an organization's information systems.

9.9.1   The role of Endpoint Protection Platforms

Endpoint Protection Platforms (EPPs) are comprehensive security solutions designed to detect, investigate, block, and contain malicious activities on end-point devices. EPPs integrate a variety of security technologies, including antivirus, anti-malware, personal firewalls, and intrusion prevention systems, to provide a multi-layered defense mechanism. By leveraging signature-based detection to identify known threats, along with heuristic and behavior-based detection for unknown or emerging threats, EPPs can effectively protect end points from being compromised by phishing attacks.

One of the critical functions of EPPs in combating phishing is their ability to scan email attachments and links for malicious content before these can cause harm. Given that phishing attacks often employ deceptive emails as their delivery mechanism, having a robust EPP in place ensures that malicious payloads are intercepted and neutralized at the end point. Moreover, EPPs can enforce security policies that prevent users from accessing known phishing sites or downloading suspicious files, further reducing the risk of compromise.

9.9.2 Application control and whitelisting

Another dimension of end-point delivery protection involves application control and whitelisting, which ensure that only trusted applications are allowed to execute on end-point devices. This approach is particularly effective against phishing attacks that rely on social engineering to persuade users to install malicious software. By maintaining a curated list of approved applications, organizations can prevent the execution of unauthorized or malicious programs, thereby safeguarding end points from being exploited by attackers.

9.9.3 Endpoint detection and response

EDR systems represent an evolution of end-point security, focusing on the detection of and response to cyber threats in real time. EDR solutions continuously monitor end-point devices for suspicious activities, leveraging advanced analytics and ML to identify patterns indicative of a phishing attack. Upon detecting a potential threat, EDR systems can automatically respond by isolating the affected device, preventing the spread of the attack, and facilitating the investigation and remediation process.

EDR's emphasis on continuous monitoring and real-time response is critical in the context of phishing attacks, which often require swift action to mitigate potential damage. Additionally, the forensic capabilities of EDR solutions enable organizations to analyze phishing attacks post-incident, gathering insights that can inform future security strategies and enhance resilience against similar threats.

9.9.4 User and entity behavior analytics

Incorporating user and entity behavior analytics (UEBA) into end-point delivery protections offers another layer of defense against phishing. UEBA tools analyze the normal behavior patterns of users and devices, establishing baselines that can be used to identify anomalous activities indicative of a phishing attack or compromise. For instance, unusual login attempts, data exfiltration patterns, or deviations in email usage could signal a successful phishing breach, triggering alerts and remediation actions.

UEBA's strength lies in its ability to detect subtle signs of compromise that may elude other security measures, providing a nuanced view of end-point security that complements traditional protection methods.

9.9.5 Challenges and future directions

While end-point delivery protections offer robust defenses against phishing attacks, they are not without challenges. The dynamic nature of phishing tactics necessitates continuous updates and adaptations of security solutions. Moreover, the growing sophistication of attacks, including polymorphic malware and advanced social engineering techniques, underscores the need for ongoing innovation in end-point security technologies.

Looking ahead, the integration of AI and ML into end-point delivery protections promises to enhance their effectiveness further. These technologies can improve detection accuracy, reduce false positives, and enable adaptive responses to evolving threats. As end-point devices continue to proliferate and serve as gateways to sensitive information, the strategic implementation of end-point delivery protections will remain a cornerstone of comprehensive cybersecurity strategies, essential for safeguarding organizations in the face of persistent phishing threats.

9.10   Multi-factor authentication

MFA has emerged as a critical security measure in the fight against phishing attacks, offering an added layer of defense that can significantly mitigate the risk of unauthorized access to sensitive information. Phishing attacks, which often aim to steal usernames and passwords, can be substantially less effective when MFA is in place. This security strategy requires users to provide two or more verification factors to gain access to an account, system, or network, making it much more challenging for attackers to breach defenses with stolen credentials alone.

9.10.1   The mechanism of multi-factor authentication

MFA protects against phishing by requiring additional proof of identity beyond just a password. These authentication factors are typically categorized into something the user knows (like a password or PIN), something the user has (such as a mobile device or security token), and something the user is (incorporating biometrics such as fingerprints or facial recognition). By combining these different categories of authentication, MFA creates a multi-layered security barrier that significantly enhances account security.

When a phishing attack successfully deceives a user into revealing their password, the presence of MFA can still prevent unauthorized access. The attacker, not in possession of the second factor—be it a physical token or a biometric feature—is blocked from further action. This critical step in the authentication process serves as a robust deterrent against the most common goal of phishing attacks: the exploitation of stolen credentials.

9.10.2   Enhanced security with adaptive multi-factor authentication

Advancements in MFA technology have led to the development of adaptive or risk-based MFA, which adjusts the authentication requirements based on the perceived risk of the access request. For example, a login attempt from an unfamiliar device or geographic location might trigger additional authentication challenges.

This adaptive approach is particularly effective against sophisticated phishing schemes that may employ stolen device information or manipulate IP addresses to appear more legitimate.

Adaptive MFA utilizes contextual information—such as time of access, device used, and network security—to dynamically assess the risk level of each login attempt. This means that in situations where there is a high risk of a phishing attack, the system can enforce stricter authentication measures, adding an extra layer of protection against unauthorized access.

9.10.3   *The role of MFA in a comprehensive security strategy*

While MFA is a powerful tool in the fight against phishing, it should not be relied upon as a single line of defense. Cybercriminals are continually evolving their tactics, and there have been instances where MFA methods were bypassed through sophisticated attacks, such as real-time phishing and SIM swap scams. Therefore, MFA is most effective when integrated into a comprehensive security strategy that includes regular security awareness training, encrypted communications, secure password practices, and the use of anti-phishing tools.

Educating users on the importance of MFA and how to correctly use it is also crucial. Users should be made aware of the types of information they should never share, including MFA codes, and be trained to recognize the signs of phishing attempts. This education, combined with a culture of security within the organization, significantly reduces the likelihood of successful phishing attacks.

9.11   Challenges and future directions

The implementation of MFA is not without its challenges. User convenience is often cited as a concern, as the additional authentication steps can be viewed as cumbersome or time-consuming. However, advancements in technology are continuously improving the user experience, making MFA more seamless and integrated into the user's workflow. Biometric authentication, in particular, offers a balance between strong security and user convenience.

As phishing attacks become more sophisticated, the development of MFA technologies will continue to evolve in response. Future directions may include the integration of AI and ML to further refine adaptive MFA systems, making them more intuitive and effective at detecting and responding to phishing threats.

MFA stands as a crucial component of modern cybersecurity defenses, offering a significant impediment to phishing attacks. By requiring multiple forms of verification, MFA effectively protects against the unauthorized use of stolen credentials, providing a critical security layer in the increasingly complex digital landscape. As part of a holistic security strategy, MFA plays a vital role in safeguarding sensitive information and systems against the pervasive threat of phishing.

9.12   Segmentation of user functions

User segmentation, a cybersecurity strategy that involves dividing network users into distinct groups based on their roles, access needs, and risk profiles, emerges as a potent mechanism for enhancing an organization's defense against phishing attacks. By applying the principles of least privilege and network segmentation, organizations can significantly minimize the impact of successful phishing attempts, limiting attackers' access to sensitive information and critical systems. This tailored approach to network access and rights management not only bolsters security but also aligns with best practices for data protection and regulatory compliance.

9.12.1   *Implementing user segmentation for enhanced security*

At its core, user segmentation involves categorizing users into segments or groups and assigning them access rights strictly based on the necessities of their job functions. This strategy prevents the over-provisioning of access rights, a common vulnerability that attackers exploit in phishing campaigns. For instance, a user in the marketing department would have access to social media tools and marketing data but not to financial systems or HR records. This compartmentalization ensures that if a phishing attack compromises a user's credentials, the breach's impact is contained within the confines of that user's access rights, significantly reducing the potential damage.

Building on the foundation of user segmentation, dynamic access controls and risk-based authentication processes can further refine the security posture. Dynamic access controls adjust users' access rights based on contextual factors such as the user's current location, the device being used, and the sensitivity of the requested data. Similarly, risk-based authentication may require additional verification steps under circumstances deemed high risk, such as login attempts from unusual locations or times. These adaptive security measures ensure that even if a user segment is targeted in a phishing attack, the authentication and access protocols dynamically adjust to maintain a heightened security level.

The zero trust security model, which operates under the assumption that threats could originate from anywhere and thus, no user or device should be automatically trusted, complements user segmentation effectively. By applying zero trust principles, organizations require continuous verification of all users' credentials and privileges, regardless of their network location. This model dovetails with user segmentation by enforcing strict access controls and minimizing lateral movement within the network, a common tactic used by attackers post-phishing breach. Zero trust architectures make it significantly more difficult for attackers to exploit compromised accounts, further safeguarding sensitive information and critical infrastructure.

While user segmentation significantly enhances an organization's defensive posture against phishing attacks, its effectiveness is amplified when combined with comprehensive user awareness and training programs. Educating users about the risks associated with phishing, the importance of safeguarding their credentials, and the procedures for reporting suspicious emails can drastically reduce the likelihood of successful attacks. Training should be tailored to the specific risks and responsibilities associated with each user segment, emphasizing the role of each user in the organization's broader cybersecurity strategy.

9.12.2   Challenges and considerations

Implementing user segmentation requires a thorough understanding of the organization's data flows, user roles, and access needs, which can be complex and time-consuming. It also necessitates ongoing management and review to adapt to changes in organizational structure, user roles, and emerging threats. Moreover, ensuring a seamless user experience while maintaining strict access controls presents a challenge, as overly restrictive policies can hinder productivity and user satisfaction.

As organizations continue to evolve and the threat landscape becomes increasingly sophisticated, automating user segmentation processes with the aid of AI and ML will become more prevalent. AI and ML can analyze user behavior, identify anomalies, and automatically adjust access controls in real time, offering a dynamic and proactive approach to minimizing the risk of phishing attacks. This evolution toward intelligent, automated user segmentation represents a significant advancement in cybersecurity, promising enhanced protection for organizations in an ever-connected world.

User segmentation stands as a critical element in a comprehensive strategy to protect against phishing attacks. By limiting access to sensitive information and critical systems based on users' specific roles and needs, organizations can significantly mitigate the impact of potential breaches. When integrated with broader cybersecurity measures, including zero trust architectures, risk-based authentication, and user training, user segmentation provides a robust defense mechanism against the pervasive threat of phishing.

9.13   Bringing it together – Microsoft Office 365 Advanced Threat Protection service

Microsoft is actively enhancing security through its Office 365 Advanced Threat Protection (ATP) service. This service leverages ML to monitor email traffic patterns and identify anomalies for each domain. If unusual email traffic is detected, the system can block these potentially harmful emails. The service offers several key features:

Microsoft Office 365 ATP analyzes email attachments and links in a secure sandbox environment. Here, ML models evaluate the content for any malicious elements and conduct a thorough inspection of links to identify and block harmful attachments and links.

The Defender for Office 365 includes features such as edge protection, sender intelligence, content filtering, and post-delivery protection. These features use AI to scrutinize content for suspicious elements and identify attack patterns. This helps in recognizing attack campaigns designed to slip past typical security measures.

To ensure the authenticity of the sender, Microsoft's tools verify the identity of the person, brand, and domain sending the email. This process involves authenticating the source to safeguard against spoofing or Business Email Compromise attacks, ensuring the sender is genuinely who they claim to be.

Microsoft applies these security measures to both internal and external emails. For emails within an organization's domains, anti-spoofing technology is used to confirm that the emails are genuinely originating from within the organization.

For emails from external domains, Microsoft's spoof intelligence assesses whether the domain adheres to SPF, DKIM, and DMARC standards. If discrepancies are found, the system learns the domain's usual message-sending patterns to identify spoofed emails.

SPF records are utilized to specify which mail servers and domains are authorized to send emails on behalf of the organization. This is a part of the DNS settings for the organization's domain.

To prevent impersonation, especially of high-profile users within an organization, Microsoft employs mailbox intelligence. This uses an ML model to create a contact graph, outlining normal communication patterns and thereby detecting any impersonation attempts of trusted contacts.

The service also includes Safe Links, which offers real-time protection by checking the reputation of links at the time they are clicked. This feature extends to internal messages, ensuring comprehensive protection against malicious links.

Microsoft's approach, combining advanced ML and AI technologies, offers a robust defense mechanism against phishing attacks, protecting users from malicious emails, links, and attachments. This multi-layered defense strategy underscores the importance of sophisticated security measures in today's digital age, where phishing threats continue to evolve.

9.14   A word on real-time threat intelligence in email security

Real-time threat intelligence has emerged as a critical component in the realm of email security, providing organizations with the ability to detect, analyze, and respond to cyber threats as they occur. This dynamic approach to cybersecurity leverages advanced technologies, including AI and ML, to offer a proactive defense against the ever-evolving landscape of email-based attacks. The benefits of real-time threat intelligence in email security are manifold, encompassing enhanced threat detection, improved incident response, and the ability to stay ahead of sophisticated cyber threats.

One of the most significant benefits of real-time threat intelligence in email security is its ability to enhance threat detection capabilities. Traditional email security measures often rely on static rules and signature-based detection methods, which can be insufficient in identifying new and sophisticated threats. Real-time threat intelligence, on the other hand, utilizes AI and ML to analyze vast amounts of data from various sources, including threat intelligence feeds, user behavior, and email content. This allows for the identification of patterns and anomalies that may indicate malicious activity.

For instance, AI-powered email security solutions can analyze email content, sender behavior, and contextual information to detect phishing attempts. By continuously learning from new data and adapting to emerging threats, these systems can identify subtle indicators of phishing that might be missed by traditional security measures. This proactive approach ensures that even the most sophisticated phishing emails are detected and blocked before they reach end users, significantly reducing the risk of successful attacks.

Real-time threat intelligence also plays a crucial role in improving incident response capabilities. In the event of a security breach or cyberattack, having access to real-time threat data and actionable insights enables organizations to swiftly identify the nature and source of the attack. This allows for the implementation of effective countermeasures to minimize the impact on operations and reputation.

When an AI-powered email security system detects a suspicious email, it can automatically quarantine the email, block any associated attachments or links, and trigger alerts to the security team. This automated response reduces the burden on security teams and accelerates the mitigation of threats. Real-time threat intelligence can provide detailed logs and alerts to SIEM systems, allowing security analysts to correlate email-based threats with other security events occurring within the organization. This comprehensive view enables a more effective and coordinated response to security incidents.

The dynamic nature of real-time threat intelligence allows organizations to stay ahead of emerging threats. By integrating threat intelligence data from multiple sources and correlating it with internal telemetry data, AI-powered email security systems can proactively identify and block new attack vectors. This proactive defense approach is particularly important in the context of rapidly evolving cyber threats, where traditional security measures may lag behind.

Real-time threat intelligence can help organizations stay informed about the latest phishing scams, malware, and other potential cyber threats targeting email systems. By analyzing data from various sources, including dark web monitoring, industry-specific threat reports, and real-time analysis of suspicious activities, organizations can anticipate and neutralize threats before they can cause substantial harm. This proactive approach not only enhances the security of email communications but also helps maintain the trust of stakeholders.

Real-time threat intelligence is designed to seamlessly integrate with broader security ecosystems, allowing organizations to orchestrate cohesive defense-in-depth strategies. This integration enables the correlation of email security events with broader security incidents and the orchestration of synchronized response actions across multiple security layers. AI-powered email security solutions can integrate with endpoint security solutions, network firewalls, and SIEM platforms. When an email-based threat is detected, the system can communicate with endpoint security tools to ensure that any malicious attachments or links are blocked at the device level. Similarly, network firewalls can be updated to block any related malicious domains, preventing users from inadvertently visiting harmful websites.

This integrated approach ensures that all security measures are working together to protect the organization.

One of the challenges of traditional email security measures is the balance between false positives and false negatives. False positives occur when legitimate emails are mistakenly classified as threats, while false negatives occur when malicious emails slip through the defenses. Real-time threat intelligence, powered by AI and ML, significantly reduces the incidence of both false positives and false negatives.

By continuously learning from user interactions and feedback, AI algorithms can refine their classification models to improve accuracy over time. This enhanced classification accuracy ensures that legitimate emails are not disrupted by false positives, while malicious emails are effectively identified and blocked. This balance is particularly beneficial in high-stakes environments where the cost of a missed threat or a false positive can be substantial.

Real-time threat intelligence leverages behavioral analysis to detect anomalies in email communication patterns and user behavior. By establishing a baseline of normal activity, AI-powered systems can identify deviations that may indicate a potential security breach. For example, if a user suddenly starts receiving emails from unfamiliar sources or exhibits unusual email usage patterns, the system can flag this activity for further investigation. This capability has proven invaluable in detecting and responding to compromised accounts and other security incidents. For instance, a multinational corporation implemented an AI-powered email security solution that monitored user behavior and detected an unusual spike in email forwarding activity. Upon investigation, the security team discovered that a phishing attack had compromised several employee accounts. The AI system's ability to detect this anomaly allowed the organization to respond swiftly, mitigating the impact of the attack and preventing further damage.

Crowdsourced threat intelligence is another powerful aspect of real-time threat intelligence in email security. By leveraging insights from a global community of security analysts and users, organizations can gain real-time visibility into emerging threats and attack vectors. This collective intelligence enhances the ability of AI-powered systems to detect and block new threats.

Platforms like VirusTotal allow users to contribute potential threat indicators, including files and URLs, to offer real-time insight into emerging attacks. This crowdsourced data is then analyzed and integrated into AI-powered email security systems, providing a more comprehensive view of the threat landscape. By incorporating this collective intelligence, organizations can stay informed about the latest threats and implement effective countermeasures to protect their email communications.

Integrating real-time threat intelligence into email security programs is essential for ensuring compliance with industry-specific standards and regulations. Data protection regulations are becoming increasingly stringent, and organizations must demonstrate a proactive approach to threat monitoring and mitigation to maintain compliance.

Real-time threat intelligence provides organizations with the tools and insights needed to meet regulatory requirements and protect sensitive data from unauthorized access or exposure. By leveraging advanced threat intelligence, organizations can strengthen their regulatory compliance posture and avoid potential fines and penalties associated with data breaches.

The dynamic nature of real-time threat intelligence ensures that email security systems are continuously improving and adapting to new threats. AI-powered systems learn from each detected threat, refining their algorithms and enhancing their ability to identify similar threats in the future. This continuous learning process ensures that email security systems remain effective against evolving threats.

Real-time threat intelligence allows organizations to review post-incident reports and identify areas for enhancement. By analyzing past incidents and patterns, organizations can strengthen their security infrastructure and prevent future incidents. This continuous improvement process is crucial for maintaining a robust and resilient email security posture.

To wrap it up, teal-time threat intelligence has revolutionized email security by providing organizations with the ability to detect, analyze, and respond to cyber threats as they occur. The benefits of real-time threat intelligence in email security are extensive, encompassing enhanced threat detection, improved incident response, proactive defense against emerging threats, seamless integration with broader security ecosystems, reduction of false positives and false negatives, enhanced user behavior analysis, leveraging crowdsourced threat intelligence, compliance with regulatory requirements, and continuous improvement and adaptation.

By leveraging advanced technologies such as AI and ML, real-time threat intelligence offers a proactive and dynamic approach to email security, ensuring that organizations can stay ahead of sophisticated cyber threats. As the threat landscape continues to evolve, the integration of real-time threat intelligence into email security programs will remain a critical component of effective cybersecurity strategies. Organizations that embrace this approach will be better equipped to protect their email communications, safeguard sensitive data, and maintain the trust of their stakeholders.

Key takeaways

- Technology plays a key role in an organization's phishing protection strategy.
- There is no one singular technology that can provide comprehensive protection against phishing attacks.
- Technologies work in concert with each other.
- A wide selection of technologies offers protections at several layers.
- Technologies include protections involving the isolation of nefarious traffic, advanced logging, detection, response and recovery as well as adaptive learning through AI and ML.
- Organizations should consider architectures that support protection concepts such as zero trust.

Chapter 10

Case studies

The following case studies were conducted between the years 2016 and 2019. The case studies not only include information generated by the case study but also include an analysis based upon the human factors data points discussed throughout this book. Additionally, each section contains a discussion around mitigation strategies and possible artificial intelligence (AI)/ML angles that could apply to a given case.

10.1 FACC phishing case study (2016)

In January 2016, Fischer Advanced Composite Components AG (FACC), an Austrian aerospace parts manufacturer, fell victim to a sophisticated business email compromise (BEC) scam, resulting in a significant financial loss of approximately €42 million (around $47 million). The incident is a classic example of CEO fraud, where cybercriminals impersonate high-level executives to deceive employees into transferring large sums of money.

The attack began when an employee in FACC's finance department received an email that appeared to come from the company's CEO, Walter Stephan. The email requested the transfer of €42 million for an alleged acquisition project. The attackers had meticulously studied the CEO's writing style and crafted a convincing email that created a sense of urgency and legitimacy. The employee, unable to discern the fraudulent nature of the email, complied with the request and transferred the funds to an account controlled by the attackers [64].

Upon realizing the fraud, FACC's management immediately involved the Austrian Criminal Investigation Department and engaged in a forensic investigation. Despite their efforts, the company was only able to block the transfer of €10.9 million, while the remaining €32 million had already disappeared into accounts in Slovakia and Asia. The financial impact of the scam was severe. FACC's share price plummeted, and the company reported a substantial decrease in earnings for the fiscal year. The incident led to an outflow of liquid funds totaling €52.8 million, resulting in an operating loss of €23.4 million [65].

The fallout from the scam led to significant changes in FACC's management. In May 2016, the company's board decided to terminate CEO Walter Stephan, citing his severe violation of duties related to the incident. The CFO and the finance department employee who executed the transfer were also dismissed.

FACC sought €10 million in legal damages from the former CEO and CFO, alleging their failure to implement adequate internal security controls. However, the Austrian courts dismissed both lawsuits.

In connection with the scam, a 32-year-old Chinese man, who was an authorized signatory of a Hong Kong-based firm that received around €4 million from FACC, was arrested on suspicion of money laundering.

In the aftermath of the attack, FACC implemented new security measures and conducted a thorough review of internal processes to prevent similar incidents in the future. The company also emphasized cybersecurity training for employees to enhance vigilance in handling sensitive communications.

10.1.1   *Likely mitigation strategies*

It is unlikely FACC published the technical mitigation strategies deployed to their enterprise infrastructure following this attack. Therefore, the mitigation strategies discussed here are our best guess and information we gathered about FACC's mitigation strategies from open sources.

After the devastating €42 million phishing attack in 2016, FACC implemented several new security measures to prevent similar incidents from occurring in the future. The company conducted a thorough review of its internal processes and made significant changes to strengthen its cybersecurity posture.

One of the key measures taken by FACC was to increase cybersecurity awareness and training for employees at all levels of the organization. The company recognized that human error played a crucial role in the successful execution of the phishing attack, as an employee in the finance department fell victim to the fraudulent email impersonating the CEO. To address this vulnerability, FACC emphasized the importance of vigilance when handling sensitive communications and provided comprehensive training programs to help employees identify and respond appropriately to potential phishing attempts.

In addition to employee training, FACC likely implemented technical controls and security solutions to enhance its email security. This may have included the deployment of advanced email filtering and authentication mechanisms, such as Domain-based Message Authentication, Reporting, and Conformance (DMARC), to detect and block spoofed or fraudulent emails. The company also likely reviewed and strengthened its access controls, password policies, and multi-factor authentication requirements to prevent unauthorized access to sensitive systems and data.

Furthermore, FACC likely conducted regular security audits, penetration testing, and vulnerability assessments to identify and address potential weaknesses in its cybersecurity defenses. These measures would help the company stay ahead of emerging threats and ensure that its security measures remained effective and up-to-date.

By taking a multi-layered approach that combined technical controls, employee awareness, and continuous monitoring and improvement, FACC aimed to rebuild trust and confidence in its cybersecurity practices after the devastating phishing attack. The company's response underscored the importance of a proactive and comprehensive approach to cybersecurity, as well as the need for constant vigilance in an ever-evolving threat landscape.

10.1.2   Will user training be effective?

FACC's implementation of user training following their significant phishing incident is a crucial step in enhancing cybersecurity awareness and reducing susceptibility to phishing attacks. However, as discussed earlier in this book, the effectiveness of such training in overcoming users' tendencies to anthropomorphize technology, their ability to regulate threat perception, and their contextual awareness when receiving unknown emails is multifaceted and complex.

To recap, anthropomorphism, the attribution of human characteristics to non-human entities, can significantly influence how users interact with technology. Research indicates that anthropomorphizing technology can lead to increased trust and perceived reliability, which might make users more susceptible to phishing attacks if they perceive the technology as inherently secure. For instance, users might trust an email from a seemingly familiar source without scrutinizing it for potential threats.

To mitigate this, training programs need to emphasize critical thinking and skepticism, even towards seemingly trustworthy sources. This involves educating users about the risks of anthropomorphism and encouraging them to verify the authenticity of communications, regardless of their perceived familiarity.

Effective security awareness training should also focus on helping users regulate their threat perception. This includes recognizing the signs of phishing emails, such as urgent calls to action, unexpected requests, and suspicious links or attachments. Training programs like those offered by Phin Security and KnowBe4 incorporate real-time simulations and progressive difficulty levels to help users develop a keen eye for identifying phishing attempts.

Moreover, training should be continuous and adaptive, ensuring that users remain vigilant and up-to-date with the latest phishing tactics. Regular simulated phishing tests can reinforce learning and help users practice their skills in a controlled environment, thereby improving their ability to detect and respond to real threats.

Contextual factors, such as being in a hurry or distracted, can significantly impact a user's ability to recognize phishing attempts. Studies have shown that users are more likely to fall for phishing scams when they are under time pressure or not fully attentive.

To address this, training programs should include scenarios that mimic real-life situations where users might be rushed or distracted, helping them develop strategies to manage these contexts effectively. For example, training can teach users to take a moment to pause and carefully examine emails, especially those that create a sense of urgency or come from unfamiliar senders. Encouraging a culture of caution and verification can help users avoid making hasty decisions that could lead to security breaches.

While user training is a vital component of cybersecurity, it must be comprehensive and multifaceted to address the various factors that influence susceptibility to phishing attacks. By incorporating elements that tackle anthropomorphism, threat perception regulation, and contextual awareness, training programs can better equip users to recognize and respond to phishing attempts effectively. Continuous education, real-time simulations, and a culture of vigilance are essential to overcoming these challenges and enhancing overall cybersecurity resilience.

10.1.3   The AI effect

Had AI been used in the FACC phishing campaign in 2016, the attack could have been significantly more sophisticated and potentially even more damaging. Leveraging AI, especially the advanced capabilities of generative AI and large language models (LLMs) that have been developed in recent years, the attackers could have crafted highly personalized and convincing emails, voice messages, or even video communications. Here's a speculative overview of how the attack might have unfolded with the use of AI:

Using AI, attackers could have analyzed vast amounts of publicly available data on social media, company websites, and professional networks to gather detailed information about FACC's organizational structure, its key personnel, and their communication habits. This would allow for the creation of highly personalized phishing emails or messages that closely mimic the writing style, tone, and typical content of communications from trusted sources within the company, such as the CEO or other senior executives.

Generative AI could have been used to produce content that is contextually relevant to FACC's business operations, current projects, or internal discussions. For instance, the phishing email could reference specific aerospace projects, use industry-specific terminology, and even include fabricated updates or requests that seem plausible to the recipient. The use of natural language generation (NLG) techniques would ensure that the messages are grammatically correct and stylistically similar to genuine communications from the impersonated executives.

With advancements in AI-powered voice and video cloning technologies, the attackers could have created fake audio or video messages from the CEO or other executives. These deepfake messages could be used to request urgent actions, such as transferring funds or providing sensitive information, adding another layer of authenticity to the phishing attempt. The use of such technology would make it extremely difficult for employees to distinguish the fake communications from real ones.

AI-driven phishing campaigns can generate unique and varied phishing content for each target, making it harder for traditional security measures to detect and block the emails. Additionally, AI can be used to optimize the timing of the phishing attempts, sending them during periods when employees are more likely to be distracted or under pressure, further increasing the chances of success.

Should the initial phishing attempts be detected or fail to elicit the desired response, AI could enable rapid adaptation of the campaign. Based on feedback from the initial attempts, the AI could alter the phishing strategy, refining the content, and targeting to overcome defenses and successfully deceive the targets.

In summary, the use of AI in the FACC phishing attack could have resulted in a far more deceptive and difficult-to-detect campaign. The personalized and contextually relevant content, combined with the potential use of deepfake technology, would significantly increase the likelihood of deceiving employees and achieving the attackers' objectives. This speculative scenario underscores the importance of advanced cybersecurity measures, including AI-driven defenses, and comprehensive employee training to combat the evolving threat of AI-powered phishing attacks [66].

10.2 Crelan Bank phishing case study (2016)

In January 2016, Crelan Bank, a prominent Belgian financial institution, fell victim to a sophisticated phishing attack known as BEC or CEO fraud. The attack resulted in a staggering financial loss of €70 million (approximately $75.8 million) for the bank.

The incident began when cybercriminals gained unauthorized access to the email account of Crelan Bank's CEO. Leveraging this access, the attackers meticulously studied the CEO's communication style and crafted a convincing email that appeared to originate from the CEO himself. The fraudulent email was sent to an employee in the bank's finance department, requesting an urgent transfer of €70 million, likely under the pretext of a legitimate business transaction or acquisition.

Unaware of the deception, the employee, trusting the seemingly authentic email from the CEO, complied with the instructions and transferred the substantial sum of money to an account controlled by the attackers. The attackers had carefully mimicked the CEO's writing style, tone, and typical communication patterns, making it extremely difficult for the employee to detect the phishing attempt.

Upon discovering the fraud during an internal audit, Crelan Bank promptly reported the incident to the Belgian authorities and initiated an investigation. Despite the significant financial loss, the bank assured its customers and partners that its reserves were sufficient to absorb the impact without affecting their operations or financial stability.

In the aftermath of the attack, Crelan Bank implemented additional security measures to prevent similar incidents from occurring in the future. The bank recognized the importance of enhancing its cybersecurity posture and strengthening its defenses against sophisticated phishing attacks.

The Crelan Bank phishing attack serves as a stark reminder of the risks posed by BEC scams and the importance of robust cybersecurity measures, including employee training, multi-factor authentication, and rigorous verification processes for financial transactions, especially those originating from high-level executives or external parties.

10.2.1 Likely mitigation strategies

After falling victim to the devastating €70 million phishing attack in January 2016, Crelan Bank implemented several mitigation strategies to strengthen its cybersecurity posture and prevent similar incidents from occurring in the future. Here are the key measures taken by the bank, presented in prose format:

Crelan Bank recognized the critical importance of enhancing employee awareness and training to combat the human element that enabled the successful phishing attack. The bank launched comprehensive cybersecurity training programs to educate employees at all levels on identifying and responding appropriately to potential phishing attempts. These training initiatives aimed to foster a culture of vigilance and caution when handling sensitive communications, particularly those involving financial transactions or requests from high-level executives.

In addition to the human aspect, Crelan Bank also focused on implementing robust technical controls and security solutions to fortify its email infrastructure. The bank likely deployed advanced email filtering and authentication mechanisms, such as DMARC, to detect and block spoofed or fraudulent emails attempting to impersonate trusted sources.

Furthermore, Crelan Bank conducted a thorough review of its internal processes and access controls, strengthening policies and procedures related to sensitive financial transactions. The bank likely implemented multi-factor authentication requirements and rigorous verification processes, ensuring that high-value transactions or requests from executives undergo multiple layers of scrutiny and authentication before execution.

To maintain a proactive stance against evolving cyber threats, Crelan Bank likely established regular security audits, penetration testing, and vulnerability assessments. These measures would enable the bank to identify and address potential weaknesses in its cybersecurity defenses, ensuring that its security measures remained effective and up-to-date.

Moreover, Crelan Bank recognized the importance of collaboration and information sharing with relevant authorities and industry partners. By reporting the incident to Belgian authorities and engaging in investigations, the bank aimed to contribute to the collective understanding of phishing tactics and facilitate the development of more effective countermeasures.

Through this multi-layered approach, combining technical controls, employee awareness, continuous monitoring, and industry collaboration, Crelan Bank sought to rebuild trust and confidence in its cybersecurity practices. The bank's response underscored the importance of a comprehensive and proactive approach to mitigating the ever-evolving threat of phishing attacks and safeguarding the financial integrity of its operations [67].

10.2.2   *Will user training be effective?*

Following the significant phishing attack in January 2016, Crelan Bank implemented user training as a key mitigation strategy to enhance cybersecurity awareness and prevent similar incidents in the future. However, as with the FACC phishing training program, while training is a crucial component of cybersecurity, it alone may not fully address certain human factors that contribute to the success of phishing attacks, such as the anthropomorphizing of technology, the regulation of threat perception, and contextual awareness.

To recap, anthropomorphizing technology involves attributing human-like characteristics to non-human entities, such as email systems or software. This can lead to a misplaced sense of trust and reliability in these systems, making users more susceptible to phishing attacks. For instance, an employee might trust an email simply because it appears to come from a familiar source, without scrutinizing it for potential threats. While training can raise awareness about the risks of phishing, it may not eliminate the tendency to anthropomorphize technology. As with the FACC case, to mitigate this, training programs should emphasize critical thinking and skepticism, encouraging users to verify the authenticity of communications, regardless of their perceived familiarity.

Effective security awareness training should help users regulate their perception of threats by teaching them to recognize the signs of phishing emails, such as urgent calls to action, unexpected requests, and suspicious links or attachments. However, the success of this training depends on the user's ability to apply this knowledge consistently. Studies have shown that even with training, users may still struggle to accurately assess the severity and likelihood of threats, especially under pressure or when distracted. Continuous and adaptive training, including real-time simulations and feedback, can help reinforce these skills and improve users' ability to detect and respond to phishing attempts.

Contextual factors, such as being in a hurry or distracted, can significantly impact a user's ability to recognize phishing attempts. For example, an employee who is rushing to meet a deadline may not take the time to carefully examine an email, increasing the likelihood of falling for a phishing scam. Training programs need to address these real-world scenarios by incorporating context-based learning and simulations that mimic the pressures and distractions users might face in their daily work. By practicing in realistic settings, users can develop strategies to manage these contexts effectively, such as taking a moment to pause and verify the authenticity of emails before taking action.

While user training is an essential component of cybersecurity, it must be comprehensive and multifaceted to address the various human factors that contribute to phishing susceptibility. By combining technical controls with robust training and awareness initiatives, organizations like Crelan Bank can better protect themselves against the evolving threat of phishing attacks.

10.2.3   The AI effect

As with the FACC phishing attack, had AI been utilized in the Crelan Bank phishing attack, the campaign could have been significantly more sophisticated and challenging to detect. Again, the integration of AI technologies, particularly generative AI and LLMs, would have enabled the attackers to craft highly personalized and convincing phishing emails, potentially leading to an even more devastating outcome.

With AI, the attackers could have leveraged advanced data analysis techniques to gather detailed information about Crelan Bank's organizational structure, communication patterns, and ongoing projects. By scraping data from public sources, such as social media profiles and company websites, the AI system could have identified key individuals, their writing styles, and typical email content. This information would have allowed the attackers to generate phishing emails that closely mimicked the tone, language, and formatting used by Crelan Bank's executives and employees.

Furthermore, generative AI could have been employed to create contextually relevant content for the phishing emails. The AI system could have incorporated industry-specific terminology, references to ongoing projects or internal discussions, and even fabricated updates or requests that would seem plausible to the recipients. This level of personalization and attention to detail would have made it extremely difficult for employees to distinguish the fraudulent emails from legitimate communications.

In addition to crafting convincing email content, AI could have been used to generate deepfake audio or video messages purportedly from Crelan Bank's CEO or other high-level executives. These deepfake messages could have been used to reinforce the urgency and authenticity of the phishing campaign, further increasing the likelihood of employee compliance.

Moreover, AI-driven automation could have enabled the attackers to rapidly adapt and refine their phishing strategy based on the initial responses received. If the initial attempts were detected or failed to elicit the desired response, the AI system could have analyzed the feedback and adjusted the content, tone, or timing of the phishing emails to overcome defenses and successfully deceive the targets.

While the Crelan Bank phishing attack was already a significant incident, the integration of AI technologies could have amplified its sophistication, personalization, and overall impact. The ability to generate highly convincing and tailored phishing content at scale, combined with the potential use of deepfake technology, would have made the attack even more challenging to detect and mitigate, potentially resulting in even greater financial losses for the bank.

10.3   The case study on the Google and Facebook phishing scams (2013–15)

In one of the most audacious and lucrative phishing scams in recent history, a Lithuanian man named Evaldas Rimasauskas orchestrated an elaborate scheme that defrauded tech giants Facebook and Google out of over $100 million between 2013 and 2015.

Rimasauskas's scheme hinged on impersonating Quanta Computer, a legitimate Taiwanese electronics manufacturer that counted Facebook, Google, and Apple among its clients. He established a company in Latvia with an identical name to Quanta Computer, complete with forged corporate stamps, email accounts, and invoices designed to mimic those of the real firm.

Over the course of two years, Rimasauskas and his accomplices sent meticulously crafted emails to employees in the accounting departments of Facebook and Google. These emails, purportedly from Quanta Computer, contained fraudulent invoices, contracts, and letters bearing forged signatures and seals, requesting payments for goods and services rendered.

The level of detail and authenticity in the fraudulent communications was remarkable. Rimasauskas had studied Quanta Computer's operations, communication styles, and even created fake corporate stamps to lend an air of legitimacy to the scheme. The emails instructed the employees to wire payments to bank accounts controlled by Rimasauskas in Latvia and Cyprus, rather than Quanta Computer's actual accounts in Asia.

Unaware of the deception, employees at Facebook and Google complied with the requests, believing they were conducting routine business transactions with a trusted vendor. Over the course of two years, Google transferred $23 million, while Facebook wired a staggering $98 million to the fraudulent accounts controlled by Rimasauskas.

The scheme eventually unraveled, leading to Rimasauskas's arrest in Lithuania in March 2017 and subsequent extradition to the United States. In March 2019, he pleaded guilty to one count of wire fraud and agreed to forfeit $49.7 million in ill-gotten gains.

Both Facebook and Google acknowledged falling victim to the scam but stated that they had recovered most of the funds shortly after the incident. However, the case highlighted the sophistication and potential impact of BEC scams, even against tech-savvy companies with substantial resources.

Rimasauskas's meticulous planning, attention to detail, and exploitation of human vulnerabilities allowed him to perpetrate one of the most successful phishing scams in history. The case serves as a stark reminder of the importance of robust cybersecurity measures, employee training, and vigilance in an era where cybercriminals are becoming increasingly sophisticated in their tactics.

While the financial losses were significant, the incident also underscored the need for enhanced security protocols, multi-factor authentication, and a heightened awareness of social engineering tactics employed by cybercriminals. The Facebook and Google phishing case study stands as a cautionary tale, reminding organizations of all sizes to remain vigilant against the ever-evolving threat landscape of cyberattacks.

10.3.1 Likely mitigation strategies

Following the sophisticated phishing attack that defrauded Google and Facebook of over $100 million, both companies likely implemented a series of robust mitigation strategies to prevent similar incidents in the future. These strategies would have focused on enhancing their cybersecurity posture, improving employee awareness, and strengthening internal controls.

One of the primary mitigation strategies would have been to enhance employee training and awareness programs. Both companies likely intensified their efforts to educate employees about the risks of phishing and social engineering attacks. This would have included regular training sessions, workshops, and simulated phishing exercises to help employees recognize and respond appropriately to suspicious emails and requests. The training would emphasize the importance of verifying the authenticity of emails, especially those requesting financial transactions or sensitive information.

To add an extra layer of security, Google and Facebook likely implemented or reinforced the use of multi-factor authentication (MFA) for accessing sensitive systems and approving financial transactions. MFA requires users to provide two or more verification factors, such as a password and a one-time code sent to a mobile device, making it more difficult for attackers to gain unauthorized access even if they obtain login credentials through phishing.

Both companies would have reviewed and strengthened their email security protocols to detect and block phishing attempts more effectively. This likely included the deployment of advanced email filtering and authentication mechanisms such as DMARC, Sender Policy Framework, and DomainKeys Identified Mail. These technologies help verify the legitimacy of incoming emails and prevent email spoofing, a common tactic used in phishing attacks.

To prevent unauthorized financial transactions, Google and Facebook likely implemented more rigorous verification processes for approving large transfers of funds. This could involve requiring multiple levels of approval, cross-checking with known contact information, and verifying requests through alternative communication channels, such as phone calls or in-person meetings, to ensure their legitimacy.

Both companies would have enhanced their continuous monitoring and incident response capabilities to detect and respond to phishing attempts and other cyber threats in real time. This likely included the use of advanced threat detection systems, security information and event management (SIEM) solutions, and dedicated incident response teams to quickly identify and mitigate potential security incidents.

Following the attack, Google and Facebook likely strengthened their collaboration with law enforcement agencies and industry partners to share information about phishing threats and best practices for mitigating them. This collaboration would help improve the overall cybersecurity landscape and enable quicker responses to emerging threats.

To ensure the effectiveness of their mitigation strategies, both companies likely conducted regular security audits and assessments. These audits would help identify potential vulnerabilities and areas for improvement in their cybersecurity defenses, allowing them to stay ahead of evolving threats.

The phishing attack on Google and Facebook highlighted the need for comprehensive and multi-layered cybersecurity strategies. By implementing enhanced employee training, multi-factor authentication, strengthened email security protocols, rigorous verification processes, continuous monitoring, and collaboration with law enforcement, both companies aimed to fortify their defenses against future phishing attacks. These measures underscore the importance of a proactive and holistic approach to cybersecurity in protecting against sophisticated social engineering threats.

10.3.2   *Training efficacy and the human factors*

When considering the human factors involved in the Google and Facebook phishing attack of 2017, there are several impediments that can hinder the effectiveness of training programs aimed at making a workforce more resistant to phishing attempts.

Despite the sophistication and resources of tech giants like Google and Facebook, the human element remains a significant vulnerability in the realm of cybersecurity. The 2017 phishing attack, which defrauded the companies of over $100 million, underscored the importance of addressing the human factors that contribute to the success of such attacks. While employee training is a crucial component of building a phishing-resistant workforce, several impediments can impede its effectiveness.

One of the primary obstacles is the inherent trust that employees place in technology and established communication channels. Many individuals assume that if an email lands in their inbox, it must be legitimate, leading to a false sense of security. This misplaced trust can make employees less vigilant in scrutinizing emails for potential phishing attempts, even after receiving training. Overcoming this ingrained belief requires a concerted effort to instill a culture of skepticism and vigilance, which can be challenging.

Another impediment is the tendency for employees to make rushed decisions in fast-paced work environments. The urgency and pressure of meeting deadlines or responding to time-sensitive requests can lead individuals to act impulsively, without taking the necessary precautions or verifying the authenticity of communications. Training programs must address this issue by providing strategies for managing time effectively and prioritizing security, even in high-pressure situations.

Emotional manipulation is another significant hurdle in combating phishing attacks. Cybercriminals often exploit emotions such as fear, curiosity, and a sense of urgency to compel recipients to take immediate action. Employees may react impulsively when faced with such emotional triggers, overriding the knowledge and awareness gained through training. Addressing this challenge requires a comprehensive approach that not only educates employees about recognizing emotional manipulation tactics but also equips them with techniques for managing their emotional responses.

Furthermore, the lack of awareness and knowledge about the various tactics and strategies employed by cybercriminals in phishing attacks can leave employees ill-prepared to recognize and respond effectively to these attempts. Training programs must cover a wide range of phishing techniques, including spear phishing, clone phishing, and BEC, to ensure that employees are well-informed about the different forms of phishing they may encounter.

Behavioral change and habit formation also present significant challenges. Even if employees are aware of the risks and know what to do, they may still fail to implement critical security practices consistently. Training programs should focus on creating a culture of security within the organization, encouraging employees to adopt secure practices in their daily work through continuous reinforcement, positive reinforcement, and creating an environment where security is a shared responsibility.

Lastly, information overload can hinder the effective communication of critical security information. With the increasing amount of information that employees are expected to absorb, it can be challenging to ensure that they retain and apply the lessons learned from phishing awareness training. To overcome this, training programs should prioritize the use of interactive and engaging methods, such as simulations, gamification, and scenario-based training, to ensure that employees retain critical security information.

While employee training is a vital component of building a phishing-resistant workforce, several human factors can impede its effectiveness. Addressing these impediments requires a comprehensive and multifaceted approach that combines technical controls, continuous training, and a culture of security that permeates every aspect of an organization's operations. By understanding and mitigating these human factors, organizations like Google and Facebook can better protect themselves against sophisticated phishing attacks and empower their employees to be a formidable line of defense against cyber threats.

10.3.3   The AI effect

As with previously discussed AI hypothetical situations, had AI been utilized in the infamous phishing attack that led to Google and Facebook wiring over $100 million to a fraudster, the sophistication and potentially the scale of the deception could have been significantly amplified. As with previously discussed attacks, the use of generative AI and LLMs would have allowed the attacker to craft even more convincing and personalized communications, potentially deceiving more individuals or securing even larger sums of money.

With AI, the attacker could have analyzed vast amounts of publicly available data on social media, company websites, and professional networks to gather detailed information about the organizational structure, communication habits, and ongoing projects of both Google and Facebook. This information could then be used to design highly personalized phishing emails that mimic the writing style, tone, and typical content of communications from trusted sources within the companies.

Generative AI could have been employed to create contextually relevant content for the phishing emails, incorporating industry-specific terminology, references to ongoing projects, or internal discussions, and even fabricating updates or requests that seem plausible to the recipients. The use of NLG techniques would ensure that the messages are grammatically correct and stylistically similar to genuine communications from the impersonated executives.

Moreover, AI-powered deepfake technology could have been used to create fake audio or video messages from high-level executives. These deepfake messages could reinforce the urgency and authenticity of the phishing campaign, further increasing the likelihood of employee compliance. The ability to clone voices with near-perfect accuracy or generate AI-created faces indistinguishable from real humans would make these phishing attempts extremely difficult to detect.

AI would also allow the attacker to automate and scale their phishing campaign efficiently. By generating numerous unique phishing emails in a short amount of time and targeting a wide range of individuals within Google and

Facebook, the attacker could increase their chances of success. AI's capability to adapt and refine the phishing strategy based on initial responses would enable the attacker to overcome defenses and successfully deceive the targets.

In response to the heightened threat posed by AI-powered phishing attacks, organizations like Google and Facebook would need to implement advanced AI-based solutions capable of detecting and blocking such sophisticated attempts. These solutions could analyze the content, context, and sender's behavior to identify suspicious or abnormal elements indicative of phishing.

Additionally, continuous employee training on recognizing and responding to phishing attempts, coupled with robust security policies and procedures, would be crucial in mitigating the risk of falling victim to such advanced attacks.

In short, had AI been used in the attack against Google and Facebook, the phishing campaign could have been significantly more deceptive and challenging to detect, underscoring the need for advanced cybersecurity measures and continuous vigilance in the face of evolving cyber threats.

10.4 Case study of spear phishing in academic research teams (2019)

In 2019, a detailed case study was conducted at Pennsylvania State University to understand the dynamics and impact of spear phishing attacks on academic research teams. The study, led by researchers Aiping Xiong, Sian Lee, Zekun Cai, Ephraim N. Govere, and Harish Kolla from the College of Information Sciences and Technology, provided valuable insights into the vulnerabilities and responses of academic environments to such targeted cyber threats.

The spear phishing attack targeted two research groups within the university. The attackers sent emails that appeared to come from trusted colleagues or advisors, exploiting the inherent trust within academic teams. The emails initially seemed benign but gradually escalated to requests for gift cards, a common tactic in phishing scams.

10.4.1 Key findings

1. **Use of mobile devices**: The study found that the victim students primarily used mobile devices to communicate with the phishers. This reliance on mobile devices played a significant role in the success of the attack. Mobile interfaces often display only the sender's name, not the full email address, making it easier for attackers to impersonate trusted individuals.
2. **Lack of prior knowledge**: The victim students did not have prior knowledge about gift card phishing scams, which contributed to their susceptibility. This lack of awareness underscores the importance of targeted cybersecurity education and training for students and faculty members.
3. **Delayed reporting**: Neither the students nor the faculty members initially reported the phishing attempts to their IT department. This delay in reporting

allowed the attackers to continue their efforts without immediate intervention, highlighting a critical gap in the incident response process.

4. **Email processing habits**: The study revealed that students paid less attention to the email address when opening emails, especially on mobile devices. This oversight was a significant factor in the success of the phishing attack. On desktop interfaces, students could see both the name and the email address of the sender. With the mobile interface, the student could only see the name of the sender but not the email address.

5. **Group interviews and analysis**: Semi-structured group interviews were conducted with two research teams involved in the phishing attacks. The participants included two faculty members and three PhD students. They shared their experiences and elaborated on their email processing habits, particularly the cues they considered important for detecting phishing emails.

10.4.2   Recommendations

Based on the findings, the researchers proposed several recommendations to enhance the detection and mitigation of spear phishing attacks in academic settings:

1. **Enhanced user interface**: Improving the email interface to make it easier for users to detect discrepancies between names and email addresses. This could involve displaying the full email address more prominently, even on mobile devices.

2. **Keyword alerts**: Implementing warnings or alerts for keywords commonly associated with phishing attacks, such as "gift cards." These alerts could prompt users to exercise additional caution when encountering such terms in emails.

3. **Immediate reporting**: Emphasizing the importance of reporting phishing attacks immediately to the IT department. Quick reporting can help contain the threat and prevent further damage.

4. **Targeted training**: Providing extra training on phishing attacks for international students, who often constitute a substantial proportion of academic research teams. This training should equip them with the knowledge and skills to detect and respond to phishing attempts effectively.

5. **Awareness campaigns**: Conducting regular awareness campaigns to educate students and faculty members about the latest phishing tactics and how to recognize them. These campaigns should include real-world examples and practical tips for staying safe online.

The spear phishing attack on academic research teams at Pennsylvania State University highlighted several critical vulnerabilities in the university's cybersecurity posture. The reliance on mobile devices, lack of prior knowledge about phishing scams, and delayed reporting were significant factors that contributed to the success of the attack. By implementing the recommended measures, such as enhancing the email interface, providing targeted training, and promoting immediate reporting, academic institutions can better protect their research teams from similar threats in the future. This case study underscores the importance of a comprehensive and proactive approach to cybersecurity in academic environments, where the stakes are high, and the potential impact of successful phishing attacks can be severe [68].

10.4.3   The human factor

The recommendations proposed to enhance the detection and mitigation of spear phishing attacks in academic settings are well-intentioned and address critical areas of concern. However, their efficacy could be diminished due to the inherent nature of human factors and the challenges associated with influencing human behavior.

Enhanced user interface: While improving the email interface to prominently display full email addresses, even on mobile devices, can aid in detecting discrepancies, the success of this recommendation relies on users actively scrutinizing the displayed information. Human factors such as inattention, cognitive biases, and the tendency to take mental shortcuts could undermine the effectiveness of this measure. Users may still overlook or disregard the displayed information, particularly when multitasking or under time constraints.

Keyword alerts: Implementing alerts for keywords commonly associated with phishing attacks, such as "gift cards," can be a useful tool in raising awareness. However, the efficacy of this recommendation could be limited by factors like alert fatigue and desensitization. If users are bombarded with too many alerts or if the alerts are perceived as disruptive, they may start ignoring or dismissing them without careful consideration.

Immediate reporting: Emphasizing the importance of immediate reporting is crucial for containing and mitigating the impact of phishing attacks. However, human factors such as fear of consequences, lack of trust in the reporting process, or a perceived lack of urgency could hinder timely reporting. Users may hesitate to report incidents due to concerns about potential repercussions or a belief that the incident is not severe enough to warrant immediate action.

Targeted training: Providing targeted training on phishing attacks for international students is a valuable recommendation, as they may face additional cultural or language barriers. However, the effectiveness of training can be influenced by factors such as motivation, learning styles, and the ability to transfer knowledge into practice. Additionally, the transient nature of student populations may necessitate continuous training efforts to ensure sustained awareness.

Awareness campaigns: Regular awareness campaigns that include real-world examples and practical tips can be effective in educating users about the latest phishing tactics. However, the success of these campaigns can be impacted by factors such as information overload, competing priorities, and the perceived relevance of the information to individual users. Sustaining engagement and ensuring the retention of knowledge from these campaigns can be challenging.

To mitigate the impact of human factors on the efficacy of these recommendations, a comprehensive approach that addresses both technical and human aspects is necessary. This could include fostering a strong cybersecurity culture through continuous awareness and training programs, promoting open communication and trust-building initiatives, and implementing incentives or accountability measures to encourage compliance and active participation in cybersecurity efforts. Additionally, involving human factors experts and incorporating user-centered design principles in the development and implementation of cybersecurity measures can help mitigate resistance and improve adoption rates.

By acknowledging and addressing the human factors that can limit the effectiveness of cybersecurity recommendations, academic institutions can enhance their overall resilience against phishing attacks and ensure a more robust and sustainable cybersecurity posture.

Chapter 11

Conclusions

Cybersecurity today stands at a critical juncture, characterized by rapid technological advancements and an ever-evolving threat landscape. The digital transformation of businesses, the proliferation of Internet of Things (IoT) devices, and the widespread adoption of remote work have collectively expanded the attack surface for cybercriminals. Among the myriad of cyber threats, phishing remains one of the most pervasive and damaging forms of cybercrime.

11.1 The current state of cybersecurity

The cybersecurity landscape in 2024 is marked by significant advancements and persistent challenges. Organizations are increasingly recognizing the importance of robust cybersecurity measures, driven by the rising frequency and sophistication of cyberattacks. According to a report by Forbes, the digital landscape in 2023 saw a significant surge in cybersecurity threats, with data breaches continuing to rise and attacks becoming more complex and intense. This escalation is largely attributed to hackers leveraging AI tools to enhance their attack strategies.

One of the most notable trends in cybersecurity is the adoption of artificial intelligence (AI) and machine learning (ML). These technologies are being used to enhance threat detection and response capabilities. AI's advanced data analysis capabilities enable the identification and prediction of cyber threats, improving early detection systems. ML algorithms are evolving to better recognize and respond to new threats, enhancing defensive measures over time. However, the adoption of AI in cybersecurity is a double-edged sword. While it offers significant benefits, it also presents challenges, as cybercriminals are increasingly using AI to develop more sophisticated attacks.

The shift to remote work has also had a profound impact on the cybersecurity landscape. The expansion of the attack surface has created new vulnerabilities, making it more challenging for organizations to secure their networks. Companies have had to adapt quickly, implementing new security measures to protect remote workers and their devices. This shift has underscored the importance of a comprehensive cybersecurity strategy that includes robust endpoint security, secure access controls, and continuous monitoring.

Phishing remains one of the most common and effective forms of cybercrime. It involves cybercriminals sending fraudulent messages, often via email, to trick individuals into revealing sensitive information or downloading malicious software. Despite advancements in cybersecurity, phishing attacks continue to be a significant threat. According to AAG IT Support, phishing is the most common form of cybercrime, with an estimated 3.4 billion spam emails sent every day. Google alone blocks around 100 million phishing emails daily, highlighting the sheer volume of these attacks.

Phishing attacks have evolved over the years, becoming more sophisticated and harder to detect. Cybercriminals use various tactics to make their phishing emails appear legitimate, such as spoofing email addresses, using official logos, and crafting convincing messages. The rise of AI has further enhanced the capabilities of cybercriminals, enabling them to create highly personalized and convincing phishing emails. According to a report by Harvard Business Review, AI-enhanced phishing emails employ advanced techniques such as personalization, optimal timing, and adaptability, making them increasingly difficult to detect and prevent.

To combat the persistent threat of phishing, organizations are employing a multi-faceted approach that includes technology, training, and policy measures. One of the primary technological defenses against phishing is the use of advanced email filtering systems. These systems use AI and ML algorithms to analyze email content and identify potential phishing attempts. By examining various features such as domain, favicon, title, and HTML properties, these filters can block many phishing emails before they reach the user's inbox.

Despite these technological advancements, human error remains a significant vulnerability. Many phishing attacks rely on social engineering tactics to exploit human behavior. As such, cybersecurity awareness training is a critical component of any anti-phishing strategy. Organizations are increasingly investing in training programs to educate employees about the dangers of phishing and how to recognize and respond to suspicious emails. Regular phishing simulations are also used to test employees' awareness and reinforce good security practices.

However, there are challenges in implementing effective training programs. According to the 2024 State of the Phish report by Proofpoint, more than 70 percent of employees admit to risky behavior that leaves them vulnerable to phishing attacks. This highlights the need for continuous education and reinforcement of security best practices. Additionally, there is often a gap between security awareness and actual behavior, with many employees feeling uncertain about their security responsibilities.

Another critical aspect of addressing phishing is the implementation of multi-layered security solutions. These solutions include multi-factor authentication (MFA), which adds an extra layer of security by requiring users to provide two or more verification factors to gain access to a system. MFA can significantly reduce the risk of phishing attacks, as it makes it more difficult for cybercriminals to gain access to accounts even if they have obtained the user's credentials.

AI and ML are playing an increasingly important role in the fight against phishing. These technologies are being used to enhance threat detection and response capabilities, making it possible to identify and mitigate phishing attacks more quickly and accurately. AI-driven threat detection systems can analyze vast amounts of data in real time, identifying patterns and anomalies that may indicate a phishing attempt. ML algorithms can also adapt and improve over time, learning from past attacks to better recognize and respond to new threats.

One of the key benefits of AI in cybersecurity is its ability to provide real-time threat analysis. This enables organizations to respond to phishing attacks more quickly, reducing the potential damage. AI algorithms can also be used to automate certain aspects of cybersecurity, such as monitoring network traffic and identifying suspicious activity. This can help to reduce the burden on human security teams and improve overall efficiency.

However, the use of AI in cybersecurity is not without its challenges. One of the main concerns is the potential for AI to be used by cybercriminals to enhance their attacks. AI-enhanced phishing emails, for example, can be highly personalized and convincing, making them more difficult to detect and prevent. Additionally, there is a risk that AI systems could be manipulated or exploited by cybercriminals, leading to new types of attacks.

11.2 The need for a comprehensive cybersecurity strategy

Addressing the problem of phishing requires a comprehensive cybersecurity strategy that includes technology, training, and policy measures. Organizations must invest in advanced email filtering systems and other technological defenses to block phishing emails before they reach users. They must also provide regular training and education to employees to help them recognize and respond to phishing attempts.

In addition to these measures, organizations must implement robust security policies and procedures. This includes establishing clear guidelines for reporting and responding to phishing attempts, as well as implementing MFA and other security measures to protect sensitive information. Organizations must also continuously monitor their networks for signs of suspicious activity and be prepared to respond quickly to any potential threats.

The role of leadership is also critical in addressing the problem of phishing. According to Ivanti's 2024 State of Cybersecurity Report, board-level attention to cybersecurity is essential for positioning it as a critical business risk. This involves ensuring that cybersecurity is a key consideration in strategic decision-making and that there is alignment between the CIO, CISO, and other leaders within the organization. By fostering a culture of vigilance and prioritizing cybersecurity at the highest levels, organizations can better protect themselves against phishing and other cyber threats.

In short, the current state of cybersecurity is characterized by significant advancements and persistent challenges. Phishing remains one of the most common and damaging forms of cybercrime, with cybercriminals continually evolving their tactics to exploit human behavior and technological vulnerabilities. To address this problem, organizations must adopt a multi-faceted approach that includes advanced technological defenses, continuous training and education, robust security policies, and strong leadership.

The integration of AI and ML into cybersecurity offers significant benefits, enhancing threat detection and response capabilities. However, it also presents new challenges, as cybercriminals increasingly use these technologies to develop more sophisticated attacks. As such, organizations must remain vigilant and continuously adapt their cybersecurity strategies to stay ahead of emerging threats.

Ultimately, the fight against phishing requires a comprehensive and proactive approach. By investing in advanced technologies, educating employees, implementing robust security measures, and prioritizing cybersecurity at the highest levels, organizations can better protect themselves and their stakeholders from the persistent threat of phishing.

11.3   The human in the loop

With phishing's continued prevalence in the cybersecurity landscape we can largely attribute its success to its adaptability and the sophisticated exploitation of human psychology. The foundation of phishing lies in deception and manipulation, techniques that have been refined over time to exploit trust and prompt urgent, often reckless, action from victims. This reliance on social engineering underscores the inadequacy of relying solely on traditional, technology-focused security paradigms. Instead, a comprehensive, multifaceted approach is imperative—one that evolves alongside phishing tactics and addresses both technological vulnerabilities and human factors.

At the heart of phishing's success is its exploitation of human vulnerabilities. Cybercriminals craft scenarios that invoke fear, urgency, or curiosity, manipulating normal human responses for malicious gain. For example, an email masquerading as a message from a trusted institution might warn of an unauthorized login attempt, urging immediate action. The urgency and fear of potential loss can cloud judgment, leading to hasty decisions like providing sensitive information or clicking on a malicious link.

Understanding the multifaceted influence of user context on susceptibility to phishing attacks is essential for crafting defenses that are not only technically robust but also empathetically designed to accommodate human variability. Today we know that in many instances our very own biology is working against us when it comes to developing phishing resistant skills.

Human biology, with its inherent cognitive biases, emotional responses, and physiological limitations, plays a significant role in hindering our ability to become phishing resistant. Phishing attacks exploit these biological and psychological vulnerabilities, making it challenging for individuals to consistently recognize and avoid deceptive emails.

Cognitive biases are systematic patterns of deviation from norm or rationality in judgment, which often lead to perceptual distortion, inaccurate judgment, or illogical interpretation. These biases are deeply rooted in human cognition and are a result of the brain's reliance on heuristics, i.e. mental shortcuts that simplify decision-making processes. While heuristics can be useful in many situations, they can also make individuals more susceptible to phishing attacks.

One prominent cognitive bias that affects phishing resistance is over-confidence. Overconfidence bias leads individuals to believe they are less likely to be deceived by phishing attempts than they actually are. This false sense of security can result in a lowered guard, making individuals more likely to fall for phishing scams. For example, an employee who has undergone basic cybersecurity training might feel overly confident in their ability to identify phishing emails, leading them to overlook subtle red flags that indicate a phishing attempt.

Another common bias is confirmation bias, which occurs when individuals give more weight to information that confirms their existing beliefs or expectations. In the context of phishing, this means that if an email appears to align with what an individual expects from a trusted source, they may overlook signs that it is frau-dulent. A common attack vector involves sending emails that look like they came from a trusted source. If an email purports to be from a bank and contains familiar branding and language, the recipient may be more inclined to trust it, even if there are subtle indicators of phishing.

The recency effect, another cognitive bias, causes individuals to give undue weight to the most recent information they have encountered. Phishers exploit this by referencing current events or recent transactions in their emails, making the fraudulent messages seem more relevant and urgent. This can lead to hasty deci-sions, as individuals may prioritize responding to what appears to be a timely and pertinent issue without thoroughly scrutinizing the email.

Emotions play a crucial role in decision-making processes, and cybercriminals are adept at exploiting this aspect of human biology to their advantage. Phishing emails often use emotional triggers such as fear, curiosity, and urgency to manip-ulate recipients into taking actions they might otherwise avoid.

Fear is one of the most powerful emotions exploited in phishing attacks. Emails that create a sense of fear or panic can prompt individuals to act quickly and irrationally. For example, a phishing email might claim that the recipient's account has been compromised and that immediate action is required to prevent further damage. The fear of losing access to important accounts or sensitive information can override rational thought, leading the recipient to click on malicious links or provide personal information without proper verification.

Curiosity is another emotion that phishers frequently exploit. Humans have a nat-ural inclination to resolve curiosity, even if it may have negative consequences. Phishing emails that offer exclusive content, rewards, or insider information can pique recipients' curiosity, prompting them to click on links or download attachments without considering the potential risks. This exploitation of curiosity can be particularly effec-tive when the email content is tailored to the recipient's interests or recent activities.

Urgency is a common tactic used in phishing emails to create a sense of immediate action. By imposing unrealistic deadlines or suggesting that immediate action is required, phishers can pressure recipients into making hasty decisions. For instance, an email might claim that a limited-time offer is about to expire or that urgent verification is needed to avoid account suspension. The perceived urgency can cloud judgment, leading individuals to act quickly without thoroughly evaluating the legitimacy of the email.

Stress and fatigue are physiological states that can significantly impair cognitive function and decision-making abilities, making individuals more vulnerable to phishing attacks. The constant vigilance required to identify and avoid phishing emails, combined with the potential consequences of falling victim, can contribute to increased stress levels and decreased productivity in the workplace.

Stressful situations can impair judgment and critical thinking, making individuals more susceptible to deception. Under stress, the brain's capacity to meticulously analyze an email or request diminishes, leading to hasty decisions that might not be made under normal circumstances. An employee who is under pressure to meet a tight deadline may be more likely to click on a phishing link without thoroughly scrutinizing the email, simply to clear their inbox and focus on their work.

Fatigue, particularly at the end of the workday, can also lower vigilance and increase susceptibility to phishing attacks. When individuals are tired, their mental resources are depleted, and their ability to process information and make sound decisions is compromised. Phishers often time their attacks to coincide with these periods of low vigilance, increasing the likelihood that their targets will respond without the usual level of scrutiny they might apply when they are more alert.

Habits and automatic behaviors are other aspects of human biology that can hinder phishing resistance. Many individuals develop routine behaviors when interacting with emails, such as quickly scanning the content and clicking on links without thorough evaluation. These automatic responses can be exploited by phishers, who design their emails to mimic legitimate communications and trigger habitual actions. For example, an employee who routinely receives and processes invoices via email may develop a habit of quickly opening and approving these emails without scrutinizing the details. A phisher who sends a fraudulent invoice that closely resembles the legitimate ones can exploit this habit, increasing the likelihood that the employee will fall for the scam. The reliance on habitual behaviors can reduce the cognitive effort required to process each email, but it also increases the risk of falling victim to phishing attacks.

Trust and authority are fundamental aspects of human social interactions, and phishers often exploit these elements to deceive their targets. Trust is built through familiarity and positive experiences, and individuals are more likely to trust communications that appear to come from known and reputable sources. Phishers capitalize on this by impersonating trusted entities, such as banks, government agencies, or senior executives within an organization. A phishing email that appears to come from a company's CEO, requesting sensitive information or urgent action, can leverage the principle of authority to compel compliance. Employees may be more inclined to follow instructions from someone in a position of authority without questioning the legitimacy of the request. This exploitation of trust and authority can be particularly effective in hierarchical organizations where employees are accustomed to following directives from senior leaders.

While human biology presents significant challenges to phishing resistance, there are strategies that organizations can implement to mitigate these vulnerabilities. One effective approach is to incorporate mindfulness training into cybersecurity programs. Mindfulness training helps individuals develop greater awareness of their thoughts, emotions, and behaviors, enabling them to recognize and manage emotional triggers that phishers exploit. By practicing mindfulness techniques, such as focused breathing and body scans, employees can learn to stay present and attentive, reducing the likelihood of impulsive actions driven by fear, curiosity, or urgency. Mindfulness training can also help individuals manage stress and fatigue, improving their overall cognitive function and decision-making abilities.

Another strategy is to implement continuous and adaptive training programs that address the evolving nature of phishing threats. Regular simulated phishing exercises can provide employees with hands-on experience in recognizing and responding to phishing attempts, reinforcing good cybersecurity habits and reducing reliance on automatic behaviors. These simulations should be designed to mimic real-world scenarios and incorporate the latest phishing tactics to ensure that employees are prepared for current threats.

Organizations can also leverage technology to support phishing resistance. AI-driven email filtering systems can analyze email content and identify potential phishing attempts, reducing the burden on employees to manually scrutinize each email. Additionally, implementing MFA can provide an extra layer of security, making it more difficult for phishers to gain access to sensitive accounts even if they obtain login credentials.

Human biology, with its inherent cognitive biases, emotional responses, and physiological limitations, presents significant challenges to phishing resistance. Cognitive biases such as overconfidence, confirmation bias, and the recency effect can skew judgment and decision-making, making individuals more susceptible to phishing attacks. Emotional manipulation, through fear, curiosity, and urgency, further exploits these vulnerabilities, prompting hasty and irrational actions. Stress and fatigue impair cognitive function and decision-making abilities, increasing susceptibility to deception. Habits and automatic behaviors reduce cognitive effort but also increase the risk of falling victim to phishing scams. Trust and authority, fundamental aspects of human social interactions, are exploited by phishers to deceive their targets.

To mitigate these vulnerabilities, organizations can implement mindfulness training, continuous and adaptive training programs, and leverage technology to support phishing resistance. By addressing the biological and psychological factors that hinder phishing resistance, organizations can enhance their overall cybersecurity posture and reduce the risk of falling victim to phishing attacks.

11.4 At the organizational level, the need for commitment

Pinning a price to a cybersecurity program can indeed be challenging. The paradox of cybersecurity is that its success is often measured by the absence of incidents—

when nothing happens, it means the defenses are working. This lack of visible outcomes can make it difficult for organizations to justify their investment in cybersecurity. However, the necessity of robust cybersecurity measures cannot be overstated. In today's digital age, the risks associated with cyber threats are too significant to ignore. Organizations must make a firm commitment to cybersecurity, developing comprehensive strategies, policies, procedures, and integrating these into the organizational culture.

11.4.1   The necessity of cybersecurity

The digital transformation of businesses has brought about unprecedented opportunities for growth and efficiency. However, it has also introduced a myriad of cyber threats. Cyberattacks can result in significant financial losses, reputational damage, and legal liabilities. The increasing sophistication of cyber threats means that no organization is immune, regardless of size or industry. Therefore, a proactive approach to cybersecurity is essential.

A robust cybersecurity program is not just about preventing attacks; it is about ensuring the resilience of the organization. This involves protecting digital assets, maintaining customer trust, and complying with regulatory requirements. The cost of a cybersecurity breach can far exceed the investment in preventive measures. According to a report by IBM, the average cost of a data breach in 2021 was $4.24 million, a figure that underscores the financial impact of inadequate cybersecurity measures.

11.4.2   Developing a comprehensive cybersecurity strategy

A comprehensive cybersecurity strategy is the cornerstone of an effective cybersecurity program. This strategy should encompass all aspects of an organization's security, including technical and operational controls, as well as employee training and awareness programs. The strategy should be periodically reviewed and updated to stay current with the evolving threat landscape and changes within the organization.

The first step in developing a cybersecurity strategy is to conduct a thorough risk assessment. This involves identifying and assessing potential risks to the organization's information systems, networks, and data. The risk assessment should consider the types of data stored, the systems and applications used, and the potential consequences of a breach. By understanding the specific risks faced by the organization, it is possible to prioritize efforts and allocate resources effectively.

Once the risks have been identified, the next step is to develop a plan to mitigate these risks. This plan should outline the measures that will be taken to prevent, detect, and respond to cyberattacks. Prevention measures might include implementing firewalls, intrusion detection systems, and anti-malware software. Detection measures could involve continuous monitoring of network activity and regular vulnerability assessments. Response measures should include a well-defined incident response plan that outlines the steps to be taken in the event of a security breach.

Cybersecurity policies are essential for translating the high-level strategy into actionable guidelines. These policies provide a framework for how the organization will protect its digital assets and respond to cyber threats. They should be clear, comprehensive, and aligned with the organization's overall security strategy.

A cybersecurity policy should cover several key areas. First, it should define the roles and responsibilities of employees, including the acceptable use of technology and the steps to take in case of a security incident. This helps to ensure that everyone in the organization understands their role in maintaining cybersecurity.

Second, the policy should outline the technical controls that will be implemented to protect the organization's systems and data. This might include access controls, data encryption, and network security measures. The policy should also specify the procedures for regularly updating and patching software to address known vulnerabilities.

Third, the policy should include guidelines for incident response. This should cover the detection and reporting of security breaches, as well as the steps to contain, eradicate, and recover from an incident. Clear communication protocols should be established to ensure that all relevant stakeholders are informed in the event of a security breach.

While policies provide the framework, procedures offer specific guidelines and instructions for implementing the policies. These procedures should be detailed and practical, ensuring that employees know exactly what steps to take to protect the organization's digital assets.

One critical procedure is access control management. This involves defining how employees, contractors, and other authorized individuals are granted, managed, and revoked access to organizational systems, applications, and data. Access rights should be granted on a need-to-know basis, and there should be a process for promptly deprovisioning access when an employee leaves the organization or changes roles.

Another important procedure is patch management. This involves identifying, testing, and deploying software updates and security patches to address vulnerabilities in the organization's systems and applications. Regularly scheduled patching cycles and guidelines for prioritizing critical patches are essential to maintaining the security of the organization's systems.

Incident response procedures are also crucial. These procedures should define the actions to be taken in case of a cybersecurity breach, including the roles and responsibilities of the incident response team, preliminary efforts to contain and mitigate the impact of the incident, and the process for documenting and reporting the incident. Predefined response playbooks for common incidents, such as malware infections or data breaches, can help ensure a swift and effective response.

Building a robust cybersecurity team is essential for implementing and maintaining an effective cybersecurity program. This requires hiring individuals with the necessary skills and expertise to manage the organization's cybersecurity efforts. Given the shortage of cybersecurity talent, attracting and retaining skilled professionals can be challenging.

Organizations should consider a multi-faceted approach to hiring cybersecurity talent. This might include offering competitive salaries and benefits, providing opportunities for professional development, and creating a positive work environment. Additionally, organizations can partner with educational institutions to develop a pipeline of future cybersecurity professionals.

Once the team is in place, it is important to provide ongoing training and development opportunities. Cybersecurity is a rapidly evolving field, and staying current with the latest threats and technologies is essential. Regular training sessions, certifications, and attendance at industry conferences can help ensure that the team remains knowledgeable and effective.

Creating a culture of cybersecurity is perhaps the most critical aspect of a successful cybersecurity program. This involves embedding cybersecurity practices and awareness into the daily activities of all employees, from the executive level to the front line. A strong cybersecurity culture ensures that everyone in the organization understands the importance of cybersecurity and their role in maintaining it.

Leadership plays a crucial role in fostering a cybersecurity culture. Executives and managers must lead by example, demonstrating a commitment to cybersecurity in their actions and decisions. This might include regularly discussing cybersecurity in meetings, participating in training sessions, and visibly supporting cybersecurity initiatives.

Employee training and awareness programs are also essential for building a cybersecurity culture. These programs should be engaging and interactive, using hands-on simulations and scenario-based learning to reinforce key concepts. By making cybersecurity training relevant and practical, organizations can help employees understand how their actions impact the organization's security posture.

Regular communication is another important aspect of building a cybersecurity culture. This might include sending out security bulletins, sharing information about recent threats, and providing tips for staying safe online. By keeping cybersecurity top of mind, organizations can help ensure that employees remain vigilant and proactive in identifying and responding to potential threats.

11.4.3   Continuous improvement and adaptation

Cybersecurity is not a one-time effort but an ongoing process that requires continuous improvement and adaptation. The threat landscape is constantly evolving, and organizations must stay current with the latest threats and technologies to remain secure. Regularly testing and updating the cybersecurity plan is essential for identifying gaps and weaknesses. This might include vulnerability assessments, penetration testing, and tabletop exercises to simulate cyberattacks. The results of these tests should be used to update the plan and improve defenses.

Organizations should conduct post-incident reviews to identify areas for improvement. By analyzing the response to a security incident, organizations can learn valuable lessons and make necessary adjustments to their cybersecurity strategy.

While a strong cybersecurity culture and well-defined policies and procedures are essential, technology also plays a critical role in protecting an organization's digital assets. Advanced security tools and technologies can help detect and respond to threats more effectively, reducing the risk of a successful cyberattack.

One important technology is security information and event management (SIEM) systems. SIEM systems collect and analyze data from various sources to identify potential security incidents. By providing real-time visibility into network activity, SIEM systems can help organizations detect and respond to threats more quickly.

Another key technology is endpoint protection. This includes anti-malware software, firewalls, and intrusion detection systems that protect individual devices from cyber threats. Regularly updating and patching these systems is essential to maintaining their effectiveness.

MFA is another important security measure. MFA requires users to provide two or more verification factors to gain access to a system, making it more difficult for cybercriminals to gain unauthorized access. Implementing MFA can significantly reduce the risk of account compromise.

11.4.4   The importance of third-party risk management

In today's interconnected world, organizations often rely on third-party vendors and partners to support their operations. However, these third parties can introduce vulnerabilities to the organization's security. Therefore, third-party risk management is a critical component of a comprehensive cybersecurity strategy.

Organizations should assess the cybersecurity posture of their third-party vendors and partners, establishing strict security requirements and regularly reviewing their compliance. This might include conducting security audits, requiring vendors to adhere to specific security standards, and including cybersecurity clauses in contracts. By managing third-party risks, organizations can reduce the likelihood of a security breach originating from a vendor or partner. This helps to ensure that the organization's digital assets remain protected, even when relying on external parties.

11.5   The bottom line

Making a commitment to cybersecurity is essential for organizations in today's digital age. While pinning a price on a cybersecurity program can be challenging, the cost of a security breach can far exceed the investment in preventive measures. Developing a comprehensive cybersecurity strategy, establishing clear policies and procedures, building a skilled cybersecurity team, and integrating cybersecurity into the organizational culture are all critical components of an effective cybersecurity program. By taking a proactive approach to cybersecurity, organizations can protect their digital assets, maintain customer trust, and comply with regulatory requirements. Continuous improvement and adaptation are essential to staying current with the evolving threat landscape, and advanced security technologies can help detect and respond to threats more effectively. Ultimately, a strong commitment to cybersecurity ensures the resilience of the organization and reduces the risk of costly security incidents.

11.6　Addressing the shortfalls in the cybersecurity workforce

Cybersecurity has evolved from a narrow, technical discipline focused on securing computer systems and networks to a multifaceted, cross-cutting field that touches every aspect of modern society. As our world becomes increasingly digitized and interconnected, the need for a comprehensive and interdisciplinary approach to cybersecurity has never been more pressing. We must rethink the way we train and educate cybersecurity professionals, fostering a workforce that can navigate the complexities of today's digital landscape and protect our most valuable assets.

The traditional view of cybersecurity as a purely technical domain is no longer sufficient. While technical expertise remains crucial, the challenges we face today extend far beyond the realm of code and hardware. Cybersecurity is inextricably linked to human behavior, organizational structures, risk management, AI, the IoT, and a myriad of other disciplines. Addressing these challenges requires a holistic understanding of the intricate interplay between technology, people, and processes.

11.6.1　*The human factor: integrating psychology into cybersecurity*

One of the most significant paradigm shifts in cybersecurity is the recognition of the human factor as a critical component of any effective security strategy. Humans are often the weakest link in the cybersecurity chain, susceptible to social engineering attacks, phishing scams, and other forms of deception. However, humans can also be our greatest asset when empowered with the right knowledge and mindset. To create a truly resilient cybersecurity workforce, we must integrate principles from psychology and behavioral science into our training and education programs. This includes understanding cognitive biases, emotional triggers, and decision-making processes that can influence an individual's cybersecurity behavior. By teaching cybersecurity professionals to recognize and mitigate these human vulnerabilities, we can significantly enhance our ability to defend against cyber threats.

Effective cybersecurity training should go beyond imparting technical knowledge and focus on cultivating a culture of security awareness and vigilance. This involves developing communication strategies that resonate with diverse audiences, fostering a sense of shared responsibility, and promoting a mindset of continuous learning and adaptation.

11.6.2　*Organizational structures and cybersecurity governance*

Cybersecurity is not merely a technical challenge; it is a strategic imperative that must be woven into the fabric of an organization's culture and decision-making processes. To achieve this, we must rethink the way cybersecurity teams are structured and integrated within organizational hierarchies.

Traditional organizational models often treat cybersecurity as a siloed function, disconnected from other business units and decision-making processes. This approach is no longer tenable in today's interconnected world, where cyber threats can impact every aspect of an organization's operations.

Instead, we must embrace a model of cybersecurity governance that promotes cross-functional collaboration and elevates cybersecurity considerations to the highest levels of organizational leadership. This requires training cybersecurity professionals to understand the broader business context, communicate effectively with stakeholders across different domains, and navigate the complexities of organizational politics and decision-making processes. By fostering a culture of cybersecurity awareness and accountability at all levels of an organization, we can create a more resilient and proactive approach to cyber risk management.

11.6.3   Quantifying cyber risks: bridging the gap between technology and business

Effective cybersecurity strategies must be grounded in a comprehensive understanding of cyber risks and their potential impact on an organization's operations, reputation, and financial performance. However, translating technical vulnerabilities and threats into tangible business risks has long been a challenge for cybersecurity professionals.

To bridge this gap, we must equip cybersecurity professionals with the skills and knowledge to quantify cyber risks in a language that resonates with business leaders and decision-makers. This involves mastering risk assessment methodologies, data analysis techniques, and financial modeling tools that can translate technical risks into measurable business impacts.

By quantifying cyber risks in terms of potential financial losses, operational disruptions, and reputational damage, cybersecurity professionals can better communicate the urgency and importance of cybersecurity investments to organizational leaders. This, in turn, can facilitate more informed decision-making and resource allocation, ensuring that cybersecurity efforts are aligned with an organization's overall business objectives and risk appetite.

11.6.4   Embracing artificial intelligence and machine learning

The rapid advancement of AI and ML technologies has profound implications for the field of cybersecurity. On one hand, these technologies offer powerful tools for detecting and responding to cyber threats in real time, automating many of the labor-intensive tasks associated with cybersecurity operations. On the other hand, AI and ML also present new challenges, as adversaries may leverage these technologies to develop more sophisticated and evasive cyberattacks.

To navigate this complex landscape, cybersecurity professionals must be trained to understand the underlying principles of AI and ML, as well as their applications in cybersecurity. This includes developing skills in data science, algorithm design, and model validation, as well as cultivating an understanding of the ethical and legal implications of using AI in cybersecurity contexts.

As AI and ML systems become more prevalent in cybersecurity operations, professionals must be equipped to manage and maintain these systems, ensuring their reliability, accuracy, and fairness. This requires a deep understanding of the potential biases and limitations of AI systems, as well as the ability to interpret and communicate their outputs effectively.

11.6.5   Securing the Internet of Things

The proliferation of connected devices, collectively known as the IoT, has introduced a vast array of new cybersecurity challenges. IoT devices, ranging from smart home appliances to industrial control systems, often lack robust security features and can serve as entry points for cyberattacks.

To address these challenges, cybersecurity professionals must develop a deep understanding of IoT architectures, communication protocols, and the unique security considerations associated with these devices. This includes knowledge of embedded systems, wireless communication technologies, and the ability to assess and mitigate vulnerabilities in IoT ecosystems.

As IoT devices become increasingly integrated into critical infrastructure and industrial systems, cybersecurity professionals must be trained to navigate the complex regulatory and compliance landscapes associated with these sectors. This may involve understanding industry-specific standards, risk management frameworks, and the legal and ethical implications of securing IoT systems in sensitive environments.

11.6.6   Fostering interdisciplinary collaboration

Addressing the multifaceted challenges of cybersecurity requires a collaborative effort that transcends traditional disciplinary boundaries. Cybersecurity professionals must be equipped to work alongside experts from diverse fields, including law, policy, ethics, economics, and social sciences, to develop holistic and effective security strategies.

This interdisciplinary approach necessitates the development of strong communication and collaboration skills, as well as a deep appreciation for the unique perspectives and contributions of each discipline. Cybersecurity professionals must learn to navigate the complexities of interdisciplinary teamwork, facilitating productive dialogue, and synthesizing insights from multiple domains. Moreover, interdisciplinary collaboration in cybersecurity education and training can foster a more diverse and inclusive workforce, bringing together individuals with diverse backgrounds, experiences, and perspectives. This diversity of thought can lead to more innovative and effective solutions to complex cybersecurity challenges.

11.6.7   Continuous learning and adaptation

The cybersecurity landscape is constantly evolving, with new threats, technologies, and regulatory frameworks emerging at a rapid pace. To remain effective, cybersecurity professionals must embrace a mindset of continuous learning and adaptation, constantly updating their knowledge and skills to stay ahead of emerging challenges.

This requires a shift in the way we approach cybersecurity education and training, moving away from a one-time, static model and towards a more dynamic and ongoing process. Cybersecurity professionals must be equipped with the tools and resources to engage in self-directed learning, leveraging online courses, professional development opportunities, and industry conferences to stay current with the latest developments in their field.

Organizations must foster a culture of continuous learning and knowledge sharing, encouraging cybersecurity professionals to collaborate, share best practices, and learn from each other's experiences. This can be facilitated through mentorship programs, knowledge management systems, and communities of practice that promote ongoing professional development and knowledge exchange.

11.6.8   Developing a cybersecurity workforce for the future

To cultivate a cybersecurity workforce capable of addressing the challenges of today and tomorrow, we must rethink our approach to education and training from the ground up. This involves integrating interdisciplinary perspectives and practical, hands-on learning experiences into cybersecurity curricula at all levels, from undergraduate programs to professional certifications.

At the university level, cybersecurity programs should be designed to provide a well-rounded education that combines technical expertise with a deep understanding of the broader social, legal, and ethical implications of cybersecurity. This may involve collaborating with other departments and disciplines to develop interdisciplinary courses and research opportunities.

For professionals already working in the field, continuous education and training opportunities must be made readily available. This could include industry-led certifications, online courses, and immersive training programs that simulate real-world cybersecurity scenarios and challenges.

We must actively promote diversity and inclusion within the cybersecurity workforce, recognizing that diverse perspectives and experiences are essential for developing effective and innovative solutions. This may involve targeted outreach and mentorship programs to encourage underrepresented groups to pursue careers in cybersecurity, as well as fostering inclusive and supportive work environments.

11.7   Conclusion

In the digital age, cybersecurity is no longer a niche concern; it is a fundamental imperative that touches every aspect of our lives and societies. As we navigate the complexities of this interconnected world, we must rethink our approach to cybersecurity, embracing an interdisciplinary and holistic perspective that integrates knowledge from diverse fields and fosters collaboration across traditional boundaries.

By integrating principles from psychology, organizational theory, risk management, AI, and emerging technologies like IoT, we can cultivate a cybersecurity workforce that is equipped to address the multifaceted challenges of today and tomorrow. This requires a paradigm shift in the way we educate and train cybersecurity professionals, emphasizing practical, hands-on learning experiences, interdisciplinary collaboration, and a mindset of continuous adaptation and growth.

Ultimately, the success of our cybersecurity efforts hinges on our ability to develop a workforce that can navigate the intricate interplay between technology, people, and processes. By embracing this holistic approach, we can build a more resilient and secure digital future, safeguarding our most valuable assets and enabling the continued growth and innovation of our interconnected world.

Things in the world of "cyber" have become complicated, and as many of my good colleagues say "The Enemy has a Vote!". If we wish to continue enjoying the benefits of our digital world, cybersecurity as a whole must be a priority.

Glossary of terms

Account Manipulation

A technique used by adversaries to maintain persistent access in a system by altering the properties of compromised accounts, such as adding permissions or credentials.

Advanced Persistent Threat (APT)

A prolonged and targeted cyberattack in which an intruder gains access to a network and remains undetected for an extended period.

AI-generated Phishing

Phishing attacks that use artificial intelligence to create more sophisticated and convincing fraudulent messages.

Altered Perceptions

Changes in the way users perceive and interact with digital environments, often leading to a dream-like state of consciousness that can impact attention and emotional responses.

Anchoring Effect

A cognitive bias where individuals rely too heavily on the first piece of information they encounter (the "anchor"), influencing subsequent decisions and judgments.

Anthropomorphism

The attribution of human characteristics, emotions, or intentions to non-human entities, such as virtual assistants or chatbots, enhancing user interaction but also potentially increasing trust and vulnerability to phishing attacks.

Attention Blindness

A phenomenon where people fail to notice unexpected objects in their visual field, especially when focused on another task, making them vulnerable to well-crafted phishing attacks.

Attack Surface

The sum of all possible points of entry for an attacker to access a system and engage in illicit activities. This includes all internet-facing assets and any devices that communicate externally.

Authority Bias

A cognitive bias where individuals tend to obey authority figures, often exploited in phishing attacks that impersonate figures of authority.

Autoencoders

Neural networks used in deepfake technology designed to encode an input into a lower-dimensional representation and then decode it back into the original format.

Bayesian Methods
Statistical methods used to quantify and assess risks, including cybersecurity risks. These methods help predict the likelihood of future events based on historical data and current information.

Behavioral Setting
A concept from ecological psychology referring to the geographical, physical, and social situation as it affects relationships and behavior.

Blockchain
A decentralized ledger technology used to create secure and immutable records, potentially for securing QR codes against fraud.

Browser Extensions
Software additions to web browsers that enhance functionality and can identify and block phishing attempts by analyzing website URLs and content in real time.

Cognitive Bias
Systematic patterns of deviation from norm or rationality in judgment, exploited in phishing attacks to manipulate users into harmful actions.

Comfort Experience
The user's sense of ease and familiarity with their technology, which can lower their defenses and make them more susceptible to phishing attacks.

Consistency
A principle in HCI design that ensures uniformity in interface elements, allowing users to leverage knowledge from one context to another. This consistency can be exploited by cybercriminals to create phishing sites or emails that mimic legitimate ones.

Contextual Learning
Tailoring training programs to reflect the specific contexts and roles of employees, enhancing the relevance and effectiveness of the training.

Contextual Perception
The user's understanding of their digital environment, influencing how they assess the authenticity of a communication.

Crowdsourcing Phishing Reporting
Encouraging employees to report suspected phishing attempts, creating a collective vigilance that enhances organizational security.

Cross-Platform User Experiences
The seamless integration of user experiences across different devices and platforms, creating a unified and coherent user journey. This enhances usability but also requires vigilance to ensure security across all devices.

Data Storytelling
Presenting data in a compelling, easy-to-understand format using visualizations, interactive charts, and narratives, making complex information accessible and engaging.

Deepfake
Synthetic media in which a person in an existing image or video is replaced with someone else's likeness using AI.

Domain-based Message Authentication, Reporting, and Conformance (DMARC)	An email authentication protocol designed to help email domain owners protect their domain from unauthorized use, commonly known as email spoofing.
Ecological Psychology	The study of human behaviors relative to their experience of their environments, emphasizing the relationship between individuals and their surroundings.
Email Monitoring Tools	Tools designed to monitor and analyze email traffic to detect and prevent phishing attacks by identifying suspicious patterns and content.
Emotional Design	A design approach that focuses on creating products that elicit positive emotions and connections with users, enhancing user engagement but also potentially increasing susceptibility to phishing.
Enhanced Simulation of Physical Experiences	Using technologies like VR and AR to bridge sensory gaps in digital interactions, providing more immersive experiences that can better simulate real-world scenarios.
Endpoint Detection and Response (EDR)	Security solutions that monitor and respond to cyber threats in real time, focusing on detecting suspicious activities on endpoint devices.
Endpoint Protection Platforms (EPPs)	Comprehensive security solutions designed to detect, investigate, block, and contain malicious activities on endpoint devices.
Equalized Status	The concept that online environments level the playing field, allowing individuals to reach wide audiences without traditional barriers such as publishers.
Experiential Learning	Learning through direct experience, which has been shown to improve retention and understanding of cybersecurity principles.
Experiential Training	Training methods that involve direct experience, such as simulated phishing attacks, to enhance learning and resistance to phishing.
Familiarity Bias	A cognitive bias where individuals show a preference for familiar or well-known entities, exploited in phishing by mimicking trusted sources.
Gamification	The application of game-design elements such as point scoring, leaderboards, and competition to training programs to enhance engagement and motivation.
Generative Adversarial Networks (GANs)	A class of machine learning frameworks used to create realistic images or videos, often employed in deepfake technology.

Gophish An open-source phishing toolkit designed for businesses and penetration testers to test the susceptibility of their organizations or clients to phishing attacks. It includes crafting emails, setting up fake web pages, and detailed reporting on interactions.

HiddenEye An automated phishing tool that allows users to create phishing sites by cloning social media platforms. It includes modules for adding keyloggers and fake security pages to enhance the legitimacy of phishing sites.

Human Factors Psychological and behavioral aspects of users that contribute to their vulnerability to phishing attacks.

Human–Computer Interaction (HCI) The study and design of the interaction between humans and computers, focusing on usability, efficiency, and user satisfaction. In the context of cybersecurity, effective HCI can mitigate or exacerbate the risk of phishing attacks.

Identity Flexibility The ability to adopt different personas online, allowing users to be themselves, partially reveal themselves, or create entirely new identities.

Initial Access The phase in a cyberattack where the attacker gains the initial foothold in a target system, often through phishing.

Intuitiveness A principle in HCI design where interfaces are designed to be easily navigable and understandable, relying on ingrained habits and understandings. Intuitive designs can be exploited by cybercriminals to create familiar-looking fraudulent sites or emails.

Kismet A sniffer tool used for capturing data packets on networks, including Wi-Fi and Bluetooth communications. It can be used with phishing tools to gather sensitive information like account details and passwords.

Machine Learning (ML) A branch of artificial intelligence focused on building systems that learn from data and improve their performance over time.

Micro-targeting The use of detailed personal data to create highly personalized phishing messages that are more likely to deceive recipients.

Minimalism A design trend that focuses on simplicity, clean layouts, and reducing cognitive load, enhancing usability but requiring vigilance to ensure security is not compromised.

MITRE ATT&CK Framework	A globally accessible knowledge base of adversary tactics and techniques based on real-world observations.
Mixed Reality (MR)	A technology that blends the digital and physical worlds allows for immersive experiences. MR can enhance user interactions but also introduces new cybersecurity threats.
Monte Carlo Simulation	A probabilistic model that uses random sampling to obtain numerical results, typically used to understand the impact of risk and uncertainty in prediction and forecasting models.
Mindfulness Training	Techniques that teach individuals to dynamically allocate attention and maintain emotional regulation, potentially improving their ability to detect phishing attempts.
Multi-factor Authentication (MFA)	A security measure requiring two or more verification factors to gain access to an account or system, significantly reducing the risk of unauthorized access.
Natural Language Processing (NLP)	A field of AI that focuses on the interaction between computers and humans through natural language.
Neuro-Fuzzy Model	A hybrid system combining neural networks and fuzzy logic for more accurate phishing detection.
Organizational Culture	The shared values, beliefs, and practices within an organization that influence behaviors, including responses to cybersecurity threats.
Perception of Cyberspace	The psychological experience of entering a digital space, often described as a place or a psychological space.
Pharming	A cyberattack that redirects a user's traffic to a fraudulent website through malicious code execution.
Phishing	A type of cyberattack that uses disguised email or other forms of communication to trick recipients into revealing personal information.
PhishX	A phishing tool that streamlines spear-phishing campaigns by masquerading phishing emails as coming from trusted contacts, increasing engagement likelihood.
Predictive Coding	A theory suggesting that the brain uses hierarchical generative models to predict sensory inputs and form mental models of experiences. This process can be exploited by cybercriminals in designing phishing attacks.
Privilege Escalation	The process by which an attacker gains elevated access to resources that are normally protected from an application or user.

QR Codes Quick Response codes that store information accessible via a smartphone camera, often used in phishing by embedding malicious links.

Reconnaissance The initial phase of a cyberattack where the attacker gathers information about the target.

Recordability The ease with which online interactions can be recorded, saved, and shared, presenting both advantages and vulnerabilities in cybersecurity.

Remote Browser Isolation (RBI) A cybersecurity technique that isolates web activities through an external browser, preventing malware from reaching internal networks.

Resource Development Activities undertaken by an attacker to gather resources necessary for executing an attack, such as purchasing domains or acquiring email accounts.

Risk Management Framework (RMF) A framework developed by NIST consisting of steps including preparation, categorization, selection, implementation, assessment, authorization, and monitoring of security controls.

Scrollytelling Combining scrolling with storytelling, creating interactive narratives that unfold as the user scrolls through a webpage or digital environment, enhancing engagement but requiring security measures to prevent exploitation.

ShellPhish An easy-to-use phishing tool that offers templates for creating phishing sites for popular platforms, leveraging information from other tools like SocialFish.

SIEM (Security Information and Event Management) Systems Systems that provide real-time analysis of security alerts generated by applications and network hardware, crucial for identifying and mitigating phishing attacks.

Simplicity and Ease of Use HCI design principles that prioritize straightforward interfaces, which can facilitate usability but also make users less likely to question straightforward, yet malicious, processes.

Simulated Phishing Campaigns Training exercises that simulate phishing attacks to test and improve an organization's resilience to phishing, though they may have limitations and unintended consequences.

Skeptical but Not Paranoid Educating employees on the importance of skepticism when dealing with emails, links, and attachments, especially from unknown sources, to verify the authenticity of requests and report suspicious activities.

Smishing

Phishing attacks conducted via text messages or SMS, exploiting the immediacy and trust in mobile communication to elicit quick responses from targets.

SocialFish

A phishing tool with an intuitive interface for creating and managing phishing attacks, capable of cloning target websites for credential harvesting.

Social Learning

Learning that occurs through social interactions, discussions, and shared experiences, which can enhance cybersecurity training by leveraging peer knowledge and support.

Social Media Phishing

Phishing attacks that leverage social media platforms to trick victims into providing sensitive information or gaining control of social media accounts to facilitate further attacks.

Spear Phishing

A targeted phishing attack aimed at a specific individual or organization, often using personalized information to increase success.

Storytelling and Scenario-Based Learning

Training methods that use narratives and realistic scenarios to make cybersecurity concepts more tangible and memorable.

Texting

The act of typing/texting, which employs different mental processes than speaking and listening, affecting communication and potentially leading to misunderstandings.

Threat Assessment and Remediation Analysis (TARA)

A framework developed by MITRE for identifying and assessing cyber vulnerabilities and deploying countermeasures.

Transcended Space

The concept that online interactions reduce geographical barriers, allowing instant communication across the globe but also increasing potential targets for phishing attacks.

UEBA (User and Entity Behavior Analytics)

Tools that analyze normal behavior patterns of users and devices to identify anomalies that may indicate a phishing attack or compromise.

Uniformity Across Technologies

A principle in HCI design ensuring a seamless user experience across different platforms and technologies, which can be exploited by cybercriminals to create phishing sites or emails that mimic legitimate ones.

User Behavior Analytics (UBA)

Tools that monitor for unusual activity that could indicate a security threat, understanding normal user behavior to alert on deviations that may signify an attack, such as phishing.

User Execution

The phase of a phishing attack where the user is tricked into executing malicious actions, such as clicking on a link or opening an attachment.

Vishing

Phishing conducted via telephone calls to trick victims into divulging personal information.

Web Filters

Tools that regulate content access by blocking websites known to be malicious or unsuitable, often integrated with email systems to prevent users from accessing phishing sites.

Whitelisting

A security measure that allows access only to known safe websites or email addresses.

References

[1] Alkhalil, Z., Hewage, C., Nawaf, L., and Khan, I. Phishing attacks: a recent comprehensive study and a new anatomy. *Frontiers in Computer Science* 2021; 3:563060, doi: 10.3389/fcomp.2021.563060.

[2] MITRE Corporation. MITRE ATT&CK: MITRE Corporation. 2015–2023. Available from: https://attack.mitre.org/ [accessed: 22 May 2023].

[3] Vijayan, J. Phishing Attacks for Initial Access Surged 54% in Q1. Dark Reading. May 19, 2022.

[4] IBM Security. Cost of a Data Breach Report 2022. IBM Corporation. July 2022.

[5] Reeves, A., Delfabbro, P., and Calic, D. Encouraging employee engagement with cybersecurity: how to tackle cyber fatigue. *SAGE Open* 2021;11(1), doi:10.1177/21582440211000049.

[6] Pattison, M., Butavicius, M., Parsons, K., McCormac, A., Calic, D., and Jerram, C. The information security awareness of bank employees. *Proceedings of the Tenth International Symposium on Human Aspects of Information Security & Assurance.* 2016, pp. 189–198.

[7] Merz, T. The Cybersecurity Trained User and Successful Phishing Attacks. ProQuest. 2017.

[8] Merz, T.R., Fallon, C.K., and Scalco, A. A context-centred research approach to phishing and operational technology in industrial control systems. *Journal of Information Warfare* 2019;18(4):24–36.

[9] Vance, A., Siponen, M., and Pahnila, S. Motivating IS security compliance: insights from habit and protection motivation theory. *Information & Management* 2012;49:190–8.

[10] Jansson, K., and von Solms, R. Phishing for phishing awareness. *Behaviour & Information Technology* 2011;32(6):584–593.

[11] Thomson, M.E., and von Solms, R. Information security awareness: educating your users effectively. *Information Management & Computer Security* 1998;6:167–73.

[12] Dodge, R.C., Carver, C., and Ferguson, A.J. Phishing for user security awareness. *Computers & Security* 2007;26(1):73–80.

[13] Puhakainen, P., and Siponen, M. Improving employees' compliance through information systems security training: an action research study. *MIS Quarterly* 2010;34(4):757–778.

[14] Federal Bureau of Investigation. Internet Crime Report 2022. Internet Crime Complaint Center, Washington, DC. 2022.

[15]   Ashenden, D. Your employees: the front line in cyber security. *Centre for Research and Evidence on Security Threats*, 2016. Available from: https://crestresearch.ac.uk/comment/employees-front-line-cyber-security/.

[16]   Heartfield, R., and Loukas, G. A taxonomy of attacks and a survey of defence mechanisms for semantic social engineering attacks. *ACM Computing Surveys* 2016;48(3):1–39.

[17]   Bergholz, A., DeBeer, J., Glahn, S., Moens, M.F., Paass, G., and Strobel, S. New filtering approaches for phishing email. *Journal of Computer Security* 2010;18:7–35.

[18]   Laszka, A., Vorobeychik, Y., and Koutsoukos, X. Optimal personalized filtering against spear-phishing attacks. *Proceedings of the AAAI Conference on Artificial Intelligence* 2015;29(1):958–64.

[19]   Kim, H., and Huh, J.H. Detecting DNS-poisoning-based phishing attacks from their network performance characteristics. *The Institution of Engineering and Technology* 2011;47(11):656–8.

[20]   Li, L., Berki, E., Helenius, M., and Ovaska, S. Towards a contingency approach with whitelist- and blacklist-based anti-phishing applications: what do usability tests indicate? *Behavior and Information Technology* 2014;33 (11):1136–47.

[21]   Chiew, K.L., Chang, E.H., Sze, S.N., and Tiong, W.K. Utilisation of website logo for phishing detection. *Computers and Security* 2015;54:16–25.

[22]   Pham, C., Nguyen, L.A.T., Tran, N.H., Huh, E.-N., and Hong, C.S. Phishing-aware: a neuro-fuzzy approach for anti-phishing on fog networks. *IEEE Transactions on Network and Service Management* 2018;15(3): 1076–89.

[23]   Rebera, A.P., Bonfanti, M., and Venier, S. Societal and ethical implications of anti-spoofing technologies in biometrics. *Science and Engineering Ethics* 2013;20:155–69.

[24]   Tang, L., and Mahmoud, Q.H. A survey of machine learning-based solutions for phishing website detection. *Machine Learning & Knowledge Extraction* 2021;3(3):672–94.

[25]   The Engine Room. Case study: spear-phishing attacks. *Internews* 2020 Available from: https://www.theengineroom.org/wp-content/uploads/2020/08/OrgSec-Case-study-Spearphishing-attacks-June-2020.pdf.

[26]   Gupta, M., Akiri, C., Aryal, K., Parker, E., and Praharaj, L. From ChatGPT to ThreatGPT: impact of generative AI in cybersecurity and privacy. *IEEE Access* 2023;11:80218–45.

[27]   Brown, T.B., Mann, B., Ryder, N., *et al.* Language models are few-shot learners. arXiv:200514165. 2020.

[28]   Neupane, S., Fernandez, I.A., Mittal, S., and Rahimi, S. Impacts and risk of generative AI technology on cyber defense. arXiv:2306.13033. 2023.

[29]   Goodfellow, I.J., Pouget-Abadie, J., Mirza, M., *et al.* Generative adversarial nets. *Advances in Neural Information Processing Systems* 2014;27.

[30]   Masi, I.K., A., Mascarenhas, R., Gurumurthy, S., and AbdAlmageed, W. Do we really need to collect millions of faces for effective face recognition?

IEEE International Conference on Biometrics Theory, Applications and Systems (BTAS). 2018.

[31] Wang, Y., Skerry-Ryan, R.J., Stanton, D., *et al.* Tacotron: towards end-to-end speech synthesis. *Proc. Interspeech* 2017, pp. 4006–10, doi:10.21437/Interspeech.2017-1452.

[32] Chesney, R.C. Deep fakes: a looming challenge for privacy, democracy, and national security. *California Law Review* 2019;107(6):1753–819.

[33] Oswald, M.G., Grace, J., Urwin, S., and Barnes, G.C. Algorithmic risk assessment policing models: lessons from the Durham HART model and 'experimental' proportionality. *Information & Communications Technology Law* 2018;27(2):223–50.

[34] Li, Y., and Lyu, S. Exposing deepfake videos by detecting face warping artifacts. arXiv:1811.00656. 2018.

[35] Chesney, R., and Citron, D. Deep fakes and the new disinformation war. *Foreign Affairs* 2019. Available from: https://www.foreignaffairs.com/articles/world/2018-12-11/deepfakes-and-new-disinformation-war.

[36] Veritti, D., Rubinato, L., Sarao, V., De Nardin, A., Foresti, G.L., and Lanzetta, P. Behind the mask: a critical perspective on the ethical, moral, and legal implications of AI in ophthalmology. *Graefe's Archive for Clinical and Experimental Ophthalmology* 2024;262(3):975–82.

[37] APW Group. Phishing activity trends report. 4th Quarter. 2022. Available from: https://docs.apwg.org/reports/apwg_trends_report_q4_2022.pdf.

[38] Skowrońska, J. A range of legal obligation for operators of Artificial Intelligence Systems in the context of Criminal Law. *Humanities & Social Sciences Reviews* 2023;11(5):1–5.

[39] Charles, E.P., and Sommer, R. Ecological psychology. In *Encyclopedia of Human Behavior*, 2nd edn, pp. 7–12. Burlington, MA: Academic Press; 2012.

[40] Gibson, J.J. *The Ecological Approach to Visual Perception*. Boston, MA: Houghton Mifflin; 1979, pp. 119–135.

[41] Michaels, C.F., and Carello, C. *Direct Perception*. Englewood Cliffs, NJ: Prentice-Hall; 1981. pp. 2, 43, 68, 127.

[42] Suler, J. *The Psychology of Cyberspace*. Lawrenceville, NJ: Rider University; 1996. Available from: https://www.johnsuler.com/pdfs/psycyber.pdf.

[43] Hipp, D., Olsen, S., and Gerhardstein, P. Mind-craft: exploring the effect of digital visual experience on changes to orientation sensitivity in visual contour perception. *Perception* 2020;49(10):1005–25.

[44] Wright, R.T., and Marett, K. The influence of experiential and dispositional factors in phishing: An empirical investigation of the deceived. *Journal of Management Information Systems* 2010;27(1):273–303.

[45] Jakobsson, M., and Ratkiewicz, J. Designing ethical phishing experiments: a study of (ROT13) rOnl query features. *Proceedings of the 15th International Conference on World Wide Web*. ACM. 2006, pp. 513–522.

[46] Weinschenk, S.M. *100 Things Every Designer Needs to Know about People*. Berkeley, CA: New Riders; 2011.

[47]   Verplanken, B., and Orbell, S. Reflections on past behavior: a self-report index of habit strength. *Journal of Applied Social Psychology* 2006;33(6):1313–30.

[48]   Li, M., and Suh, A. Anthropomorphism in AI-enabled technology: a literature review. *Electronic Markets* 2022;32(4):2245–75.

[49]   Konya-Baumback, E., Biller, M., and von Janda, S. Someone out there? A study on the social presence of anthropomorphized chatbots. *Computers in Human Behavior* 2023;139:107513.

[50]   Rhim, H., Ha, S., and Lee, Y. Application of humanization to survey chatbots: Change in chatbot perception, interaction experience, and survey data quality. *Computers in Human Behavior* 2021;121:107034.

[51]   Go, E., and Sundar, S.S. Humanizing chatbots: The effects of visual, identity and conversational cues on humanness perceptions. *Computers in Human Behavior* 2019;97:304–316.

[52]   Texas Medical Liability Trust. Cyber fraud case: failure to recognize phishing email. 2023. Available from: https://hub.tmlt.org/case-studies/cyber-fraud-case-study-failure-to-recognize-phishing-email.

[53]   Lange, R.D., and Haefner, R.M. Characterizing and interpreting the influence of internal variables on sensory activity. *Current Opinion in Neurobiology* 2017;46:84–89.

[54]   Hubbard, D., and Seiersen, R. *How to Measure Anything in Cybersecurity Risk*. Hoboken, NJ: Wiley; 2016.

[55]   Lain, D., Kostiainen, K., and Capkun, S., (eds.) Phishing in Organizations: Findings from a Large-Scale and Long-Term Study. *2022 IEEE Symposium on Security and Privacy (SP)*. IEEE; 2022, pp. 842–59.

[56]   Jayatilaka, A., Beu, N., Baetu, I., *et al.* Evaluation of security training and awareness programs: Review of current practices and guideline. arXiv:2112.06356. 2021.

[57]   Orunsolu, A.A., Sodiya, A.S., and Akinwale, A.T. A predictive model for phishing detection. *Journal of King Saud University – Computer and Information Sciences* 2022;34(2):1437–1445.

[58]   Visier. (n.d.). Predictive HR analytics: What it is and how to use it. Available from: https://www.visier.com/blog/predictive-hr-analytics/ [retrieved 20 July 2024].

[59]   Bada, M., and Sasse, A. *Cyber Security Awareness Campaigns: Why Do They Fail to Change Behavior?* Global Cyber Security Capacity Centre: Draft Working Paper. 2014.

[60]   Sumner, A., Yuan, X., Anwar, M., and Mcbride, M. Examining factors impacting the effectiveness of anti-phishing trainings. *Journal of Computer Information Systems* 2022;62(5):975–97.

[61]   Jensen, M.L., Dinger, M., Wright, R.T., and Thatcher, J.B. Training to mitigate phishing attacks using mindfulness techniques. *Journal of Management Information Systems* 2017;34(2):597–626.

[62]   Rizzoni, F., Magalini, S., Casaroli, A., Mari, P., Dixon, M., and Coventry, L. Phishing simulation exercise in a large hospital: a case study. *Digital Health* 2022;8:13.

[63] Schuman-Olivier, Z., Trombka, M., Lovas, D.A., *et al.* Mindfulness and behavior change. *Harvard Review of Psychiatry* 2020;28(6):371–94.

[64] Erwin, L. IT Governance European Blog [Internet]. Available from: https://www. itgovernance.eu/blog/en/author/lirwin. County Louth, Ireland: IT Governance. 2022 [cited 2024].

[65] Trend Micro. Austrian Aeronautics Company Loses over €42 Million to BEC Scam; 2016. Available from: https://www.trendmicro.com/vinfo/in/security/ news/cybercrime-and-digital-threats/austrian-aeronautics-company-loses-42m-to-bec-scam.

[66] Vanderbilt, M. *Bad Actors Using AI to Target Businesses via Business Email Compromise (BEC)*. Los Angeles, CA: Baker Tilly; 2023.

[67] Zorz, Z. Belgian bank Crelan losses €70 million to BEC scammers. Help Net Security; 2016. Available from: https://www.helpnetsecurity.com/2016/01/ 26/belgian-bank-crelan-loses-e70-million-to-bec-scammers/.

[68] Xiong, A., Lee, S., Cai, Z., Govere, E.N., and Kolla, H. You Received an Email from Your Advisor? A Case Study of Phishing Scam in a University Setting. Pennsylvania State University; 2019. Available from: https://www. usenix.org/system/files/soups2023-poster11_xiong_final.pdf.

Index